Microbial biotechnology in the laboratory and practice

WYDAWNICTWO
UNIWERSYTETU
ŁÓDZKIEGO

Microbial biotechnology in the laboratory and practice

Theory, exercises and specialist laboratories

edited by Jerzy Długoński

WYDAWNICTWO
UNIWERSYTETU
ŁÓDZKIEGO

JAGIELLONIAN
UNIVERSITY
PRESS

Łódź–Kraków 2021

Jerzy Długoński – University of Łódź, Faculty of Biology and Environmental Protection
Institute of Microbiology, Biotechnology and Immunology
Department of Industrial Microbiology and Biotechnology, 90-237 Łódź, Banacha St. 12/16, Poland

Published by Łódź University Press & Jagiellonian University Press
First edition, Łódź–Kraków 2021

ISBN 978-83-8220-150-5 – paperback Łódź University Press
e-ISBN 978-83-8220-420-9 – electronic version Łódź University Press
ISBN 978-83-233-4984-6 – paper back Jagiellonian University
e-ISBN 978-83-233-7215-8 – electronic version Jagiellonian University

Łódź University Press
Lindleya St. 8, 90-131 Łódź, Poland
www.wydawnictwo.uni.lodz.pl
e-mail: ksiegarnia@uni.lodz.pl
phone +48 (42) 665 58 63

Jagiellonian University Press
Editorial Offices, Michałowskiego 9/2, 31-126 Kraków, Poland
Phone: +48 12 663 23 80, Fax: +48 12 663 23 83
Distribution: Phone: +48 12 631 01 97, Fax: +48 12 631 01 98
Cell Phone: + 48 506 006 674, e-mail: sprzedaz@wuj.pl
Bank: PEKAO SA, IBAN PL 80 1240 4722 1111 0000 4856 3325

www.wuj.pl

The book is available in the Columbia University Press catalog:
https://cup.columbia.edu

Table of contents

Foreword

Life sciences, including biotechnology, microbiology, and related disciplines, are among the fastest growing areas of science since the second half of the 20th century. The discovery of the DNA structure in 1953 by Watson and Crick, as well as the development of modern research techniques and methods that allow the analysis and description of the processes taking place in organisms with great accuracy in a short period of time, have contributed to this. This is reflected in the ever-increasing number of scientific publications in the international literature, as well as the emergence of new fields of study at universities, often of an interdisciplinary nature, enabling students to acquire knowledge from different areas of science and facilitating their pursuit of employment in the ever-changing jobs market. Such a role is fulfilled, among others, by biotechnology, which combines the latest achievements of the intensely developing natural sciences with their use in various areas of human activity, as shown in the figure below. The areas of knowledge and practice that are considered both in the upper and lower fields, constitute the main topics that are discussed in this handbook.

The first two chapters discuss the principles of obtaining strains of microorganisms used in practice, the methods of their cultivation, improvement in terms of properties useful from the application point of view, and the most important issues related to storage in conditions conducive to the survival of the microorganisms and maintenance of their desired properties. This part of the handbook omits basic microbiological techniques, assuming that they are known from general microbiology courses. The focus is on familiarising the reader with the latest analytical techniques used in microbiology, biotechnology, and related sciences – confocal and fluorescence microscopy, spectrofluorimetry, isotope techniques, chromatography, atomic absorption spectrometry. The principles of aerial and satellite remote sensing are also presented here, enabling rapid detection and objective assessment of threats caused by bacterial, viral, and fungal pathogens in agricultural, forest, and urban green areas (Figure 2.6.1.4).

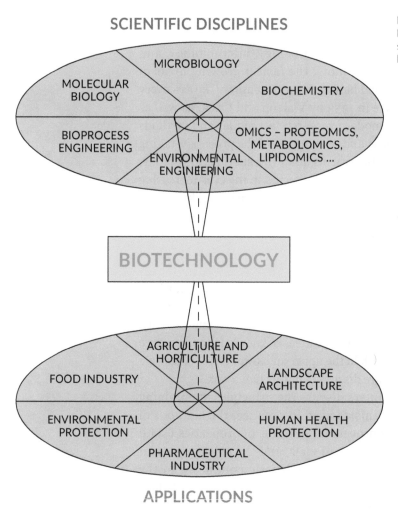

Fig. 1. Relationships between selected natural sciences and areas of human activity.

The third chapter is devoted to the identification of previously isolated bacteria and fungi, considering the use of the latest techniques, including MALDI-TOF/TOF and LC-MS/MS methods already presented in the previous chapter.

The fourth chapter presents the possibilities of using microorganisms on an industrial scale. This part of the publication is largely based on the previously published handbook Microbiological Biotechnology. Exercises and specialist laboratories edited by Jerzy Długoński, which was published in 1997 by the University of Lodz Publishing House. Due to the fact that the layout of that handbook gained the approval of both teachers and students, the division of the discussed issues into biosynthesis, fermentation, and biotransformation processes was maintained, at the same time updating its contents and introducing additional subsections concerning the production of biosurfactants and food produced with the participation of microorganisms.

The rationale for the development of chapter five, the most extensive part and at the same time reflecting the interdisciplinary approach of the authors of individual subchapters to solving problems related to the protection of human health and the environment. These issues were explored in earlier publications, including an extensive monograph Microbial Biodegradation. From Omics to Function and Application (edited by Jerzy Długoński, Caister Academic Press, Norfolk 2016) as well as cooperation with scientists from other fields of science in Poland and abroad.

Chapter six is a collection of examples demonstrating the benefits of using the latest analytical techniques in scientific research and in the protection of health and the environment in a broad sense.

The final part of the handbook (chapters seven and eight) is a list of media and buffers with their composition, used in the practical part of individual subchapters, as well as macroscopic and microscopic images of fungal strains used by the authors of the handbook in research and didactics.

The handbook was prepared on the basis of many years of didactic and scientific experience of the authors, conducting classes in various fields and specialties, which has bestowed a multilateral character on the work. For this reason, the authors of the handbook are convinced that it will be useful for students of biotechnology, microbiology, and ecology, as well as interdisciplinary faculties such as eco-town, environmental protection, biomonitoring and ecological biotechnology or urban revitalisation.

The authors of the handbook would like to express their heartfelt thanks to the Reviewers, Prof. Grażyna Płaza, Prof. Jerzy Falandysz, Prof. Dominik Kopeć and Dr. Alwyn Fernandes whose valuable comments and remarks contributed fruitfully to the preparation of its final version. Special thanks are also addressed to Aleksandra Góralczyk-Bińkowska, MSc, Dr. Anna Jasińska, Dr. Katarzyna Zawadzka, Dr. Andrzej Długoński and Małgorzata Krokocka, MSc for taking care of the careful editorial and graphic design of the handbook.

With a view to a possible next edition of the handbook, the authors will be grateful if any critical comments were sent to the scientific editor (jerzy.dlugonski@biol.uni.lodz.pl).

Łódź, 12 December 2019

Jerzy Długoński

Foreword to the handbook *Microbiological Biotechnology*. Exercises and specialist laboratories from 1997

According to the definition adopted by the European Federation of Biotechnology, "biotechnology is an integrated application of natural and engineering sciences with the aim of using living organisms, cells and their component parts for products and services". One of the sciences that is part of biotechnology is industrial microbiology, which deals with the use of microorganisms in practice. The most important factors that favour the widespread use of microorganisms in biotechnological processes include relatively easy, large-scale culturing techniques, the ability to use various raw materials as a source of carbon and energy, high metabolic rates, the well-known genetic structure of some species and the ease of genetic manipulation. It should be noted, however, that to take full advantage of the properties of microorganisms, it is necessary to know the physiology of industrial strains and the mechanisms regulating the changes taking place inside the cell. These issues are the main subject of specialist laboratories and exercises in industrial microbiology for students of the 4th year of biotechnology and microbiology at the Faculty of Biology and Environmental Protection, University of Łódź. Students, when they join classes conducted in semesters 8 and 9, and already have a considerable amount of knowledge in the fields of chemistry, general microbiology, biochemistry, mycology and microbial genetics. Therefore, the handbook does not discuss the basic techniques used in laboratory work, but focuses mainly on issues concerning the physiology of industrial microorganisms and their use in biotechnological processes.

The handbook has been developed on the basis of many years of didactic and scientific experience of the employees of the Department of Industrial Microbiology of the University of Łódź. The authors would like to express their deep gratitude to the organiser and long-term manager of the Department of Industrial Microbiology, Prof. Leon Sedlaczek, without whose contribution to the didactic and scientific achievements of the Department, it would not be possible to develop this handbook.

Equally sincere thanks are due to the Reviewer of this study, Prof. Aleksander Chmiel, whose valuable remarks and suggestions have helped to create the final version of the handbook.

The handbook was written primarily for students specialising in biotechnology and microbiology. Nevertheless, it will be useful for students of other specialisations interested in microbiology and for various people involved in the use of microorganisms in practice.

Łódź, 27 June 1996

Jerzy Długoński

1.
Methods of screening, culturing, improvement and storage of microorganisms of industrial importance

1.1. General characteristics of microorganisms used in biotechnological processes

Although more than a century and a half has passed since Louis Pasteur created microbiology and biotechnology, it is estimated that only 5% of microfungi (filamentous fungi and yeast) and 12% of eubacteria and archaeon species, capable of growing in the laboratory, are now known. This creates a large field of action for biologists, especially microbiologists, especially since microorganisms are characterised by a number of valuable properties favourable for their use in technological processes:

- rapid, in relation to other organisms, course of metabolic processes;
- well-known genomes of numerous species of microorganisms and ease of genetic manipulation;
- well-developed methods of culturing on a laboratory and industrial scale;
- cheap and widely available ingredients for culture media;
- diversity of metabolism (different products can be obtained from the same raw materials, using different strains of microorganisms);
- adaptability of microorganisms (using the same microorganism, the same product can be obtained from different raw materials, e.g., in alcoholic fermentation – ethanol production).

The features limiting the use of microorganisms include:

- production of large biomass and thus the consumption of medium components and the costs of disposing of biomass and/or post-culture liquid;
- unprofitable production of most of the low molecular weight substances (except for the food industry and compounds used in medical treatment).

When isolating microorganisms from different environments and then assessing their suitability for use in different technological processes, not only the ability to synthesise, transform or degrade various compounds is considered, but also:

- nutritional properties (low nutritional requirements);
- temperature properties (thermophilic strains are preferred, due to the lack of need to cool the culture in large-volume bioreactors);
- process type (continuous cultures);
- no harmful effects of the microorganism on the equipment components (biological corrosion);
- stability of morphological and biochemical characteristics;
- resistance to bacteriophages and bacterial infections;
- susceptibility to genetic manipulation (acquisition of mutants and GMMs based on the starting strain);
- high performance (based on substrate conversion);
- high productivity (product performance per unit of time);
- product recovery (cost of extraction and disposal of by-products present in the extract).

Table 1.1. Examples of negative and positive manifestations of microbial activity (from the point of view of human health and activity)

Microorganisms (dominant)	Practical application	Adverse activity
Claviceps purpurea	Biosynthesis of psychotropic drugs	Ergot poisoning ("Fire of St. Anthony")
Numerous species of *Bacillus, Pseudomonas, Micrococcus, Flavobacterium, Streptomyces, Aspergillus*	Production of numerous complexes, agri-food industry waste treatment, composting, sewage treatment	Decomposition of raw materials and food products
Bacillus, Pseudomonas, Acinetobacter Micrococcus, Sphingomonas, Mucor, Phanerochaete	Bioremediation of areas contaminated by petroleum substances, production of biosurfactants	Mucosalisation and degradation of petroleum products (oils, lubricants, gasoline)
Bacillus, Pseudomonas, Trichoderma harzianum, Paeciliomyces marquandii Cunninghamella echinulata Umbelopsis isabellina Nectriella pironii Myrothecium roridum	Biodegradation of toxic xenobiotics, including EDCs (Endocrine Disrupting Compounds), industrial sewage decolorisation, biosynthesis of hydrolases, laccases and other enzymes	Biological corrosion of construction materials, library resources, interior design elements, plant diseases

When analysing the properties of isolated microorganisms and their suitability for use in specific technological processes, it should also be remembered that the same microorganism (or groups of interacting microorganisms) can be useful in several different branches of industry or environmental protection, but they can also cause serious damage. Examples are the so-called acetic acid bacteria of the genera *Acetobacter* and *Gluconobacter*, used for the production of acetic acid or sorbose, and concurrently posing a serious threat in the production of wine or beer. Other similar examples are presented in Table 1.1.

1.2. Screening of microorganisms useful in microbiological processes

1.2.1. Suitability of various environments for isolation of microorganisms used in industrial processes

Microorganisms with specific metabolic characteristics are widely used in modern biotechnological processes. The main source of microorganisms is the natural environment, especially the soil. The soil contains bacteria, actinomycetes and micromycetes that are capable of biosynthesis, decomposition and biotransformation of many compounds important for humans (antibiotics, vitamins, organic acids, steroid compounds), as well as elimination of pollutants, including toxic xenobiotics. Genetic diversity of soil microorganisms can be an excellent source of strains with recognised biotechnological properties, used in modern technologies.

The exploration of strains useful in industrial processes, from less typical (extreme) microbial habitats, e.g. saline water reservoirs, waters of polar and high mountain lakes, hot springs, geysers, etc. should also be considered.

By isolating microorganisms from sites polluted with xenobiotics, such as industrial landfills, strains capable of detoxifying and degrading them can be obtained. Most ecological niches are useful reservoirs of microorganisms with unique properties, often very valuable from the biotechnological point of view. In particular, microorganisms capable of surviving in unusual, often extreme conditions (extremophiles), can be a source of enzymes used for industrial, therapeutic, scientific purposes (Table 1.2.1).

Thermophiles (organisms adapted to live above 80°C) are successfully used in various industries. They are usually isolated from volcanic areas (soils). A classic example is *Thermus aquaticus* – a species that produces thermostable DNA polymerase, used for the PCR reaction. A lot of enzymes obtained with the use of this group of microorganisms (e.g. protease, lipase) are used in industry, e.g. food, pharmaceuticals and dyes. Furthermore, microorganisms inhabiting cold environments, such as the polar and tundra biomes, cold ocean waters, glaciers or underground caves (psychrophilic microorganisms) are a source of enzymes, commonly used as an additive to washing powders that are active at low temperatures. In turn, proteases and lipases, obtained from psychrophiles, are also used in the food industry, e.g. for cheese production.

Another group of microorganisms, inhabiting unusual environments, are acidophilic microorganisms living in acidic environments. They were identified for the first time in the waters created in the process of mine drainage. This group of bacteria oxidises inorganic sulphur compounds, playing an important role in the geochemical cycle of this element. They have also found practical use in the mining industry for desulphurisation of hard coal or for the recovery of rare metals, including those of strategic importance, such as uranium. The discussed group also includes lactic acid bacteria that have found use in the food and pharmaceutical industries, e.g. probiotics (Section 4.2.2). Alkaliphiles – microorganisms living in alkaline environments – are a very valuable source of proteases used in the food industry, as well as in the production of cleaning products and amylases used in the brewing, food and baking industries. Moreover, the treatment of highly saline wastewater can be considerably assisted by the presence of NO_2 and NO_3 reductases.

1.2.2. Soil as a source of potential producers of biologically active compounds

The soil is a rich and inexhaustible source of microorganisms. There are about 10^6–10^8 bacterial cells, 10^4–10^6 actinomycete conidia, 10^2–10^4 fungal spores in 1 g of arable soil. They are very important in the course of biogeochemical cycles and the mineralisation of organic remains of plant and animal origin, improving at the same time the fertility (humus formation) and productivity of soils. Soil microorganisms also affect the functioning of ecosystems and the condition of plants. Moreover, microorganisms with properties that make them useful in biotechnology have been isolated for years, from the soil environment.

Soil is a natural habitat for many species of microorganisms. It consists of mineral compounds and organic substances (50% of the soil composition), soil gas (about 35%) and soil solution (15%). The solid part of the soil consists mainly of soil colloids. The most important function is performed by organic colloids (humic compounds), which are formed as a result of the decomposition of organic matter by microorganisms. Mineral colloids, on the other hand, determine water-air relations. Soil water, containing dissolved organic and mineral substances, is called soil solution. This soil fraction has buffer properties and determines the pH of the soil. The space between soil particles (unoccupied by soil solution) is filled by the soil gas. The soil gas consists mainly of NH_3, CO_2, N_2 and O_2, also of smaller quantities of H_2S and CH_4. Soil structure is composed of mineral grains, which are glued together by humic compounds and mucous, forming 0.5–5 mm size particles. The microorganisms are located on the surface of particles in humic substances, as well as in organo-mineral complexes.

Soil is rich in microbiota in partially decomposed organic matter (humus). Larger clusters, the so-called soil aggregates, are built from soil particles. The exterior of these clusters is inhabited mainly by spore-forming bacteria, actinomycetes and filamentous fungi. The inner part is colonised by bacteria assimilating mineral nitrogen compounds, other (mainly Gram-negative) bacteria, and *Fusarium* moulds.

Soil microbiota are abundant and depend on the type of soil, including chemical composition (inorganic and organic compounds), pH, oxygen availability, humidity, temperature, depth, geographical zone, presence of plants. In general, soil microbiota are divided into: autochthonous microbiota (constantly present in the soil) and zymogenous microbiota (introduced periodically). The source of carbon and energy for the autochthonous microbiota are humus substances (compounds formed from the decomposition of organic matter of plant and animal origin). The development of the zymogenous microbiota depends on the inflow of easily assimilable organic matter.

Filamentous fungi and a significant proportion of bacteria prefer moist environments. Microorganisms with low water requirements develop in dry soils – they are xerophilic microorganisms. Actinomycetes and nitrifying bacteria of the *Arthrobacter* genus also inhabit dry soils. Wet soils, which have a limited amount of nutrients and oxygen, are abundant in anaerobic and facultative microorganism. Biodiversity of soil microbiota also depends on temperature. The best time of year for the microorganism's isolation is spring. In winter, the soil microbiota is significantly reduced. An important parameter that determines the development of microorganisms in the soil is pH. Soils with alkaline or neutral pH contain several times more microorganisms than acidic soils (e.g. peatlands). Fertile soil, rich in humus substances, is also rich in microbiota compared to compacted meadow soil.

Microorganisms are very sensitive to sunlight (UV solar radiation) and wind (drying), so very few dwell on the soil surface.

Most microorganisms are present in the root zone (rhizosphere) and on the root surface (rhizoplane). On the other hand, the number of microorganisms is much smaller in deeper soil layers (at the depth of 1–2 m).

1.2.3. Microbial screening

Special screening methods have been developed in order to obtain new strains from the environment. The screening can be defined as a set of selective procedures aimed at isolating from a large number of microorganisms, only those that meet specific technological requirements. Screening also includes a preliminary assessment of their suitability for a given process, e.g. assessment of their biodegradation potential. In the initial stage of screening, we are dealing with microorganisms with undefined systematic affiliation, which may also include pathogenic strains. Usually strains isolated directly from the environment produce small amounts of metabolites of interest. Only after they have been enhanced, using genetic and microbiological techniques (see Section 1.6), can strains be used on an industrial scale. An effective and simple technique for the isolation of microorganisms is the selective multiplication of cultures. In this case, strictly defined media and specific culture conditions are used, i.e., medium composition, temperature, pH, osmotic pressure, light availability, oxygen availability, which allow the growth of a specific group of microorganisms (Table 1.2.1).

Culture conditions	Type of isolated microorganisms
Extremely acidic pH (pH 2–4)	Acidophiles
Low temperature (4–15°C)	Psychrophiles
High temperature (42–100°C)	Thermophiles
NaCl at great concentration	*Nocardia*, halophiles
Presence of atmospheric nitrogen (N$_2$)	Anaerobics
Chitin as growth substrate	*Lysobacter*
Tree bark, roots	Myxobacteria
Grain pollen	*Actinoplanes*
Media with heavy metals addition	Microrganisms accumulating heavy metals

Table 1.2.1. Methods of stimulating the growth of selected groups of microorganisms by application of appropriate culture conditions

The search for microorganisms with desired characteristics can also be accelerated by inoculation of the samples taken from prepared extracts, directly onto differential media plates, allowing the isolation of microorganisms with specific metabolic activity (Table 1.2.2).

Table 1.2.2. Examples of methods that favour the acquisition of microorganisms with specific metabolic activities

Producers	Pre-selection rules
Antibiotics	Inoculation on agar plates with strains of test microorganisms, e.g. *Staphylococcus aureus, Proteus vulgaris, Candida albicans, Penicillium avellaneum.* Growth inhibition zones are an indicator of activity.
β-lactamase-resistant antibiotics	Inoculation on agar plates with β-lactamase addition.
Proteases	Inoculation on agar plates with casein. Colonies that produce lightening zones in the medium are selected.
Amylases	Inoculation on agar plates with starch. Selection of colonies after staining with Lugol's iodine (J/KJ).
Lipases	Inoculation on agar plates with emulsified oil. Selection of colonies after precipitation of free fatty acids with calcium (Ca) ions.
Phosphatases	Inoculation on agar plates with phenolphthalein diphosphate. Selection based on the colour change of the substrate around the colony.
Cellulases	Inoculation on agar plates with phosphocellulases. Selection of colonies that produce transparent zones of enzyme activity.
Pectinases	Inoculation on agar plates with pectins. Selection of colonies producing bright zones on an opalescent background – reaction with a solution of CTAB (cetyltrimethylammonium bromide).
Xenobiotic decomposing microorganisms	Inoculation on agar plates with selected xenobiotic, being the only or main source of carbon and energy. Analysis of the growing colonies in terms of their degradation activity.

The screening stage includes the selection of the collection site and the method of isolating microorganisms using appropriate media. If we are looking for protease-producing bacteria, a convenient place to collect samples may be, among others, the areas around industrial plants producing significant amounts of waste containing proteins, e.g. dairies, meat plants. At a further stage, pure microbial cultures are isolated, on the basis of which liquid cultures are prepared. A preliminary assessment of the suitability of the isolated strains for use in a specific technological process is made. If a specific product is desired, process optimisation is carried out, often combined with improving the isolated strain and increasing the scale of culture. The final stage of screening includes purification of the product (if necessary), patenting of the relevant (key) production steps and commercialisation of the obtained research results. The key stages of screening are shown in Figure 1.2.1.

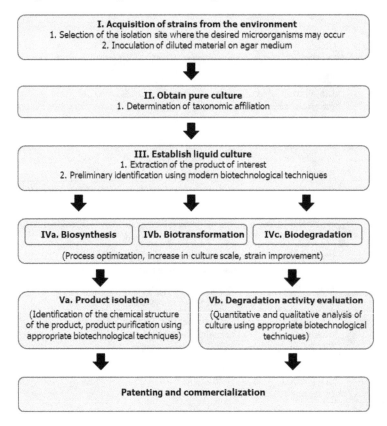

Fig. 1.2.1. Stages of screening microorganisms that are useful in biotechnological processes.

1.2.4. Soil of polluted environments as a source of microorganisms used in environmental protection processes

A significant increase in environmental pollution has been observed as a result of intensive industrial development, which has a very high impact on soil composition and biodiversity. A large part of the species of microorganisms inhabiting the soil have the ability to biodegrade xenobiotics. The degradation potential of microorganisms depends primarily on their environmental biodiversity but also on physical and chemical parameters (temperature, oxygenation, pH, availability of nutrients) of soil. Of all the pollutants, aromatic hydrocarbons are particularly dangerous. They exhibit strong toxic and carcinogenic properties. Contamination with petroleum products is greatest in the vicinity of refineries, petrol stations, airports, freeways, railroads and diagnostic stations, repair shops and car washes. Nevertheless, the contamination of green areas of cities by aromatic hydrocarbons and other harmful substances, and the related revitalisation of degraded areas, are important problems. These issues are discussed in more detail in Section 5.1. Petroleum products, especially polycyclic compounds, are resistant to degradation and some may remain in the environment for many years. Elimination of these compounds is mainly caused by microorganisms. Therefore, the search for strains capable of using oil derivatives as carbon and energy sources or of converting them into environmentally non-toxic products is a very important task of modern environmental biotechnology.

Chemical structures of selected aromatic hydrocarbons are presented in Figure 1.2.2.

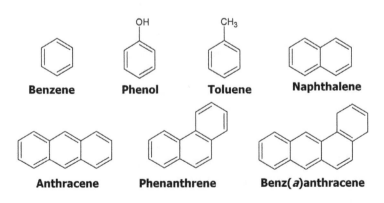

Benzene **Phenol** **Toluene** **Naphthalene**

Anthracene **Phenanthrene** **Benz(*a*)anthracene**

Fig. 1.2.2. Structure of selected aromatic hydrocarbons.

The rate of degradation of these compounds depends on the number of aromatic rings in the molecule. Today, numerous microorganisms capable of degrading low molecular weight hydrocarbons are known, but not many that can degrade high molecular weight polycyclic aromatic hydrocarbons (PAHs). Aromatic hydrocarbons can be partially or completely degraded (mineralised) by consortia of microorganisms and less frequently by individual species or strains.

Prokaryotic microorganisms used in the degradation of aromatic hydrocarbons mainly use dioxygenases, which are capable of introducing two oxygen atoms into the aromatic ring. Then, in the dehydrogenation stage, dihydroxy derivatives (catechols) are formed. The catechol ring, a key intermediate metabolite formed, is further degraded via enzymes of *ortho* and *meta* cleavage pathways. Subsequent reactions produce metabolites that are included in the Krebs cycle (Figure 1.2.3).

Some microorganisms, such as filamentous fungi, are capable of biotransforming aromatic hydrocarbons into intermediate products. These organisms use monooxygenases containing cytochrome P-450 that introduce one oxygen atom into the aromatic ring. The resulting epoxide is then converted to phenol or dihydrodiol and these can be conjugated with polar compounds such as glucose, glucuronic acid and xylose (Figure 1.2.4).

In addition, PAH biotransformation processes in both fungi and bacteria can be catalysed by oxidising and reducing enzymes such as laccases and peroxidases. The mechanism of these reactions is described in Section 5.11.

The susceptibility of hydrocarbons to biochemical degradation is determined not only by the chemical structure of petroleum products. Other factors influencing this process include substrate solubility and production of surfactants by microorganisms, hydrocarbon adsorption processes, number of microorganisms, concentration of nitrogen and phosphorus biogenic compounds, concentration of hydrocarbons, pH, humidity, temperature, oxygen content, presence of carbon sources for microorganisms other than hydrocarbons and the presence of toxic compounds.

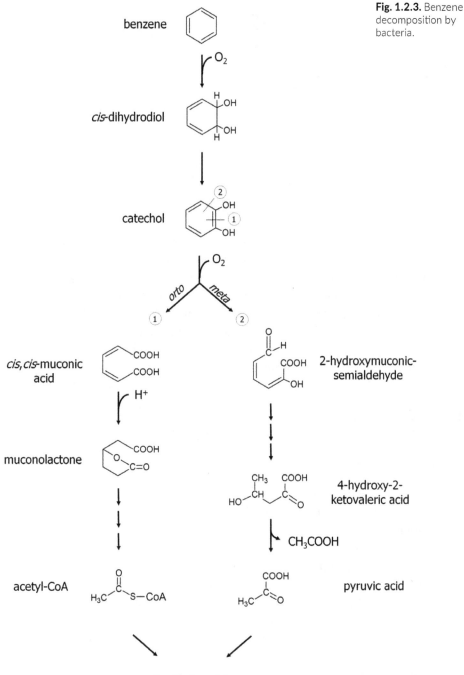

benzene

cis-dihydrodiol

catechol

cis, cis-muconic acid

muconolactone

acetyl-CoA

2-hydroxymuconic-semialdehyde

4-hydroxy-2-ketovaleric acid

pyruvic acid

the Krebs cycle

Fig. 1.2.3. Benzene decomposition by bacteria.

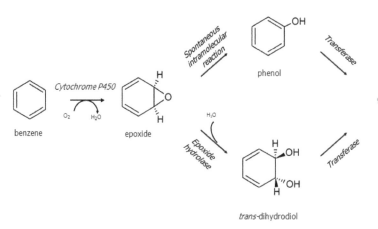

phenol

Conjugates with polar
compounds (e.g. glucose,
glucuronic acid, sulfate, GSH)

benzene   epoxide

trans-dihydrodiol

Only a small number of soil bacteria (0.01–0.3%) are capable of using hydrocarbons as sole source of carbon and energy. Screening methods are very useful in isolating microorganisms capable of degrading xenobiotics. The easiest way to do this is to take a soil sample and perform an inoculation from appropriate dilutions on selective media with the addition of xenobiotics as a source of carbon and energy. Growing colonies are then isolated and checked for their degradation activity.

Fig. 1.2.4.
Biotransformation
of benzene by
filamentous fungi.

PRACTICAL PART

Materials and media

1. Source of isolated bacteria: garden and meadow soil, as well as soil samples taken from the area of a dairy, a gas station or freeway.
2. Microorganisms:
   a) Strains isolated from soil samples;
   b) Reference bacterial strains: *Staphylococcus aureus* ATCC 6538, *Escherichia coli* ATCC 25922, *Pseudomonas aeruginosa* ATCC 15442.
3. Solid media: agar, ZT medium, universal, maltose, universal with anthracene and phenanthrene (0.1 g/l), medium with milk, with starch, Tween 80.
4. Reagents: anthracene, phenanthrene.

5. Apparatus: gas chromatograph Agilent GC7890 with mass spectrometer MS5975C.

Aim of the exercise
Obtaining strains of microorganisms capable of biosynthesis of biologically active substances and decomposition of petroleum derivatives.

Procedure

1. Prepare a soil extract.
   Weigh 2 g of soil, transfer to a 100 ml flask containing 20 ml of 0.85% NaCl solution and add sterile glass beads. Incubate on a shaker for 0.5 h at room temperature. Once the incubation is complete, wait 2–3 min until the larger soil particles have settled.
2. Make a series of dilutions of the soil extract in 0.85% NaCl solution according to the following scheme:

Fig. 1.2.5. Series of dilutions of the soil extract

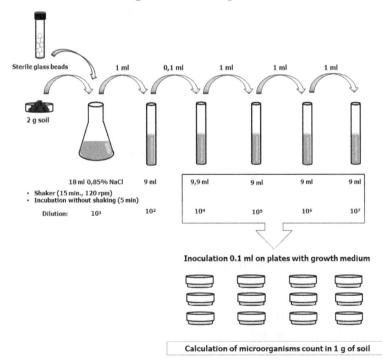

Calculation of microorganisms count in 1 g of soil

3. Inoculate from appropriate dilutions of the soil extract (Figure 1.2.5) 0.1 ml per plate in the following media: agar, ZT, agar with anthracene or phenanthrene (0.1 g/l), medium with milk, with Tween 80, with starch.

   Inoculate duplicates from each dilution. Incubate the plates 48 h for 7 days at 28°C.
4. Perform macroscopic observation of the cultures, count the number of microorganisms in 1 g of soil.
5. Isolation of single microbial colonies:
   a) producing proteases (lightening around the colonies on plates with milk containing medium); single colonies should be transferred to plates with milk and a streak culture should be carried out;
   b) Actinomycetes; single colonies should be transferred to maltose plates and inoculated using the loop. After 7 days, examine the antagonistic properties of the selected soil actinomycetes in relation to the test bacteria: *Staphylococcus aureus, Escherichia coli, Pseudomonas aeruginosa*;
   c) microorganisms potentially capable of decomposing petroleum-based compounds (anthracene and phenanthrene); single colonies should be transferred to a universal medium.

      Additionally, a microscopic preparation of the isolated microorganism should be made.

      Isolates of microorganisms capable of producing proteolytic enzymes that produce a substance that is antagonistic to the tested bacteria should be secured for further analyses.
6. Evaluation of proteolytic enzyme activity of microorganisms isolated from the soil (see Section 4.1.3).
7. Evaluation of the activity of substances that are antagonistic to the bacteria being tested, produced by microorganisms isolated from the soil (see Section 4.1.5).
8. Evaluation of the ability of isolated strains (bacteria and fungi) to degrade petroleum-based compounds:
   a) streaking of selected colonies from plates on universal (bacteria) and ZT slants (fungi);

b) streaking from liquid medium (10% inoculum) of strains isolated from the soil – up to 20 ml of universal medium with addition of anthracene or phenanthrene at a concentration of 0.5 g/l.

Incubation for 96 h at 28°C (fungi) and 30°C (bacteria);

c) extraction should be performed with ethyl acetate. Prepare the evaporated extract for chromatographic analyses (see Section 5.6.6);

d) chromatographic analysis for samples extracted with ethyl acetate and concentrated under reduced pressure is performed using a gas chromatograph coupled to a mass spectrometer GC/MS (Agilent GC 7890 and MS 5975C) with HP 5 MS column (30 m × 0.25 mm × 0.25 mm) at a temperature range of 80–300°C (see Section 2.3) at a helium flow of 1 ml/min.

Analysis of results

On the basis of the conducted analyses, evaluate the suitability of the isolated microorganisms for:

1) production of proteolytic enzymes;

2) production of substances of antagonistic activity;

3) degradation of petroleum-based compounds.

1.3. Methods of culturing and stabilisation of microorganisms under aerobic conditions (including bioreactors and immobilisation in gels)

INTRODUCTION

Knowledge of microbial physiology, including growth physiology, is essential for proper control of technological processes in which microorganisms are used. In biotechnology, surface or submerged culturing methods can be used for industrial strains. Due to the need to have relatively large production facilities and usually lower process efficiency, surface methods are rarely used today. Submerged cultures use bioreactors (fermenters) which are hermetically closed metal and/or glass vessels (laboratory fermenters) that allow full control of the growth and production of products by microorganisms. On a laboratory scale, fermenters ranging from several dozen millilitres to several dozen litres are used. In the industry, bioreactors of even 200 thousand litres or more, are used. In the laboratory, at the stage of inoculum multiplication, glass flasks of up to several litres are also used for submerged cultivation. In biotechnological processes, apart from microorganisms slowly suspended in the medium, cells "trapped" in gels are often used. The use of immobilised cultures allows for their use in a few or even several production cycles, which significantly reduces the cost of the technological process. The culture of microorganisms in aerobic conditions in liquid media can be divided into three basic types: batch, fed-batch with continuous dosing of medium and continuous culture.

1.3.1. Batch culture

Batch culture is a closed system. It consists of preparing a sterile culture medium with added inoculum and conducting the culture in a fermenter for a specific period of time. Changes in the amount of microbial biomass are observed from the onset of incubation (Figure 1.3.1).

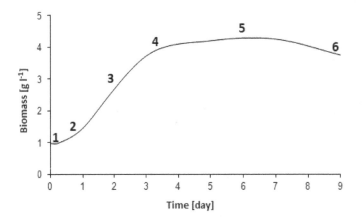

Fig. 1.3.1. Example of the batch culture of microorganisms in time.

Microbial growth is described through 6 stages. During the stagnation phase (lag-phase) no biomass growth is observed; this time is used by the microorganisms to adapt to the new environment. In the acceleration phase, there is an increasing growth in the amount of biomass until it reaches an exponential rate (exponential phase or logarithmic growth).

During this stage the growth of the microorganism reaches its highest value. After some time the growth substrate in the culture system is exhausted and toxic metabolic products may appear, so the multiplication of the microorganism slows down (growth inhibition phase), and in the next stage it is completely inhibited (stationary phase). During this period of growth, the largest amount of biomass in the system is observed. With the lack of nutrients and increasing accumulation of toxic metabolites, the culture enters a phase of death associated with a decrease in the amount of biomass in the system. The growth rate of the microorganism is described by the equation:

$$r_x = \mu X$$

where:
r_x – actual growth rate,
μ – specific growth rate,
X – culture biomass content.

The μ factor, also called the specific growth factor, can be defined as the ratio of the biomass growth rate to the existing one. Most often its unit is h^{-1}.

Growth phase	Specific growth rate
Lag-phase	$\mu = 0$
Acceleration growth	$\mu < \mu_{max}$
Logarithmic	$\mu = \mu_{max}$
Growth inhibition	$\mu < \mu_{max}$
Stationary	$\mu = 0$
Death	$\mu < 0$

Table 1.3.1. Growth rate depending on growth phase.

Since the growth of the microorganism in the exponential phase is at its highest rate, the substrate is easily accessible and the resulting metabolites do not adversely affect the growth (optimal conditions prevail), the biomass growth reaches its highest value of μ_{max}.

The biomass balance for this type of culture is described as follows:

$$\frac{dx}{dt} = \mu X$$

In the exponential phase, for the value $\mu = \mu_{max}$:

$$X = X_0 e^{\mu_{max}(t - t_0)}$$

The growth of the microorganism can also be expressed by the parameter
– time to double the biomass (td). Assumption that:

$$2\,X_0 = X_0 e^{\mu_{max}t_d}$$

leads to the relationship $\ln 2 = \mu\, t_d$, i.e., $t_d = \dfrac{\ln 2}{\mu_{max}}$

The easiest way to determine μmax in the culture is to observe the changes in the amount of biomass over time, preparing a graph in lnX – time of culture system. For the identified exponential phase, we determine two points, $t_1 X_1$ and $t_2 X_2$ and using the formula:

$$\mu_{max} = \frac{\ln X_2 - \ln X_1}{t_2 - t_1}$$

calculate μ_{max} [h⁻¹].

The growth of the microbial biomass is associated with the disposal of the substrate in the culture medium. The parameter describing the ratio of biomass growth to the amount of used substrate is the mass yield factor Y_{XS}

$$Y_{xs} = \frac{\Delta X}{-\Delta S} = \frac{X - X_0}{-(S - S_0)}$$

for X_0, S_0 describing the initial concentration of biomass and substrate.

Advantages of batch culture:
• short culture time facilitates the maintenance of sterile conditions and prevents mutations;
• Simple culture preparation.

Disadvantages:
- No substrate concentration regulation and low process efficiency.

1.3.2. Fed-batch culture – batch with continuous dosing of nutrient to the fermenter

In this process, unlike batch culture, the active volume of the bioreactor increases during incubation. This type of process has been designed for cultures where low substrate concentration is necessary to obtain a large amount of product (e.g. synthesis of antibiotics) and this process can occur during the growth inhibition or stationary phase. These types of cultures are often used under industrial conditions.

Advantages:
- Possibility of regulating the culture parameters according to the type of microorganism and the product obtained as well as culture medium composition.

Disadvantages:
- Possibility of medium infection and strain degeneration.

1.3.3. Continuous culture

After the initial multiplication of the microorganisms, continuous cultures are carried out in a flow bioreactor fed with medium. The bioreactors are continuously fed with sterile medium at flow rate F and at the same time, the culture liquid is collected from the tank at the same flow rate, maintaining a constant volume of the liquid (Figure 1.3.2).

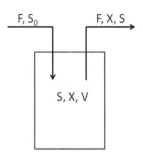

Fig. 1.3.2. Chemostat scheme
Explanations:
F – volumetric flow rate; S_0 – substrate concentration in the dosed medium; S – substrate concentration in the culture medium; X –biomass content in the culture; V – culture volume

There are two basic types of continuous culture:

- turbidostat – in this model the microorganisms can grow at a maximum rate, because the biomass content is kept constant and the amount of supplied substrate depends on the growth rate;
- chemostat – the parameter that plays a decisive role in the chemostat is the dilution rate D, depending on the volume flow rate F and the culture volume V:

$$D = \frac{F}{V}$$

Under steady conditions, when the rate of biomass removal from the bioreactor is equal to the rate of its growth in the apparatus $\mu = D$. This condition can be derived from the biomass balance of the chemostat.

$$\frac{d(VX)}{dt} = FX_0 - FX + r_x V$$

$$X\frac{dV}{dt} + V\frac{dX}{dt} = FX_0 - FX + r_x V / : V$$

No change in tank volume:

$$\frac{dX}{dt} = \frac{F}{V}(X_0 - X) + r_x$$

where $X_0 = 0$ and there is no change in biomass content:

$$\frac{dX}{dt} = -DX + \mu X$$

$$D = \mu$$

In the case of $\mu < D$, the biomass is leached from the tank and in the case of $\mu > D$, the biomass is stored in the tank. The relationship between the growth rate and the concentration of the growth limiting substrate is described by the Monod equation:

$$\mu = \frac{\mu_{max} S}{K_S + S}$$

where K_S – a constant of substrate saturation for a substrate concentration at which $\mu = \mu_{max}/2$.

For the condition $\mu = D$ the Monod equation is modified and the concentration of the substrate in the tank is:

$$S = \frac{DK_S}{\mu_{max} - D}$$

Taking this relationship into account and based on the constant value of the mass yield factor Y_{XS}:

$$Y_{XS} = \frac{\dfrac{dX}{dt}}{\dfrac{dS}{dt}}$$

the following relationship between substrate concentration and biomass content and dilution rate is obtained:

$$X = Y_{XS} \left(S_0 - \frac{DK_S}{\mu_{max} - D} \right)$$

where S_0 – concentration of the substrate in the sterile medium introduced into the culture.

The growth rate of the microorganism and the amount of biomass in the outlet stream is related to the flow rate, therefore its maximum value for a situation where $S = S_0$ is described by the equation:

$$D_{max} = \frac{\mu_{max} S_0}{K_S + S_0}$$

If $D > \mu_{max}$, a wash-out process occurs. It is also worth noting that for most cultures, K_S has much lower values than the substrate concentration introduced with sterile medium, i.e., $D_{max} = \mu_{max}$.

Another parameter describing continuous culture is productivity, expressed as $Qx = DX$. Maximum productivity occurs at the dilution rate:

$$D_{opt} = \mu_{max}\left(1 - \sqrt{\frac{K_S}{K_S + S_o}}\right)$$

Advantages of continuous cultures:
- Preferred for obtaining biomass and primary metabolites;
- Use smaller bioreactors compared to batch cultures.

Disadvantages:
- Risk of infection and possibility of mutation in the microorganisms used due to long working time;
- Cannot be used for filamentous fungi.

1.3.4. Bioreactors for submerged cultures

The basic device used for the cultivation of microorganisms in industrial conditions is a bioreactor. Most often it is a cylindrical vessel with different capacities, aeration and/or mechanical mixing (Figure 1.3.3). The capacity of bioreactors varies greatly, ranging from several millilitres to several thousand litres depending on the application. Bioreactors are used to culture microbial cells as well as plant, animal, and subcellular structures.

Bioreactors for anaerobic cultures are usually very simple constructions without aeration and often mixing. More complicated devices are bioreactors for aerobic cultures. These mainly include bioreactors with mechanical mixing. The use of turbine (most common), paddle or propeller stirrers allows similar culture parameters to be maintained throughout the volume of the tank. The exact type of stirrer used depends on the type of biocatalyst. Baffles located on the sides of the tank prevent the formation of vortexes, and oxygen and air are usually supplied by using a bubbler, jet nozzle or a porous disc. On the other hand, such baffles may have a negative impact on cells strongly susceptible to mechanical stress, e.g. animal cells.

Usually, the tank is designed so that the culture medium occupies up to 80% of the tank volume. The remaining part is designed for the emerging foam. If the excess foam is a significant problem during culturing, then mechanical or chemical means are used to reduce it. However, it should be remembered that the use of chemical methods may limit the access of oxygen during the culture of microorganisms.

Tanks with gas mixing are an alternative to bioreactors with mechanical mixing. In the case of the bubble column, air bubbles, pressed from the bottom, affect the movement of the liquid while heading upwards. A more complex version, with a clear division of the tank into ascending and descending parts, is the air-lift device. Bioreactors of these types are usually high columns, so that air bubbles travel a long way before they reach the surface. Their use is often associated with cell cultures that are easily affected by the shear stress resulting from stirrer operation.

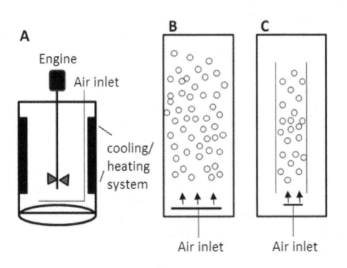

Fig. 1.3.3. Bioreactor types: a) with mechanical mixing, b) barbotage column, c) air-lift.

Another type of bioreactor is a tank with a stationary packed bed, through which the growing medium flows continuously. The particle size of the carrier is particularly important here. Very large particles slow down the mass exchange and diffusion resistance, and particles that are too small make the flow through the bed difficult.

In addition to classic bioreactors, disposable bioreactors with capacities ranging from several millilitres to several thousand litres are increasingly popular. These tanks are prepared for aeration, samples collection and mixing. Thanks to their advantages, such as low cost, sterility, no cleaning requirement, they are increasingly used in industrial conditions.

1.3.5. Stabilisation of microorganisms by immobilisation

Immobilisation consists of adsorption, chemical binding or mechanical "closure" of microbial cells or free enzymes in/on properly selected carrier substances. An important feature of the carrier used for immobilisation (in processes carried out in aqueous solutions) must be its insolubility in water, as well as a significant mechanical strength, and these are important in various technological operations. An important advantage of using immobilised enzymes or whole cells of microorganisms in biotechnology, is the possibility of multiple repetitions of the process using the same biological material, without the necessity of repeatedly multiplying the microorganisms or preparing the enzyme. In addition, the use of immobilised biological material makes it possible to carry out biotechnological processes in continuous, economically important systems.

Historically, the oldest method of microorganisms immobilisation is the deposition of acetic acid bacteria on beech chips and the use of such immobilised microorganisms for the production of food vinegar using the rack method (18th century), and then, after significant technological modifications, the development of a highly efficient generator method (early 20th century). In the 1960s, work on immobilisation of various enzymes obtained from microorganisms, useful in obtaining specific products, developed very intensively. For example, it was possible to successfully immobilise invertase and aspartase and obtain fructose from glucose and aspartic acid from ammonium fumarate.

However, if the product of interest is based on the action of enzyme complexes, or if several enzymes have to act in successive reactions, it is far more beneficial to immobilise the entire cells of the microorganisms. An additional advantage of such a procedure is that the optimum pH of the enzymatic reactions and the optimum temperature of the process is often different from the optimum conditions required for the growth of the microorganisms used for immobilisation. Thus, by using immobilised microorganisms, it is possible to become independent of the conditions of their culture.

The methods of trapping (closing) in appropriate matrix polymers are most commonly used in the immobilisation of whole microbial cells. Both specific chemicals and gels of natural origin are used as carriers. The most common ones are: polyacrylamide, polyurethane, polystyrene, and natural ones: collagen, agar, sodium alginate and χ-carrageenan.

Alginate is a polysaccharide obtained from brown marine algae, containing β-D-mannopyranosyl-uronate residues bound *via* 1–4 bonds with α-L-gulopyranosyl-uronate residues in regular sequences. In turn, χ-carrageenan, also of marine origin, is a mixture of polysaccharides, in which the main component is the α-D-galactopyranosyl residue in the form of sulphate ester.

Intensive research on the application of whole cell immobilisation in a variety of biotechnologies, developed in the 1970s, mainly in Japan and the USA.

A particularly successful solution is the method of obtaining aspartic acid from fumaric acid and ammonia using *E. coli* cells immobilised in carrageenan with active aspartase. The *E. coli* immobilizate used in this process proved to be extremely durable, its half-life was about two years.

Several systems for obtaining different amino acids have been developed using polyacrylamide gels, polyurethane, calcium alginate or other carriers. Enzymatic production of L-arginine and L-cytrulin by *Pseudomonas putida*; L-alanine from aspartic acid by *P. dacunhae*; tryptophan from indole and L-serine – by *E. coli*; glutamic acid by *Micrococcus glutamicus*, etc., are well known.

There are also systems for the production of organic acids using immobilised microorganisms, e.g. using bacteria of a genus *Brevibacterium* immobilised in a polyacrylamide gel or carrageenan, 80% of the conversion efficiency of fumaric acid to malic acid was achieved on an industrial scale.

Immobilised conidia of *Aspergillus niger* in alginine have found a wide use, e.g. for the production of citric acid in a continuous system.

Immobilised cell preparations are also used in the bioconversion of steroids (Sections 1.6 and 4.3.2). Spores, mycelium and protoplasts of *Cunninghamella elegans* immobilised in alginate gel were used to convert cortixolone (by 11α- and 11β-hydroxylation) into epihydrocortisone and hydrocortisone. *Curvularia lunata* mycelium trapped in polyacrylamide or calcium alginate was also used in the process of cortixolone 11β-hydroxylation. The cells of *Arthrobacter globiformis* immobilised in polyacrylamide were used in the bioconversion of hydrocortisone to prednisolone (dehydrogenation). Further transformation of prednisolone was also possible using immobilised microorganisms.

The technique of immobilisation of *Gluconobacter* bacteria for bioconversion of sorbitol to sorbose, raw material for the production of vitamin C, is successfully applied.

The cells of *Saccharomyces cerevisiae* and *Kluyveromyces marxianus* immobilised in alginates were also used to obtain (for running purposes) ethanol from glucose and inulin.

PRACTICAL PART

A. Practical part concerning batch and continuous cultures (on example of yeast is presented in Section 4.1.1)

B. Examination of filamentous fungi growth. Determination of specific growth factor

Materials and media

1. Microorganisms: *Cunninghamella echinulata*.
2. Media: Sabouraud liquid medium.

Aim of the exercise
Determination of the specific growth rate of the microorganism in submerged and surface cultures.

Procedure

1. Comparison of growth rate of *Cunninghamella echinulata* in submerged and surface culture in Sabouraud medium.

2. SUBMERGED CULTURE. 7-day culture of C. echinulata on a slanted Sabouraud medium. Wash off 3 ml of sterile distilled water and transfer the culture suspension to 20 ml of sterile Sabouraud liquid medium placed in a 100 ml flask. Use 6 flasks separately for the experiment. Streaking is carried out immediately after washing the slants, culture is carried out on a shaker at 28°C, then after 0, 6, 9, 12, 18 and 24 h of incubation:

   a) describe the growth of microorganisms in the culture;

   b) filter the culture from each flask through the filter paper under vacuum, wash with sterile distilled water and dry at 105°C to constant weight;

   c) determine the maximum specific growth rate.

3. SURFACE CULTURE. 7-day culture of *C. echinulata* on a slanted Sabouraud medium. Wash off 3 ml of sterile distilled water and transfer the culture suspension to 20 ml of sterile Sabouraud medium placed in a 100 ml flask. Use 6 flasks separately for the experiment. Streaking is carried out immediately after washing the slants, then after 12, 24, 36, 48, 60 and 72 h of culture it is necessary to:

   a) describe the growth of microorganisms in the culture;

   b) filter the culture from each flask through the filter paper under vacuum, wash with sterile distilled water and dry at 105°C to constant weight;

   c) determine the maximum specific growth rate.

Analysis of results

On the basis of the obtained values of the specific growth rate, explain the differences in the rate of biomass growth in both types of culture.

C. Stabilisation of microorganisms by immobilisation

Materials and media

1. Microorganism: *Saccharomyces cerevisiae*.
2. Medium: wort 12°Blg and 8°Blg.
3. Reagents: sodium alginate, carrageenan, 0.2 M solution of $CaCl_2$.

Aim of the exercise
Comparison of fermentation capacity of free yeast cells and yeast immobilised in gels.

Procedure

1. Immobilisation of *S. cerevisiae* cells in alginate.
   a) Centrifuge the 48-h shaken *S. cerevisiae* culture in 12°Blg malt wort from the medium, wash with a solution of sterile saline and then centrifuge again. Thoroughly disperse yeast precipitate in sterile distilled water (in 0.1 original volume). In the obtained suspension, count the yeast cells in the Thoma chamber and give their density per 1 ml of the immobilizate. Divide the yeast precipitate obtained from the culture into 2 equal parts;
   b) prepare a 4% sterile solution of sodium alginate;
   c) mix the yeast suspension with the alginate solution at a volume ratio of 1:1;
   d) put the cell suspension in the alginate in a sterile syringe with a needle 1 mm in diameter and a cut end; drop the suspension into the sterile 0.2 M $CaCl_2$. Leave the beads in solution for 20 min. Wash the hardened beads twice with sterile distilled water;
   e) place the beads in a flask containing 100 ml of 8°Blg of malt wort; transfer the flask to a thermostat at 28°C;
   f) carry out an identical fermentation test with S. cerevisiae free cells at the same initial density;

g) after 3 days of fermentation, determine the ethanol content in the sample with the immobilised yeast and compare it with the control sample;

h) carry out the next fermentation cycle. After removing the fermented wort and washing the beads of the immobilizate with sterile distilled water, introduce fresh medium in the same volume as in the first cycle. Repeat the fermentation test and ethanol content analysis after 3 days of fermentation. Compare the results with the first and the control sample. The following cycles can be repeated several times, the mechanical strength of the beads as well as the enzymatic activity of the immobilised yeast is at least 12 weeks.

2. Immobilisation of *S. cerevisiae* cells in χ-carrageenan. Prepare the yeast suspension as described in point 1a.

a) Preparation of the χ-carrageenan solution. Prepare hot (60°C) 4% carrageenan solution in 0.9% NaCl;

b) Thoroughly mix the warm carrageenan solution (40°C) with the yeast suspension at a ratio of 9:1; maintain at 40°C;

c) Place the cell suspension in the carrageenan in a syringe (as in point 1d) and drop it in a sterile 2% KCl solution. Leave the beads in the solution for 20 minutes. Rinse the beads created twice with sterile distilled water;

d) Continue as in 1 e-h.

1.4. Microorganisms culture in anaerobic conditions

INTRODUCTION

Anaerobic bacteria play a key role in the biogeochemical cycle of many important elements (e.g. carbon (C), nitrogen (N), phosphorus (P), sulphur (S), iron (Fe)), contributing to the biological balance in both natural and man-made (anthropogenic) ecosystems.

Many anaerobic species are pathogens that are the basic etiological agent of many human and animal infectious diseases.

An essential condition for anaerobe cultivation is to create and then maintain an atmosphere without molecular oxygen. The classification of anaerobic culture methods is based primarily on the type of procedures used to create and maintain an oxygen-free growth environment, with particular emphasis on aspects such as: (1) obtaining a vacuum, (2) replacing air with gases that do not contain molecular oxygen, (3) oxygen reduction, (4) absorption of oxygen by chemical agents, (5) restricting access to atmospheric oxygen using physical principles or mechanical devices. The techniques most commonly used to culture microorganisms under anaerobic conditions are divided into: chemical, biological, physical and mixed techniques.

Chemical methods are mainly based on removing oxygen by means of a reducing agent. For this purpose, chemical compounds are placed in jars or other vessels in which anaerobic cultures are carried out. These compounds include pyrogalol or KOH. The cultures also use agents such as dithiothreitol, L-cysteine hydrochloride, 2-mercaptoethane sulfonate (coenzyme M), sodium sulphite, sodium thiosulphate, glutathione or titanium (III) citrate, which can be introduced directly into liquid or solidified agar media. However, due to different chemical and biological properties of the reducing agents used (e.g. redox potential, toxicity), they may inhibit the growth of certain groups of anaerobes.

Biological techniques of anaerobic microorganisms culturing are mainly based on the Fortner's method (the so-called Fortner's plates). It is based on the common culturing (side by side) of anaerobes and aerophilic microorganisms. After some time, aerobic microorganisms, using oxygen during their development, create the conditions conducive to the growth of anaerobes.

The most frequently used physical and mixed (physicochemical) methods of obtaining anaerobic conditions include culturing in specially prepared devices called anaerostats.

Anaerobic conditions are obtained in these vessels by: (1) pumping out the air using a vacuum pump and then introducing the anaerobic gas or gases through a system of appropriate valves and/or (2) using chemical methods to generate an anaerobic atmosphere based on simple chemical reactions and metallic catalysts. The following are mainly used in procedures with chemical agents: sodium boride, palladium catalyst and CO_2

tablet containing additionally sodium bicarbonate and citric acid, which supplement the anaerobic atmosphere with hydrogen. Disposable, commercially available chemical anaerobic gas generators such as GasPak, AnaeroGen or Anaerocult are widely used and they are mostly in the form of sachets or envelopes placed inside the culture jars. In order to control the anaerobic conditions in the anaerostate, appropriate indicators are often introduced, e.g. based on the reaction of methylene blue, which occurs in oxidised form in blue and becomes colourless when hydrogen is added.

One of the standards of obtaining anaerobic conditions by physical methods is the ANOXOMAT system, which is based on a device that allows preparation of the culture in anaerobic conditions or with low oxygen content by means of a fully automated mechanism of air pumping and atmospheric composition exchange in anaerobic jars. The process of achieving the desired growth conditions for microorganisms that are sensitive to different oxygen concentration in the environment, is controlled at each stage of the process through a microprocessor system, ensuring precise, fast and reproducible achievement and maintenance of anaerobic or low oxygen conditions.

Anaerobic chambers are also widely used in anaerobic cultures, which minimise the exposure of anaerobic microorganisms to oxygen. Anaerobic chambers are equipped with specialised systems supporting control of the process of creating and maintaining anaerobic atmosphere. They are also equipped with sets of sleeves or cuffs attached to hand access ports and a sluice for material insertion and removal.

In addition, bioreactors are also used for anaerobic culture, mainly due to the continuous feeding of substrates and removal of by-products, which allows stable conditions to be achieved for microorganism growth, while reproducing the natural environmental conditions.

Many methods of anaerobes cultivation are based on modifications of already existing techniques, which often boil down to the separation of specific groups of microorganisms. These improvements include changes in the composition of culture media (e.g. the use of antibiotics to facilitate the growth of methanogenic archaeons), the introduction of new cocultures or the use of in-situ/in vivo cultures, which by mapping living conditions almost identical to the natural ones allow the growth of anaerobes, causing many difficulties in culturing. Improved anaerobic culture procedures also include the method using 6-well plates and anaerobic gas packaging system. This technique is based on an inoculation of a specific dilution of the tested material, mixed previously with a growth medium containing gellan gum on a 6-well plate, and then placing the plate with the appropriate catalysts in a sealed bag and creating an anaerobic atmosphere through a needle inserted into the bag. This method is particularly useful for the culture of obligate anaerobes, including methanogens and sulphate-reducing bacteria.

Bacteria of the genus *Propionibacterium* belong to the Acinobacteria phylum. They are characterised by high heterogeneity in terms of morphological and biochemical features and nutritional requirements. Their cells can take different shapes (pleomorphism) depending on the growth phase and the amount of oxygen in the environment. They are mostly cylindrical, but can also take spherical or Y or V-like forms. All *Propionibacterium* species are non-sporulating, have no motion organelles and belong to Gram-positive bacteria. It should be noted, however, that some strains, in the initial stage of growth, may be Gram-variable. They require anaerobic conditions for growth, although most can tolerate low oxygen concentrations in the environment and therefore belong to microaerophilic bacteria.

The species belonging to Propionibacterium are conventionally divided into two groups: the so-called "classical", also known as lactic, isolated from cheese and other dairy products, but also from the soil or natural products such as orange juice; and the "skin" bacteria belonging to the exo- and endogenous bacterial microbiota that inhabit the skin and mucous membranes of the mouth or digestive tract. There are 13 species of this genus, of which 6 belong to the "classical" bacteria (*P. thoenii, P. freudenreichii, P. jensenii, P. cyclohexanicum, P. microaerophilum and P. acidipropionici*), and 7 to the "skin" bacteria (*P. acnes, P. granulosum, P. australiens, P. propionicus, P. humerusii, P. acidifaciens and P. avidum*).

Due to its many beneficial classical properties, *Propionibacterium* bacteria are used in many industries. The greatest potential is found in the food industry and in the general protection of human and animal health. Due to their health-promoting properties, such as production of vitamin B_{12} and B_9, propionic acid, trehalose, stimulating the growth of other probiotic strains or synthesis of bacteriocins with a wide range of antimicrobial activity, they can act as probiotic preparations or vitamin and protein supplements, positively affecting the body by acting in the digestive tract. In addition, "classical" bacteria can increase the bioavailability of many minerals and proteins, reduce symptoms of lactose intolerance and counteract intestinal complaints such as constipation and diarrhoea. On an industrial scale, dairy Propionibacterium are used in the production of vitamin B_{12}, animal nutrition as a feed and silage additive, as well as in the dairy, bakery and cheese industries because of their ability to produce propionic and acetic acid and CO_2 as a by-product of fermentation. These properties are particularly useful in the production of hard rennet cheese, fermented acidified products or vegetable salads. Carbon dioxide is responsible, among others, for the hole forming process in some cheese, while propionic and acetic acid have strong antimicrobial properties and give the products appropriate organoleptic features as well as prevent the formation of unpleasant smell and taste.

PRACTICAL PART

Materials and media

1. Microorganisms: *Propionibacterium freudenreichii* ssp. *shermanii* DSMZ 4902.
2. Media: with yeast extract, casein medium, medium with casein hydrolysate.
3. Reagents: standard substances (vitamin B_{12}, lactic acid, acetic acid, propionic acid), cobalt chloride ($CoCl_2 \times 6\ H_2O$).
4. Apparatus: ANOXOMAT device for preparation and culture of bacteria in anaerobic and microaerophilic atmosphere (6% O_2), jars for culture of bacteria in anaerobic and microaerophilic conditions, incubator, rotary shaker, liquid chromatograph coupled with mass spectrometer (LC/MS), laboratory centrifuge, mechanical disintegrator.

5. Other materials: serological pipettes, automatic pipettes, sterile tips for automatic pipettes, Erlenmeyer flasks (100 ml).

Aim of the exercise

Effects of oxygen content and the type of culture medium on the efficiency of the biosynthesis of vitamin B_{12} and organic acids (lactic, acetic and propionic) by Propionibacterium freudenreichii ssp. shermanii.

Procedure

1. Establishing the Propionibacterium cultures: establish Propionibacterium cultures on a liquid medium with yeast extract, medium with casein hydrolysate and casein in 100 ml flasks. For this purpose, inoculate 20 ml of the medium with a 24 h material collected with a loop from a 24 h bacterial culture on a medium with yeast extract. Use ANOXOMAT to create anaerobic and microaerophilic conditions. Add cobalt salts (10 mg/l $CoCl_2 \times 6\ H_2O$) to all cultures. The cultures should be conducted in anaerobic jars and under aerobic conditions for 6 days at 28°C, on a rotary shaker at 160 rpm under the following conditions:
   a) anaerobically,
   b) aerobically,
   c) 3 days anaerobically and then 3 days under aerobic conditions,
   d) constantly in an atmosphere with a reduced oxygen content of 6%,
   e) 3 days in an atmosphere with a reduced oxygen content of 6%, and then 3 days under aerobic conditions.
2. Preparation of standard curves: Standard curves of organic acids and vitamin B_{12} for the determination of samples tested by liquid chromatography (LC-MS) should be prepared according to the conditions recommended by the manufacturer of the apparatus for individual analysed substrates. The separation should be conducted using a C18 phase chromatographic column.

3. Quantitative determination of substrates in cultures: after 6 days of incubation, centrifuge all tested culture systems (6,000 × g, 15 min), filter the supernatants through 0.2 µm pore diameter filters and then determine the content of tested acids and vitamins by liquid chromatography (according to the same method as the standard curves). In the case of vitamin B$_{12}$ biosynthesis analysis, the bacterial cultures are additionally incubated 20 min. at 80°C, then homogenised using a mechanical disintegrator, then using the same procedure as used for the samples for quantitative determination of organic acids.

Analysis of results

On the basis of the quantitative results obtained from the chromatographic analysis, draw conclusions concerning the influence of oxygen content and the composition of the growth medium on the production yield of vitamin B12 and organic acids by the *P. freudenreichii* ssp. *shermanii* strain.

1.5. Fungal protoplasts: release, properties and application

INTRODUCTION

The term "protoplast" is used to describe a bacterial, fungal or plant cell completely devoid of a cell wall, but has a cell membrane as the only external shield. The term "spheroplast", on the other hand, refers to a cell on the surface of which fragments of the cell wall are present. Protoplasts and spheroplasts do not occur under natural conditions. The exceptions are strains (that are pathogenic to insects) of certain microfungi of the Entomophthoromycotina subtype, Zoopagomycota (formerly Zygomycota), where protoplastic and spheroplastic cells are present during one of the stages of development. Additional data on this group of fungi are provided in Section 5.13. Compared to the baseline cells, the physiological and biochemical properties of protoplasts are similar. The exceptions are the rate of DNA synthesis (higher in the baseline cells than in protoplasts) and the rate of processes in which the cell wall is a barrier to substrate access to intracellular enzymes. In this case, processes such as e.g. hydroxylation reactions of steroid compounds are faster using protoplasts.

The characteristic features of protoplasts include:
- Spherical shape;
- Lack of mechanical resistance – for this reason it is necessary to use low shaking values (below 90 rpm) during the process of protoplast release and delicate conditions of protoplasts separation from undigested mycelium biomass (centrifugation rate should not exceed 500 × g);
- Sensitivity to osmotic pressure – protoplasts placed in water, crack within seconds. Therefore, during the release of protoplasts and in subsequent stages of testing, it is necessary to use a hypertonic environment, provided by the use of osmotic stabilisers. Osmotic stabilisers are aqueous solutions (0.4–1.0 M) of inorganic (e.g. $MgSO_4$, $CaCl_2$, KCl) or organic (e.g. mannitol, sorbitol, sorbose) compounds. Usually, inorganic osmotic stabilisers; (e.g. 0.6–0.8 M KCl) are used to obtain protoplasts of filamentous fungi, whereas the release of yeast protoplasts is favoured by an organic compound environment of 0.8–1.2 M.

The first scientific reports on the isolation of yeast and fungi protoplasts come from the 1960s. Research conducted at that time focused on developing efficient methods of bacterial and fungal cell wall disintegration to isolate and characterise intracellular proteins and genetic material. In the following years, the scope of research using protoplasts and spheroplasts was extended.

Directions for the practical use of protoplasts and spheroplasts include:
- examination of the structure and course of cell wall synthesis – after obtaining the protoplasts, a gradual reversal (the process of cell wall reconstruction) enables characterisation of subsequent stages of cell wall reconstruction and the enzymes involved in the above process;
- isolation and characterisation of cell organelles;

- determination of genetic homology – based on comparison of similarities of large sections of DNA isolated from protoplasts; the results of the above research were used in taxonomy – to describe taxonomic units and to establish an affinity of microorganisms (especially fungi);
- studies of metabolic processes (including the biosynthesis of certain antibiotics and mycotoxins and the transformation of steroid compounds), where the cell wall may constitute a barrier that significantly limits the rate of substrate intake into the cell or secretion of product into the growth environment;
- increase in the efficiency of microbial metabolic processes of high commercial significance (so-called strain improvement), in which methods of mutagenisation, recombination and fusion of protoplasts are used (additional data are provided in Section 1.6).

The methods developed so far for removing the cell wall to obtain protoplasts are based on the use of physical or biological factors. Physical methods consist in mechanical disintegration of bacterial cells and mycelium hyphae, and were used mainly in the initial stage of protoplasts research. This was due to low productivity and also because of the adverse effect on the physiology of the obtained cells. Biological methods use enzymes capable of decomposing components of the microbial cell wall. In the case of filamentous fungi, these are so-called lytic complexes, which are a mixture of many enzymes. These enzymes are mostly inductive in nature and catalyse the decomposition of various compounds (mainly polysaccharides) that apart of the cell wall or that are included in cell wall.

The following can be used to obtain lytic enzymes:
- enzymes of the digestive juice of the edible snail (*Helix pomatia*), previously fed with biomass of such microorganisms from which protoplasts are required;
- culture liquids obtained from the culture of some microorganisms – soil bacteria (mainly Actinomycetes of *Streptomyces* and *Micromonospora* genera) and micromycetes strains of *Trichoderma* genus (including, above all, *T. viride* and *T. harzianum* species). Since, as previously indicated, lytic enzymes are mainly inductive

enzymes, culture media for the multiplication of the above bacteria and fungi must contain, as the main source of carbon and energy, the biomass of these microorganisms from which protoplasts will be obtained. Additional data on the formation of lytic complexes using microorganisms are provided in Section 4.1.3.

Figures 8.1.6. J-K show the digestion of mycelium from *Cunninghamella echinulata* IM1785 21Gp with a lytic complex obtained from *Trichoderma viride*. Both spheroplasts and protoplasts of *C. echinulata* are visible in the figure.

The yield of protoplasts obtained depends on many factors, the most important being:

- type and quality of the lytic enzyme (these features are influenced by a number of factors including the source of the enzymes, the way they are obtained and the presence of impurities, such as proteolytic enzymes);
- type, concentration and pH of the osmotic stabiliser used during cell wall digestion (affects the activity of lytic enzymes and, by ensuring adequate osmotic pressure and environmental pH, the performance and metabolic activity of released protoplasts);
- process temperature;
- age and composition of cell wall of the microorganism from which the protoplasts are obtained (this depends, among others, on the type of growth medium and the time of cultivation; usually, cultivation conducted in a medium poor in nutrients and for a time allowing the culture to reach only the growth inhibition phase allows biomass to be obtained, which is an efficient inductor of the production of the lytic complex).

Reversion of protoplasts – this term describes the process of cell wall reconstruction and the return of protoplasts to their shape and characteristics of normal cells. The frequency of reversion never reaches 100%. Depending on the origin of the protoplasts (taxonomic type) and the conditions used for cell wall reconstruction, 5 to about 85% of the protoplasts are subject to reversion. The process of cell wall reconstruction is accompanied by an increase in cell size, doubling of genetic material and loss of protoplastic characteristics – spherical shape and sensitivity to osmotic shock.

PRACTICAL PART

Materials and media

1. Microorganisms: microscopic filamentous fungal strains: *Curvularia lunata* and *Cunninghamella echinulata*.
2. Media: PL-2; PL-2 with an addition of 0.8 M KCl.
3. Reagents: lytic complex: enzymes degrading the cell wall of mycelium, for which the inductor was an autoclaved mycelium of *C. lunata* or *C. echinulata*; 0.8 M KCl.
4. Apparatus: incubator for shaking cultures, centrifuge, microscope.

Aim of the exercise
Evaluation of the course of protoplast release, description of characteristics of protoplasts, spheroplasts and initial mycelium, as well as the reversion of protoplasts.

Procedure

1. Perform microscopic observations (staining of living cells, magnification 400 times) of mycelium of *C. lunata* and *C. echinulata*, which were obtained after 24 h of screening cultures in PL-2 medium, under aeration conditions (120 rpm) and at 28°C.
2. Prepare the mycelium biomass for digestion with a lytic complex (enzymes degrading the mycelium cell wall), for which the inductor was an autoclaved mycelium of *C. lunata* or *C. echinulata*:
   a) separate the mycelium biomass by centrifugation or filtration;
   b) rinse the obtained mycelium biomass with an osmotic stabiliser (0.8 M KCl);
   c) place mycelium in flasks containing lytic complex (8–20 mg/ml) dissolved in osmotic stabiliser;
   d) incubate sample for 2–18 h under shaking conditions (70 rpm).

3. Perform microscopic observations:
   a) mycelium digested with a lytic complex and released protoplasts (after 2 and 18 h of digestion);
   b) protoplasts subjected to osmotic shock;
   c) reverted protoplasts after 6 and 18 h of incubation (90 rpm) of the protoplasts suspension in PL-2 medium with the addition of 0.8 M KCl.

Analysis of results

Draw conclusions about the effectiveness of the release and characteristics of protoplasts, as well as the course of protoplast reversion.

1.6. Strain development – mutagenesis, fusion and electroporation of protoplasts

Microorganisms obtained from the natural environment conduct biosynthesis, degradation or bioconversion processes with an efficiency usually insufficient for the technological process and causing economic losses. The metabolic activity of wild strains can be partially changed by the selection of culture conditions and an increase in the yield of the desired product can be achieved. However, fundamental metabolic and physiological changes can only be achieved through modification of the microbial genotype. In industrial laboratories, classical methods of mutagenisation and selection are most commonly used. Genetic engineering methods are also used at the cellular level (fusion of protoplasts) and subcellular level (genetic transformation, electrotransformation), allowing for recombination of genetic information from different organisms.

Improvement of industrial strains through mutagenesis

One of the basic ways of improving industrial strains is mutagenesis and selection of mutants with beneficial properties for a given biotechnological process. There is a large number of chemical and physical mutagenic factors known, such as nitrous acid, gaseous yperite, nitrogenous base analogues, alkylating compounds, acridine dyes, UV, X and gamma rays, fast neutrons. The types of mutation depend on the type of mutagen, and one or more different types of damage may occur to the DNA structure, e.g. N-methyl-N'-nitro-N-nitrosoguanidine is a methylating agent, UV radiation mainly causes dimerisation of pyrimidine bases, as well as hydroxylation of these bases and crosslinking of double threaded DNA. The effectiveness of the mutagenesis process depends on the type of factor used, its concentration, growth phase, conditions of microbial culture prior to the exposure of cells to the mutagen, the composition of the environment during the process, as well as the efficiency of repair mechanisms. Selected mutants are subjected to precise characterisation using different culture conditions. Then, optimal process conditions for the best strain are developed.

The constantly increasing demand for steroid medications (see Section 4.3.2), as well as the scarcity of traditional industrial raw material for their production – diosgenin (steroid glycoside obtained from a plant belonging to the genus *Dioscorea*), has led to interest in plant and animal sterols. Sterols are 3-monohydroxysteroids with a side chain of 8, 9 or 10 carbon atoms in C-17 (see Section 4.3.2).

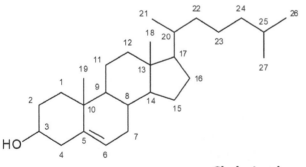

Cholesterol

Fig. 1.6.1. Carbon numbering of cholesterol molecule.

The use of sterols for the production of steroid medications first requires shortening or complete removal of the aliphatic chain associated with C-17 of the steroid molecule (Figure 1.6.1). Conventional chemical degradation is not economical and leads to the formation of numerous by-products in a multistage process. The microbiological degradation of sterols enables the desired multistage transformation to be carried out selectively during the single-step process in a non-aggressive environment. An extensive list of microorganisms capable of sterol degradation is described, and they belong among others to the following genera: *Arthrobacter, Mycobacterium, Rhodococcus, Corynebacterium, Nocardia, Streptomyces.* In particular, products with the ketone group in position 17 and the carboxyl group in position 20 are of industrial importance and can be directly applied in the treatment as well as converted by microbiological and/or chemical means to derivatives with valuable therapeutic properties (Figure 1.6.2).

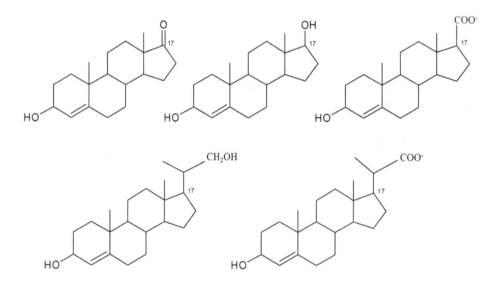

Attempts to use microorganisms for targeted degradation of sterols have been derived from research on the use of cholesterol as the only source of carbon for microorganisms. In most of the examined microorganisms, the inductive nature of enzymes responsible for sterols degradation was found. Microorganisms can use both the ring structure and aliphatic side chain as a source of carbon and energy. Degradation of the ring arrangement of sterols, for example, androst-4-ene-3,17-dione (AD) (Figure 1.6.3), is preceded by oxidation to 3-oxoderivatives and then double bond isomerisation (Δ5-Δ4).

Fig. 1.6.2. Steroid products of industrial importance.

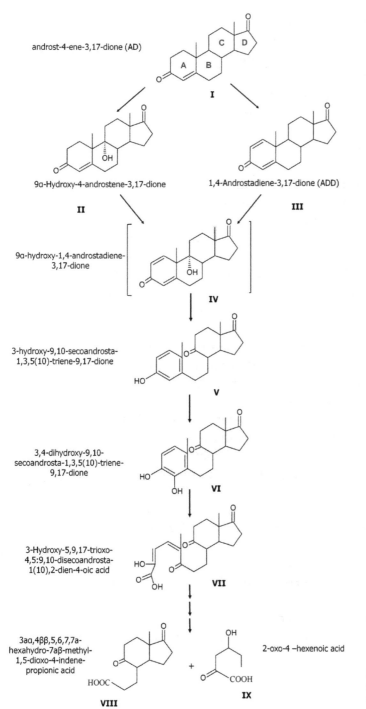

androst-4-ene-3,17-dione (AD)

I

9α-Hydroxy-4-androstene-3,17-dione

II

1,4-Androstadiene-3,17-dione (ADD)

III

9α-hydroxy-1,4-androstadiene-3,17-dione

IV

3-hydroxy-9,10-secoandrosta-1,3,5(10)-triene-9,17-dione

V

3,4-dihydroxy-9,10-secoandrosta-1,3,5(10)-triene-9,17-dione

VI

3-Hydroxy-5,9,17-trioxo-4,5:9,10-disecoandrosta-1(10),2-dien-4-oic acid

VII

3aα,4ββ,5,6,7,7a-hexahydro-7aβ-methyl-1,5-dioxo-4-indene-propionic acid

VIII

2-oxo-4 –hexenoic acid

IX

Fig. 1.6.3. Scheme of steroids ring structure distribution.

Both these reactions can be enzymatic or non-enzymatic. Reactions initiating the ring structure splitting are 9-α-hydroxylation and dehydrogenation between 1 and 2 carbon atoms. The order of action of these enzymes is different and depends on the type of microorganisms. The resulting metabolite (IV) is simultaneously aromatised in ring A and split in ring B. A 6-carbon fragment (IX) is separated, whose cleavage leads to pyruvate and propionic aldehyde. Further microbiological degradation occurs with the opening of the remaining steroid rings D and C and the release of Krebs cycle acids. The process of microbiological degradation of side chains of sterols is similar in many microorganisms to β-oxidation of fatty acids. Degradation of the side chain (bile acids, cholesterol, sitosterol) is preceded by conversion to Δ4–3-one derivatives. Stages of the side chain shortening involve splitting the bonds between C-24 and C-25, C-22 and C-23 and C-17 and C-20, with simultaneous separation of propionic acid, acetic acid and again propionic acid (Figure 1.6.4).

The processes of degradation of the ring structure of sterols and their side chain do not occur in the established order, but usually simultaneously and independently of each other.

Screening tests showed high prevalence of microorganisms breaking down cholesterol and other sterols. However, it is extremely rare to isolate microorganisms that have the ability to selectively break down alkyl side chains without breaking down the ring structure of sterols. Reactions initiating ring breakdown are 9-α-hydroxylase and 1(2)-dehydrogenase. Different ways of blocking the activity of these enzymes have become the basis for three methods enabling selective microbiological degradation of sterols side chains:

Structural modifications of the substrate to prevent enzymatic attack on the ring system are presented in Figure 1.6.5.

Application of enzyme inhibitors initiating degradation of the ring structure, e.g. 8-hydroxyquinoline, 1,10-phenyantroline, 2,2'-dipyridyl.

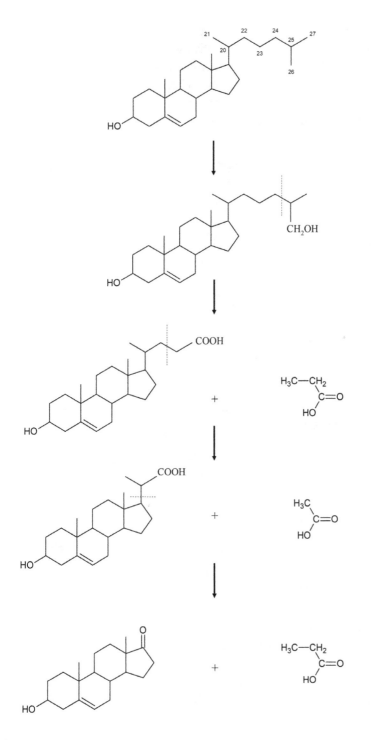

Fig. 1.6.4. Degradation of cholesterol side chain

Mutagenesis of microorganisms that completely decompose sterols and isolation of mutants that have permanently lost their susceptibility to 9-α-hydroxylation and 1(2)-dehydrogenation. Studies indicate significant potential for stimulating microbiological transformation of sterols to desired products, such as: 17-ketosteroids, 22-carbon acids and perhydroindane derivatives.

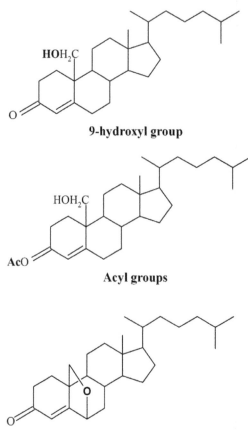

Fig. 1.6.5. Examples of structural modifications of the substrate.

9-hydroxyl group

Acyl groups

Oxygen bridge

These metabolites are potential intermediate products for further partial synthesis of steroid medications. AD is the starting material for androgens and diuretic drug, spironolactone, while ADD (androsta-1,4-dien-3,17-dione) is used for the production of estrone and 19-nor derivatives for the production of oral contraceptives.

PRACTICAL PART

Materials and media

1. Microorganisms: *Rhodococcus* sp. IM 58 isolated from the soil. This microorganism is able to grow in media with cholesterol as the only source of carbon and energy. It attacks the cholesterol molecule from both the alkyl side chain and the ring skeleton. It also metabolises other steroids without side chain (e.g. AD).

2. Media: liquid 1a – for multiplication and transformation, agar plates and slants, selective – plates with mineral medium A – with added cholesterol (0.5 g/l) as the only source of carbon and energy, B – with added AD as the only source of carbon, agar slants.

3. Reagents: mutagen N-methyl-N'-nitro-N-nitrosoguanidine (MNNG).

4. Apparatus: set HPLC Hewlett-Packard 1100 Series.

Aim of the exercise
Growth and selection of mutants that selectively degrade the side chain of cholesterol (without breaking its ring structure) with the production of desired 19 and 22-carbon steroid products.

Procedure

1. Take 2 ml of the 18 h culture of the starting strain and place in 4 sterile centrifuge tubes. Add 1, 4, 8, 12 mg of mutagen (MNNG) dissolved in 0.1 ml acetone to each. **CAUTION: Use gloves and glasses when working with the mutagen.**

   Incubate the tubes for 60 min under the conditions of the previous initial culture. After incubation, add 5 ml of sterile 0.85% NaCl solution to dilute the mutagen and then centrifuge the cells at 3,000 × g for 10 min. Remove the centrifugate, wash (twice) cells in 2 ml of saline, mix thoroughly with the centrifugate and make a series of dilutions of the resulting suspension (10^2, 10^3, 10^4, 10^5, 10^6). From the dilutions 10^4, 10^5 and 10^6, streak 0.1 ml on the agar plates (5–10 plates of each dilution) and spread with a glass spreader. Incubate the plates 72 h at 30°C.

2. After growth of the colonies on agar plates (as single colonies) draw a survival curve depending on the mutagen concentration and transfer as many colonies as possible onto selective media A and B.

   Divide the plates with media A and B into sectors and then streak each colony from the agar plate (point-by-point, using a straight loop) onto media A and B in identical numbered sectors.

3. After 2–7 days of incubation at culture temperature, select the potential mutants for further testing. Selected clones should be transferred from cholesterol media into agar slants. Mutants should be selected according to the following criteria:

   a) mutants growing on a medium with cholesterol, but not growing on a medium with AD;

   b) mutants growing less or more slowly on a medium with AD than on a medium with cholesterol.

4. Examination of transformation capabilities of selected mutants.

   a) Preparation of transformation samples.

   Rinse the culture of analysed mutant with 3 ml of medium 1a and inoculate the obtained suspension with 50 ml of the same sterile medium in a 1 litre flask. Add cholesterol at a concentration of 1 g/l of medium to the flask as a transformation substrate.

   Preparation of the substrate: 250 mg of cholesterol dissolve hot in 2 ml of ethanol, add 4 ml of 1% Tween 80 solution and micronise with ultrasound for 30 min.

   After adding the substrate, incubate the flasks for 48 h at 30°C. Conduct the control culture in the same way.

   b) After completion of the biotransformation, extract 10 ml samples 3 times with 10 ml of dichloromethane. Dehydrate the extracts with anhydrous sodium sulphate, filter and evaporate to dryness on a vacuum evaporator.

c) Chromatographic analysis of the extracts.
Dissolve the dry residue in 1 ml of chloroform:ethanol mixture (1:1 v/v) and determine steroids content using HPLC Hewlett-Packard 1100 Series, UV-VIS detector (254 nm), ODS Hypersil 5 mm × 125 × 4 mm column, mobile phase methanol:water (60:40), flow 1 ml/min.

Analysis of results

Based on the chromatograms, draw conclusions about the ability of the selected clones to degrade cholesterol in relation to the initial strain.

Improvement of industrial strains using protoplast fusion

The process of mutation and selection (often multiple times) on the one hand, allows the attainment of highly efficient mutants, on the other hand, it often contributes to the acquisition of many unfavourable features that make it difficult to use new strains on an industrial scale. The removal of undesirable changes in the microbial genotype can be done by crossing mutants with the initial wild strain or crossing different mutants with each other. Sexual or parasexual processes are relatively rare in industrial microorganisms. Therefore, protoplast fusion is used for cell combination and genetic material recombination.

Fusion of protoplasts suspended in osmotic stabilisers rarely occurs due to strong negative charge on the surface of the protoplasts. The addition of PEG (polyethylene glycol) and Ca ions favours the fusion of protoplasts. Most often 30% PEG solutions, with a molecular weight of 6000, containing 0.01–0.1 M $CaCl_2$ are used. The frequency of protoplasts fusion with PEG is 60%. The use of electric impulses (electrofusion) of 10 kV/cm, lasting several milliseconds, for protoplasts to connect, increases the frequency of fusion up to 90%.

The fusion of genetic material of combined protoplasts is much less frequently observed. In fungi, diploid nuclei are formed in at most, a few percent of the protoplasts which were subject to fusion. Recombination of DNA fragments between chromosomes (genophores) of parental forms occurs most easily when a protoplast (mutants) of the same strain or species closely related to each other is used for crossbreeding. In the case of fusion of protoplasts of taxonomically distant organisms, the resulting hybrids are most often haploidised with the production of parental genotypes.

After the fusion, the protoplasts are transferred to a medium containing, apart from the osmotic stabiliser, components enabling cell wall regeneration and reversion to cells with external shields.

Separation of recombinants from cells that have not been subject to fusion facilitates the use of strains with different morphological characteristics (e.g. colour of mycelium, conidia), dietary requirements (auxotrophic mutants) or drug resistance (antibiotics, sulfonamides).

Many industrial strains and especially of micromycetes and Actinomycetes have been improved by protoplast fusion, e.g. *Phanerochaete chrysosporum, Streptomyces* sp., and *Trichoderma reesei* used for lignin and cellulose degradation; *Streptomyces antibioticus, Cephalosporium acremonium* and *Penicillium chrysogenum* for antibiotics; (*Aspergillus sojae* for proteases and glutaminases); *Candida tropicalis* and *C. boidinii* for utilisation of n-alkanes and methanol; *Phaffia rhodozyma* for production of carotenoids, *Aspergillus niger* for citric acid biosynthesis; and *Mycobacterium* sp. and *Cunninghamella elegans* for steroids conversion.

The fusion of protoplasts was also used to recombine DNA microorganisms belonging to different species and even genera. An example is the hybridisation of industrial strains of *Saccharomyces diasticus* – hydrolysing starch, and *S. cerevisiae* using glucose, maltose and sucrose in ethanol fermentation. The recombinants obtained produced ethanol from a starch-containing medium.

PRACTICAL PART

Materials and media

1. Microorganisms: protoplasts of *Cunninghamella echinulata* (mutant A *lys$^-$*, mutant B *ala$^-$*) with density 10^8/ml.
2. Media: solid EG and Czapek-Dox in form of bars.

3. Reagents: 30% PEG-6000 in 0.1 M $CaCl_2$, 0.8 M KCl with an addition of 50 mM $CaCl_2$.

Aim of the exercise
Performing the fusion of protoplasts of the auxotrophic mutants C. echinulata lys⁻ (A) and C. echinulata ala⁻ (B) in order to obtain recombinants capable of growing on a minimal medium.

Procedure

1. Rinse the protoplasts of *C. echinulata*, clone A and B, 3 times with 0.8 M KCl with 50 mM $CaCl_2$ addition, centrifuge (1,000 × g, 5 min). Combine 1.0 ml of clone A and B protoplasts with 10^8/ml density, mix gently, centrifuge (1,000 × g, 5 min).
2. Add 1 ml of 30% PEG-6000 in 0.1 M $CaCl_2$ to protoplasts precipitate, mix gently and leave at room temperature for 15 min.
3. After that time, make a series of dilutions in 0.8 M KCl (10^1 to 10^6). From the prepared dilutions, take 0.1 ml of the protoplasts suspension and introduce it into the bars with stabilised EG medium (9.9 ml) and similarly into the bars with stabilised minimal medium (Czapek-Dox). Then pour the protoplasts suspended in the media onto sterile Petri dishes and incubate at 28°C after solidifying the media.
4. After 7 days of incubation, count the colonies grown on EG medium, give the number of reversible protoplasts (in percent, relative to the density counted in the Thoma chamber). Count also the colonies grown on the minimal medium and give the fusion efficiency (in percent, assuming the number of colonies grown on EG medium as 100%).
5. Transfer the obtained recombinants onto plates with minimal medium in order to obtain permanent recombinants able to grow on minimal medium.
6. Cheque the ability of the isolated strains to hydroxylate steroids (procedure as in Section 4.3.2).

Improvement of industrial strains using electrotransformation

Sexual reproduction, parasexual processes and protoplast fusion allow for genetic recombination (the formation of new gene combinations) only within the same species or organisms closely related to each other. The development of in vitro DNA recombination techniques, also referred to as genetic engineering methods at the subcellular level, allow introduction into the recipient's organism, DNA fragments of a specific sequence, originating from living organisms (microorganisms, plants, animals) or artificially synthesised, and result in the creation of gene combinations that do not occur naturally.

In vitro DNA recombination technology includes:
- isolation and purification of natural DNA and/or synthesis of artificial genes;
- incorporation of desired DNA sequence into the vector;
- introduction of the carrier together with the insert into the recipient's cell (genetic transformation, electrotransformation);
- cloning and detection of the expression of introduced genes.

Electrotransformation is a process in which the cytoplasmic membrane is subjected to short, strong electrical impulses, which causes its temporary destabilisation, leading to an increase in its permeability to the DNA molecules. For the destabilisation of the cytoplasmic membrane to be a reversible process, the duration of electrical impulses should be from micro to milliseconds and the intensity from 1 to 20 kV/cm. Above these values, the membrane and consequently the cell is irreversibly destroyed.

This method allows the introduction of foreign DNA into the cells of bacteria, yeast, fungi, plant and animal cells. Both protoplasts and whole vegetative cells, sprouting spores and mycelium hyphae can be electrotransformed. The highest efficiency of electrotransformation was achieved with *E. coli*. Under optimal conditions, 10^{10} transformants per μg of DNA were obtained. This process already takes place with a lower yield in Gram-positive bacteria, which is related to cell wall construction – the highest yield is in this case 10^{6} transformants per μg of DNA. Even lower yields have been reported for *Saccharomyces cerevisiae*,

Schizosaccharomyces pombe about 10^5 recombinants per μg DNA and for filamentous fungi (despite the use of protoplasts) from 2 to 100 electrotransformants per μg DNA, depending on the strain and vector used. Better results (10^3 electrotransformants per μg of DNA) were obtained for *Penicillium urticae* and *Leptosphaeria maculans*, subjecting sprouting spores of the mentioned strains to the process of electrotransformation.

When using the electrotransformation method, much better performance is always obtained compared to the traditional transformation method using polyethylene glycol (PEG). The traditional method of genetic transformation of fungi uses protoplasts to which foreign DNA and PEG of appropriate molecular weight (determined experimentally for the tested strain) are added. However, the yields obtained by this method were not satisfactory.

In the case of *Curvularia lunata* strain, which was transformed with plasmid DNA pAN7−1, only 0.1 transformant per μg of DNA was obtained. For *Cochliobolus lunatus*, on the other hand, yields ranging from 0.15−0.29 transformant per μg of DNA were obtained. Moreover, PEG is often toxic to the cells undergoing transformation, which makes it impossible to obtain the appropriate number of transformants. When electrotransformation is used, such a problem does not exist.

Electrotransformation is a convenient method to bestow new characteristics to these organisms. The obtained transformants can also be used as a starting material for studies on the organisation of genes in the chromosome, their regulation and expression, especially in the case of filamentous fungi.

PRACTICAL PART

Materials and media

1. Microorganisms: *Curvularia lunata* IM 2901.
2. Media: solid EG (with addition of osmotic stabiliser – in the form of bars), liquid EG (with addition of osmotic stabiliser), medium for protoplasts regeneration.
3. Equipment: electrotransformation apparatus.

4. Reagents: 0.8 M KCl with addition of 50 mM CaCl$_2$; 1.2 M sorbitol with added 10 mM CaCl$_2$; hygromycin in concentration 100 mg/ml, plasmid DNA pAN7–1.

Aim of the exercise

Obtaining Curvularia lunata recombinants resistant to hygromycin by introducing plasmid DNA pAN7-1 by electrotransformation method.

The pAN7–1 vector is a bifunctional (shuttle) plasmid that can multiply autonomously only in Prokaryote cells, while in Eukaryote cells it multiplies only when the vector or parts of it are integrated with the chromosomal DNA of the recipient. It has two selection markers: resistance to ampicillin and hygromycin (Figure 1.6.6).

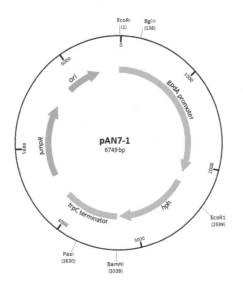

Fig. 1.6.6. Plasmid pAN7–1

Notes: green colour represent the DNA of *E. coli* and orange colour represents promoter and terminator from *A. nidulans* and hygromycin resistance gene

Procedure

1. Transfer *C. lunata* protoplasts, 10^8 cell/ml density, into an Eppendorf tube (400 μl in 1.2 M sorbitol with 10 μM CaCl$_2$) and place in an ice bath.
2. Add 10 μl of plasmid DNA pAN7–1 (1 to 5 μg DNA), mix, leave in an ice bath for 10 min.

3. After this time, transfer the sample to a cooled (electroporation) cuvette and subject it to electrical impulses in the electrotransformation apparatus.

4. Electrotransformation conditions: voltage 2.0 kV, frequency 3.0 µF.

5. Read and note the duration of the electrical pulse.

6. Transfer the cuvette immediately to an ice bath and leave at 4°C for 20 min.

7. Transfer the contents of the cuvette into a tube, add 1.6 ml of the medium for regeneration, place on a shaker, and mix gently for 1 h.

8. Prepare a series of dilutions (10^2 to 10^6) in 0.8 M KCl, take 0.1 ml of the protoplasts suspension, and introduce it into the bars with stabilised EG medium (~ 40°C). Then pour the protoplasts suspended in EG medium onto sterile Petri dishes and after solidifying the medium, incubate at 28°C for 7 days.

9. Take 500, 250, 100 µl of the protoplasts suspension (after subjecting to an electrical pulse) and introduce into the dissolved and cooled bars containing stabilised EG medium with added hygromycin at a concentration of 100 µg/ml.

10. Make a series of dilutions in 0.8 M KCl (control), continue as described in point 8.

11. After 7 days of incubation at 28°C, count the colonies grown on EG medium (control) and give the percentage of protoplasts capable of reversion, taking the density of the protoplasts counted in the Thoma chamber as 100%.

12. Count also the colonies that grown on the EG medium (after an electrical pulse), give the percentage of the protoplasts capacity of regeneration, assuming the number of output protoplasts capable of reversion, at 100%.

13. Count the colonies grown on a medium with the addition of hygromycin, give the percentage of electrotransformation efficiency.

14. Transfer the recombinants onto media with hygromycin in concentrations: 100, 500, 1,000 µg/ml. After 7 days of incubation at 28°C, assess the degree of isolated electrotransformants resistance to hygromycin.

1.7. Storage of industrial strains

INTRODUCTION

Appropriate protection of industrial microbial strains is a fundamental problem in any microbiological biotechnology. On the one hand, it is about securing microbiological purity, on the other hand, maintaining maximum life span and, above all, maintaining the desired technological features. The latter is the most important element, because even with a very low percentage of microbial survival during storage, it is important that the remaining cells have biotechnological properties at the initial culture level.

With a great variety of ways to store microorganisms, there is virtually no universal method that can be used in all cases. This calls for the development of an individual preservation method for specific strains used in a specific biotechnological process. The basic criterion for selecting the method of preservation of pure microbial cultures is the storage time, therefore these methods are divided into short- and long-term (Table 1.7.1).

Table 1.7.1. The most common methods of microorganism storage

	Preservation methods		Preservation temperature	Preservation period
Short-term	Storage on solid or liquid media		4–8°C	≤ few months
	Storage on solid or liquid media coated with mineral oil		4–8°C	≤ 3 years
Long-term	Vacuum-drying of microorganisms		4°C	≤ 4 years
	Lyophilisation		4°C	≥ 10 years
	Cryopreservation at low temperatures		–20°C	≤ 3 years
			–80°C	≤ 10 years
	Cryopreservation at ultra-low temperatures		–196°C	≤ 30 years

All methods of microorganisms preservation are based on the phenomenon of anabiosis, i.e., the organism ability to reversibly reduce its life processes. The state of anabiosis is therefore associated with the inhibition of microbial cell metabolism, increased resistance to extreme external conditions and the ability to recover basic life activity. Therefore, spore-forming microorganisms or spores are relatively easiest to store, while vegetative forms of bacteria and fragments of vegetative mycelium of filamentous fungi and actinomycetes are much more difficult. The production of capsules, e.g. dextran, promotes the storage of microorganisms. Of course, the type of the microorganism already pre-determines the choice of preservation method.

The easiest way to protect microorganisms in the short term, is storage in refrigerated conditions (4–8°C) on solid or liquid media. This method requires periodic passage of stored microorganisms to fresh portions of the medium, and its frequency depends on the type of microbiological medium used, storage conditions, as well as the type of microorganism. Additionally, tubes with stored microorganisms are sealed with a parafilm foil to limit the oxygen supply. Currently, this method is not widely used as it is high cost and labour intensive, but most of all, due to the high frequency of mutations and loss of desired microbial characteristics. Microorganisms stored on solid or liquid media can also be covered with a layer of sterile mineral oil, which can increase their life span. Mineral oil should have a high degree of chemical purity and its layer should be about 1–2 cm thick to effectively protect the microbial cells from oxygen. Although the protection of microorganisms with a layer of oil increases their survival rate and allows for longer storage, this method still has a high frequency of undesirable mutations.

The storage of microorganisms in a dry state is the preferred method of long-term preservation of pure cultures and is widely used for commercial collections of microorganisms. This method allows for long-term preservation of the collections of microorganisms, and an easy way of their further distribution. The technology of microorganisms drying consists of direct drying of their cells suspended in a protective solution (liquid-drying) or drying of a previously frozen microorganism (freeze-drying).

Freeze-drying is a commonly used method of securing microorganisms by dehydrating the cells of the microorganisms placed in deep-freeze state. This process requires the use of a vacuum pump to lower the pressure so that the frozen water in the microbial cells can sublimate. The cells of microorganisms are suspended in cryoprotective agents and then deep frozen (in liquid nitrogen (-80°C or -196°C) and lyophilised.

In the case of microorganisms sensitive to freezing, the protection process consists of vacuum-drying the cells of the microorganisms suspended in protective solutions without prior freezing (liquid-drying). The dried microorganisms are usually stored in a refrigerator (4°C). Their rehydration is a critical stage that affects their ability to reproduce life processes. Similar to the protective measures used to freeze microbial cells, there are a number of effective rehydration measures, including 10% skimmed milk solution, 10% sucrose solution and PTM liquid medium (1.5% peptone, 1% tryptone and 0.5% meat extract).

Another method of storage of microorganisms is their cryopreservation in the form of "stocks" with the addition of cryoprotectants (e.g. glycerin stocks), which are stored deep frozen (−80°C or −196°C). The consequence of low temperatures is that the biochemical and physiological activity of the cells stops, so that they enter into a state of anabiosis and can be stored for a long time. In this case, the microorganisms are stored in a deep freezer (−80°C) or in liquid nitrogen (−196°C). The method of cryopreservation of microorganisms is particularly valued due to the low risk of mutations in frozen microorganisms and their high survival rate. The key factor affecting microorganism survival is the choice of the right cryoprotectant. It has also been shown that the rate of cooling and thawing of microorganisms is another important factor influencing their ability to reproduce life processes. It was found that a controlled cooling rate (1–5°C per min) and fast thawing of protected cells (water bath with a temperature appropriate for the culture of a particular strain) are optimal for achieving maximum survival of stored microorganisms.

On the other hand, protective compounds (cryoprotectants) should exhibit several key features that allow their use in the storage of microorganisms, such as:
- increasing the survival of microorganisms;
- no toxicity to microbial cells;
- high water solubility;
- easy mixing with water and colligative properties;
- low eutectic temperature;
- prevention of salt hyperconcentration in suspension;
- stabilisation of hydrogen bonds in the crystalline network and prevention of large crystal formation;
- in the case of endocellular cryoprotectants, easy penetration through the cell membrane.

Cryoprotectants were divided into intra- and extracellular types due to the location of their activity. The most commonly used intracellular cryoprotectants are glycerol and dimethylsulphoxide (DMSO), which penetrate into the cells. These compounds produce small water crystals in the frozen cells, which protect them from damage. The cryoprotectant concentration affects the survival of microorganisms and should be selected experimentally depending on the type of microorganism being protected. Among the extracellular protective agents, polyvinylpyrrolidone (PVP), hydroxyethyl starch (HES) and dextran are most commonly used.

The aim of commercial microbial culture collections is to assemble and distribute reference strains of microorganisms used, among others, in various industries, quality control tests or scientific research. They are recognised as a way of maintaining the diversity of microorganisms ex situ. The first commercial collection of microbial cultures was founded by Professor Král in 1890 at the German University in Prague. Currently, the World Data Centre for Microorganisms (WDCM) has registered about 568 collections from 68 countries. The best-known of these are the American Type Culture Collection (ATCC), the German Collection of Microorganisms and Cell Lines (Deutsche Sammlung von Mikroorganismen und Zellkulturen GmbH – DSMZ) and the National Collection of Yeast Cultures (NCYC).

PRACTICAL PART

Materials and media

1. Microorganisms: *Lactococcus lactis, Sacharomyces cerevisiae, Trichoderma viride, Cunninghamella echinulata.*
2. Media: ZT plates, agar plates, wort plates.
3. Reagents: glycerol, DMSO, trehalose, skimmed milk.
4. Apparatus: low-temperature freezer, lyophiliser.

Aim of the exercise
Evaluation of the usefulness of freezing and lyophilisation of microorganisms as methods of their long-term preservation and storage. Evaluation of protective properties of selected cryoprotective solutions.

Procedure

1. Freezing of cultures in protective solutions:
   a) mix 500 µl *S. cerevisiae* and *L. lactis* cultures in early stationary phase with 500 µl of 50% glycerol solution in sterile Eppendorf tubes and freeze at −80°C;
   b) mix 900 µl *S. cerevisiae* and *L. lactis* cultures in early stationary phase with 100 µl DMSO (dimethylsulphoxide) in sterile Eppendorf tubes and freeze at −70°C;
   c) make a series of dilutions from the culture of yeast and bacteria and inoculate 0.1 ml each to solid wort and agar medium using the spread plate method. Determine the number of cells in the suspension before freezing. Provide the result in CFU/ml.
2. Evaluation of culture survival in protective solutions:
   a) check the survival of *S. cerevisiae* and *L. lactis* stored in protective solutions after 12 and 24 months of storage at −70°C:
      Make a series of dilutions from thawed yeast and bacteria cultures and then inoculate 0.1 ml each into solid wort and agar media using the spread plate method. Determine the number of cells in the suspension after the freezing process. Provide the result in CFU/ml.
3. Storage of lyophilised filamentous fungi:
   a) rinse the spores from 10-day cultures of *T. viride* and *C. echinulata* with 7 ml of 10% glucose and then filter through sterile glass wool funnels to remove the remains of mycelium;
   b) add a solution of 30% trehalose or 20% skimmed milk (1:1 ratio) to the suspension intended for lyophilisation;
   c) transfer 1 ml of the suspension into glass vials and freeze in liquid nitrogen and then lyophilise;
   d) make a series of dilutions from a part of the suspension and inoculate 0.1 ml on ZT plates using the spread plate method;
   e) determine the number of spores in the suspension before the lyophilisation process. Provide the result in CFU/ml.

4. Evaluation of the survival of lyophilised strains:
   a) check the survival of *T. viride* and *C. echinulata* after 12 months of lyophilised storage at 4°C:
      - add 1 ml of 10% sucrose solution to the lyophilizates tested, then make a series of dilutions and inoculate 0.1 ml (from 10^{-2}, 10^{-3}, 10^{-4}, 10^{-5} dilutions) on ZT plates using the spread plate method;
      - count the number of colonies grown on the plates. Compare CFU/ml before and after lyophilisation and on this basis determine the survival of the strains.

Analysis of results

On the basis of the obtained results, evaluate the role of DMSO, glycerol, trehalose and skimmed milk as cryoprotective agents in the process of microbial protection. Also assess the usefulness of the lyophilisation process and freezing of microorganisms in protective solutions as storage methods for industrial strains.

Literature

Books

Bednarski, W., Fiedurek, J., 2007. *Podstawy biotechnologii przemysłowej*. WNT, Warszawa.

Długoński, J., 1993. *Protoplasty Cunninghamella elegans (Lendner) jako model w badaniu 11-hydroksylacji steroidów*. Wydawnictwo Uniwersytetu Łódzkiego, Łódź.

Długoński, J. (red.), 1997. *Biotechnologia mikrobiologiczna – ćwiczenia i pracownie specjalistyczne*. Wydawnictwo Uniwersytetu Łódzkiego, Łódź.

Długoński, J. (red.), 2016. *Microbial Biodegradation From Omics to Function and Application*. Caister Academic Press, Norfolk.

Doran, P., 2013. *Bioprocess Engineering Principles*. Academic Press, London.

Kowal, K., 2000. *Wpływ czynników fizycznych i chemicznych na wzrost drobnoustrojów*, w: Libudzisz, Z., Kowal, K. (red.), *Mikrobiologia techniczna I*. Wydawnictwo Politechniki Łódzkiej, Łódź, s. 168–172.

Kusewicz, D., 2000. *Gleba*, w: Libudzisz, Z., Kowal, K. (red.), *Mikrobiologia techniczna I*. Wydawnictwo Politechniki Łódzkiej, Łódź, s. 221–223.

Ledakowicz, S., 2011. *Inżynieria biochemiczna*. WNT, Warszawa.

Narihiro, T., Kamagata, Y., 2016. *Anaerobic cultivation*, w: Yates, M.V., Nakatsu, C.H., Miller, R.V., Pillai, S.D. (red.), *Manual of Environmental Microbiology*, 4th ed. ASM Press, Washington, s. 2.1.2-1–2.1.2-12.

She, R., Petti, C., 2015. *Procedures for the storage of microorganisms*, w: Jorgensen, J., Pfaller, M., Carroll, K., Funke, G., Landry, M., Richter, S., Warnock, D. (red.), Manual of Clinical Microbiology, 11th ed. ASM Press, Washington, s. 161–168.

Strobel, H.J., 2009. *Basic laboratory culture methods for anaerobic bacteria*, w: Mielenz, J. (red.), *Biofuels. Methods in molecular biology (methods and protocols)*. Humana Press, Totowa, NJ, s. 247–261.

Review articles

Alves, P.D.D., Siqueira, F.F., Facchin, S., Horta, C.C.R., Victoria, J.M.N., Kalapothakis, E., 2014. *Survey of microbial enzymes in soil, water, and plant microenvironments*. Open Microbiol. J. 8, 25–31.

Croughan, M.S., Konstantinov, K.B., Cooney, C., 2015. *The future of industrial bioprocessing: Batch or continuous?* Biotechnol. Bioeng. 112, 648–651.

Frąc, M., Jezierska-Tys, S., 2010. *Różnorodność mikroorganizmów środowiska glebowego*. Post. Microbiol. 40, 47–58.

Giorgi, V., Menéndez, P., García-Carnelli, C., 2019. *Microbial transformation of cholesterol: reactions and practical aspects – an update*. World J. Microbiol. Biotechnol. 35, 131.

Larsen, B.B., Miller, E., Rhodes, M.K., Wiens, J.J., 2017. *Inordinate fondness multiplied and redistributed: the number of species on Earth and the new pie of life*. Q. Rev. Biol. 92, 229–265.

Li, D., Tang, Y., Lin, J., Cai, W., 2017. *Methods for genetic transformation of filamentous fungi*. Microb. Cell Fact. 16, 168.

Morgan, C.A., Herman, N., White, P.A., Vesey, G., 2006. *Preservation of micro-organisms by drying. A review*. J. Microbiol. Meth. 66, 183–193.

Pawlicka-Kaczorowska, J., Czaczyk, K., 2016. *Klasyczne bakterie propionowe – taksonomia, warunki hodowlane oraz zastosowanie*. Post. Mikrobiol. 55, 367–380.

Piwowarek, K., Lipińska, E., Hać-Szymańczuk, E., Kieliszek, M., Ścibisz, I., 2018. *Propionibacterium spp. – source of propionic acid, vitamin B12, and other metabolites important for the industry*. Appl. Microbial. Biotechnol. 102, 515–538.

Prakash, O., Nimonkar, Y., Shouche Y.S., 2013. *Practice and prospects of microbial preservation*. FEMS Microbiol. Lett. 339, 1–9.

Original scientific papers

Długoński, J., Paraszkiewicz, K., Sedlaczek, L., 1997. *Maintenance of steroid 11-hydroxylation activity in immobilized Cunninghamella elegans protoplasts*. World J. Microbiol. Biotechnol. 13, 469–473.

Długoński, J., Bartnicka, K., Chojecka V., Sedlaczek, L., 1992. *Stabilization of steroid 11-hydroxylation activity of Cunninghamella elegans protoplasts in organic osmotic stabilizers*. World J. Microbiol. Biotechnol. 8, 500–504.

Długoński, J., Bartnicka, K., Zemełko, I., Chojecka, V., Sedlaczek, L., 1991. *Determination of cytochrome P-450 in Cunninghamella elegans intact protoplasts and cell-free preparations capable of steroid hydroxylation*. J. Basic Microbiol. 31, 347–356.

Eyini, M., Rajkumar, K., Balaji, P., 2006. *Isolation, regeneration and PEG-induced fusion of protoplasts of Pleurotus pulmonarius and Pleurotus florida*. Mycobiology 34, 73–78.

Fariña, J.I., Molina, O.E., Figueroa, L.I., 2004. *Formation and regeneration of protoplasts in Sclerotium rolfsii ATCC 201126.* J. Appl. Microbiol. 96, 254–262.

Kim, B.K., Kang, J.H., Jin, M., Kim, H.W., Shim, M.J., Choi, E.C., 2000. *Mycelial protoplast isolation and regeneration of Lentinus lepideus.* Life Sci. 25, 1359–1367.

Ramamoorthy, V., Govindaraj, L., Dhanasekaran, M., Vetrivel, S., Kumar, K.K., Ebenezar E., 2015. *Combination of driselase and lysing enzyme in one molar potassium chloride is effective for the production of protoplasts from germinated conidia of Fusarium verticillioides.* J. Microbiol. Methods. 111, 127–134.

Savitha, S., Sadhasivam, S., Swaminathan, K., 2010. *Regeneration and molecular characterization of an intergeneric hybrid between Graphium putredinis and Trichoderma harzianum by protoplasmic fusion.* Biotechnol. Adv. 28, 285–292.

Tan, G.H., Mustapha, N., 2014. *A comparative analysis of preservation of functional food cultures by freeze-drying, liquid-drying and freezing methods.* Direct Res. J. Agric. Food. Sci. 2, 13–18.

Turgeon, B.G., Condon, B., Liu, J., Zhang, N., 2010. *Protoplast transformation of filamentous fungi.* Methods Mol. Biol. 638, 3–19.

Wilmańska, D., Milczarek, K., Rumijowska, A., Bartnicka, K., Sedlaczek, L., 1992. *Elimination of by-products in 11β-hydroxylation of Substance S using Curvularia lunata clones regenerated from NTG-treated protoplasts.* Appl. Microbiol. Biotechnol. 37, 626–630.

Conference materials

Walisch S., Kaczmarowicz G., 1990. *Porównanie biosyntezy glukonianu w bioreaktorach: zbiornikowym z mieszadłem oraz wieżowym bezmieszadłowym, w: IV Ogólnokrajowa Sesja Naukowa „Postępy inżynierii bioreaktorowej", Łódź 1990. Materiały sesyjne.* Zakład Poligraficzny Politechniki Łódzkiej, Łódź, s. 19–24.

2
Fundamentals of modern analytical techniques used in microbial biotechnology and related sciences

2.1. Confocal, fluorescent microscopy and spectrofluorimetry

INTRODUCTION

Fluorescence is a luminescent process in which light is emitted by an excited atom or molecule. Luminescent radiation can be caused by chemical, mechanical or physical factors (e.g. absorption of light). Photoluminescence is a type of luminescence in which the molecule is excited by photons from UV or visible light. This phenomenon is divided into two categories: fluorescence and phosphorescence. Fluorescence is the ability of certain atoms and molecules to emit light of a certain wavelength, resulting from energy changes inside the atoms or molecules of a substance. It is short-lived (10^{-10} to 10^{-6} seconds) and disappears simultaneously with the interruption of the excitation. Phosphorescence occurs in a similar way to fluorescence, but with a much longer excitation period, that can last for several days. The phenomenon of fluorescence is commonly used in biological research and is based on the use of fluorescent markers (fluorochromes), which in a specific way combine with cellular organelles and enable their visualisation. Currently, the entire spectrum of such substances is available on the market.

2.1.1. Fluorescence phenomenon

The fluorescence phenomenon was first described in 1852 by George Stokes, who discovered a shift in wavelength to longer emission spectra values. Fluorescence is a multistage process that involves a series of consecutive events:
  • excitation of a particle by photons from an external light source. The energy of the photons is absorbed by the fluorochrome electrons, creating an excited singlet state;

- the excitation state exists for a certain period of time (usually up to 10 nanoseconds) and then the fluorochrome undergoes conformational changes. As a result, the energy is partially dissipated and the electrons of the fluorochrome molecule pass to a singlet level;
- return of the fluorochrome molecule to its basic state through the emission of fluorescent radiation (electrons return to one of the oscillating levels of the basic state).

The excited electrons collide, which causes energy dissipation. Returning to the basic state, the electron emits the remaining energy in the form of fluorescent radiation, so the energy of the emitted wave is less than the energy of the fluorescent wave. Therefore, the radiation emitted during the fluorescence process is always shifted towards wavelengths longer than the excitation wavelength. This difference is called the Stokes' law (Figure 2.1.1).

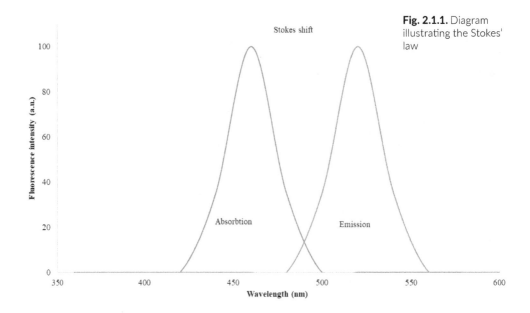

Fig. 2.1.1. Diagram illustrating the Stokes' law

2.1.2. Spectrofluorimetry

Spectrofluorimetry (fluorescent spectroscopy) is a technique that uses the phenomenon of fluorescence. The use of fluorescent markers makes this technique very sensitive and specific; it is often used in biological and biotechnological studies. The commonly used instrument is a spectrofluorimeter equipped with a light source (e.g. xenon arc lamp, emitting radiation in the range of 200 to 900 nm), two monochromators, a chamber in which the samples are placed (e.g. in a cuvette) and a detector. The first monochromator enables the selection of the excitation wavelength, while the second one enables the selection of the emitted light wavelength. Since the second monochromator only directs light of the selected wavelength to the detector, it allows the measurement of fluorescence intensity. In other devices (fluorimeters) there are optical filter sets in place of the monochromators, which transmit light of specific wavelengths. This solution is commonly used in fluorimetric microplate readers.

Usually, in biological studies, the relative amount of fluorescence is determined which is the difference between a test and a blank sample. Otherwise, the fluorescence of the starting sample is measured, to which a system disrupter (e.g. a toxic compound) is added and the measurement is repeated. The results obtained are usually expressed as percentages.

2.1.3. Fluorescent and confocal microscopy – comparison

Figure 2.1.2 shows a schematic representation of the operation of the fluorescent and confocal microscope.

Fluorescent microscopy is a variation of light microscopy. Fluorescent microscopes have a higher resolving power than light microscopes, which allows for accurate observation of intracellular structures. It is a relatively non-invasive technique that allows observation of living cells. Fluorescent microscopes are equipped with a special light source and sets of filters (excitation and barrier), in addition to the standard components of a light microscope.

Nowadays, these microscopes are often equipped with high resolution cameras, which enable the recording of images.

Light sources: the most common light source in fluorescent microscopes is a halogen lamp, although xenon, mercury or LED lamps can also be used. In confocal microscopes, lasers with specific wavelengths are used as light sources.

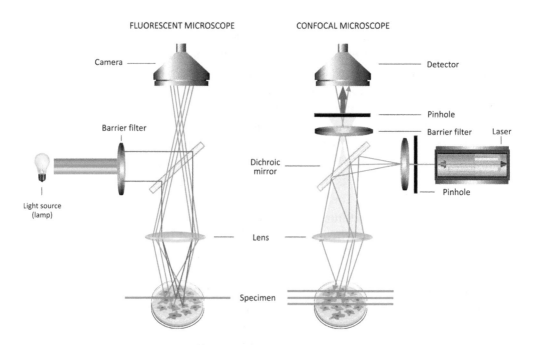

FLUORESCENT MICROSCOPE CONFOCAL MICROSCOPE

Fig. 2.1.2. Scheme of fluorescent and confocal microscope construction.

Filter sets: the essential elements of every fluorescent microscope are the excitation and barrier filter sets. The use of these first filters is to excite the sample by means of illumination with a specific wavelength adapted to the fluorochrome used. Barrier filters are designed to cut off the excitation light from the fluorescent light in order to obtain a glowing image on a dark background.

Confocal microscope is an expanded fluorescent microscope, where the traditional light source is replaced by a laser. The most commonly used lasers are argon (with wavelengths 458, 476, 488, 514) and helium-neon (543, 633).

Recently, "white lasers" have become popular, and these allow the choice of any wavelength of excitation light with the accuracy of 1 nm (in the range from 470 nm to 670 nm). This solution allows for precise adjustment of excitation and emission wavelengths to the used fluorescent markers.

Thanks to the "pinhole" aperture, built into the confocal module of the microscope, images are obtained not only from the surface of the preparation, but also from its depth. The application of this solution allows the contrast and quality of the obtained images to be increased, but primarily, it allows the recording of very thin (1 µm) layers of microscopic preparations, which are then folded together (so-called z-stack). Thanks to appropriate software, three-dimensional (3D) images are created from individual layers of the preparation, which is the greatest advantage of confocal microscopy.

2.1.4. Autofluorescence and fluorescent markers

The phenomenon of autofluorescence (primary fluorescence) is the ability of some cells or their components to emit visible light if they are excited by light of a certain wavelength. Amino acids with aromatic rings (tyrosine, phenylalanine and tryptophan), some cofactors (NAD and FAD) and e.g. purines and pyrimidines in nucleic acids show the ability to fluorescence.

As mentioned previously, fluorescent markers are used very often in biological sciences, i.e., substances which, when excited by light of a certain wavelength, are capable of emitting light. Currently, the whole spectrum of such markers is available to determine the metabolic state of a cell (fluorescein diacetate, propidine iodide, commercial Live/Dead tests), mitosis and meiosis studies (DNA staining, e.g. SYTOX Green, Hoechst 33342, DAPI) or studies of cell membrane fluidity (application of DPH, TMA-DPH).

2.1.5. Fluorescent proteins

The discovery in the last century of Green Fluorescent Protein (GFP), naturally occurring in jellyfish of the genus Aequorea, revolutionised microscopic examination. It is a small protein, with a molecular weight of 26.9 kDa, composed of 238 amino acids. Its structure is atypical and forms a so-called beta-barrel, inside of which there is an α-helical section, responsible for the formation of a chromophore built of 3 amino acids – Tyr66, Gly67, Ser65. Currently, fluorescent proteins (FP) are a basic element of biological research and new, improved versions are constantly appearing. The improved proteins are lighter than GFP, cover a wide spectral range (from blue to yellow), exhibit increased photostability, reduced oligomerisation, are not pH-sensitive and have a faster maturation rate. These proteins are introduced into living organisms through genetic engineering.

cDNA of most fluorescent proteins is widely available and can easily be cloned into many known vectors, which when introduced into living cells will cause them to fluorescence. It is much more difficult to construct a fusion fluorescent protein that is composed of FP and the protein subject to examination. Most often, the inclusion of the FP gene at the 3' or 5' end of the gene encoding the protein of interest, does not negatively affect the adoption of a native form by both proteins and leads to the formation of a hybrid. Fusion proteins enable tracking the movement of proteins in cells under the influence of various factors (e.g. temperature or pH changes). The use of fluorescent proteins in practice, includes:
- labelling of cell organelles or proteins;
- location of proteins in a cell;
- determination of the amount of the tested protein in the cell;
- testing the movement of proteins in a cell, e.g. using the FRAP technique (Fluorescence Recovery After Photobleaching);
- observations of changes taking place in cells, e.g. cell divisions.

PRACTICAL PART

Spectrofluorimetry

Materials and media

1. Microorganisms: *Metarhizium* sp. strains on ZT slants.
2. Media: Sabouraud (doubly concentrated).
3. Reagents: water with 1% Tween 60, commercially available insecticides (e.g. Karate Zeon 050 CS), sterile water, phosphate buffer pH 6.4, FDA solution (fluorescein diacetate) (0.7 mg/l).
4. Materials: sterile funnels with glass wool, Thoma chamber, 96-well plate.
5. Apparatus: light microscope, FLUOstar Omega spectrofluorometer (BMG Labtech), thermostat (temperature 28°C).

Aim of the exercise
Determination of the influence of chemical insecticides on the metabolic activity of entomopathogenic fungi.

Procedure

1. Rinse the ZT slants containing the tested microorganisms with 5 ml of water containing Tween 60 at a concentration of 0.1% and filter through a funnel with glass wool.
2. Determine the suspension density under the light microscope in the Thoma chamber. The resulting spore suspension density should be $2-5 \times 10^6$ cfu/ml.
3. Dilute the suspension of a density of $2-5 \times 10^6$ cfu/ml 10 times in Sabouraud medium (doubly concentrated), thus obtaining a suspension at a density of $2-5 \times 10^5$ cfu/ml.
4. Apply sterile water, insecticides in specific concentrations on a 96-well titration plate and make a series of dilutions. Additionally, perform abiotic tests.
5. Apply 100 µl of the spore suspension at a density of $2-5 \times 10^5$ cfu/ml to all wells of the plate. Incubate the plate for 48 h at 28°C.
6. After the incubation is complete, add a phosphate buffer mixture of pH = 6.4 and 0.7 mg/l FDA solution to each well of the plate. Then, after the specified time (30 min – 1 h) measure the plate fluorescence on a spectrofluorimeter (excitation 485 nm, emission 530 nm).

Analysis of results

The fluorescence intensity results obtained for biotic samples correspond to 100% viability. Calculate the viability of the remaining samples from the proportion. Present the results in the form of graphs. Interpret the results obtained and on this basis draw conclusions about the toxic effects of insecticides on entomopathogenic fungi.

Confocal microscopy

Materials and media

1. Microorganisms: liquid culture of *Metarhizium* sp. on medium according to Lobos.
2. Reagents: propidium iodide (1 mg/ml), 4-*n*-NP (20 mg/ml), PBS buffer pH = 7.0.
3. Materials: Eppendorf tubes, cover slides, microscope slides.
4. Equipment: centrifuge, confocal microscope LSM5Meta (Zeiss).

Aim of the exercise
Evaluation of the permeability of fungal cell shields by confocal microscopy in the presence and absence of a toxic xenobiotic – 4-nonylphenol.

Procedure

1. Collect the hyphae from liquid cultures with 4-*n*-NP and from biotic cultures (prepared according to the methodology given in Section 6.2) and place them in Eppendorf tubes.
2. Centrifuge (8,000 × g, 5 min).
3. Rinse with PBS buffer.
4. Centrifuge (8,000 × g, 5 min), suspend in 1 ml of PBS buffer.
5. Add 1 μl of propidium iodide.
6. Mix and incubate at 28°C in the dark for 5 min.
7. Centrifuge (8,000 × g, 5 min), then rinse with PBS buffer.
8. Centrifuge again (8,000 × g, 5 min) and then suspend in 1 ml PBS.
9. Place the samples on the microscope slide and cover with a cover slide.
10. Perform scans (2D and 3D) under the confocal microscope (excitation: 543 nm, emission: 620 nm).

Analysis of results

For 2D scans, use the available software to determine the total area of the hyphae surfaces (coloured and uncoloured) and the area of the hyphae surfaces that are glowing red. Present the results as a ratio of the red fluorescence surface area to the area occupied by hyphae. Presented the 3D scans as photos. Interpret the obtained results and draw conclusions about the influence of 4-n-NP on the permeability of the cell membranes.

2.2. Isotopic techniques (radioactive isotopes)

INTRODUCTION

The most commonly used radioactive isotopes in biological research are ^{14}C and ^{3}H labelled chemical compounds that emit ionizing β radiation. According to the current regulations in force in numerous countries (in Poland – Atomic Law and its implementing regulations), radioactive isotope research is carried out in isotopic laboratories. Individuals who start to work with radioactive isotopes should be thoroughly acquainted with the instructions and regulations in the isotope laboratory, developed on the basis of the above-mentioned legal regulations, in particular:

- laboratory regulations;
- technological instruction for work with individual radioactive isotopes specifying detailed radiological protection procedures for each type of work performed;
- instructions for radioactive waste;
- emergency action procedure for the isotope laboratory;
- dosimeter user manual with valid calibration certificates;
- operating instructions for the scintillation counter.

PRACTICAL PART

CAUTION: Before working in the isotope laboratory, turn on the radiometer and cheque for contamination of tables, hands, aprons and reusable equipment as necessary. Similar measurements should be taken after the work with radioactive isotopes is completed (according to isotope laboratory regulations).

Materials and reagents

1. Microorganisms: selected bacterial or fungal strains capable of 4-nonylphenol decomposition.
2. Substrates: 4-n-NP, 4-n-NP [ring-U-^{14}C] (25 µCi/ml).
3. Media: mineral media according to Lobos (with or without 2% glucose).
4. Equipment: ^{14}CO$_2$ capturing device, scintillation counter.
5. Reagents: scintillators Ultima Gold and Permafluor E, Carbosorb E (sorbent for capturing ^{14}CO$_2$).

Aim of the exercise
Determination of 4-n-NP mineralisation capacity of the examined microorganisms. The characteristics of 4-n-NP are presented in Section 5.6.4.

Procedure

1. Multiplication of selected bacterial strains on mineral medium according to Lobos with 2% glucose.
2. Before incubation, dilute the bacterial pre-cultures in Lobos mineral medium (without addition of glucose) so that the initial culture OD value is 0.1.
3. To 20 ml of prepared cultures, add 20 mg/l 4-n-NP (i.e., the so-called "cold" nonylphenol) and 4-n-NP labelled with ^{14}C isotope (i.e., the so-called "hot" nonylphenol) so that the radioactivity in the whole culture is 1 µCi/ml. The initial 4-n-NP [ring-U-^{14}C] solution contains 250 µCi/ml at 10 mg/ml. Mix the prepared cultures thoroughly.
4. Collect 1 ml each from the culture into an Eppendorf tube and centrifuge 10,000 × g for 5 min.
5. Take 0.5 ml of supernatant to the scintillation tube, add 1.5 ml of UltimaGold scintillator. Concurrently prepare

a negative control (add 1.5 ml of UltimaGold scintillator to 0.5 ml of mineral medium).

6. Rinse the precipitate with deionised water and centrifuge again at 10,000 × g for 5 min. Discard the supernatant and suspend the precipitate in 2 ml scintillator. Concurrently prepare a negative control (add 2 ml of UltimaGold scintillator to the scintillator tube).

7. Take measurements on the scintillation counter.

8. Incubate the prepared samples (point 3) at 28°C for 72 h in a specially prepared set to capture $^{14}CO_2$. Add 10 ml of Carbosorb E sorbent to each of the appropriate elements of the set. The set must be located in the isotopic laboratory.

9. Preparation of the 4-*n*-NP [ring-U-^{14}C] standard curve. To prepare a standard curve (the output radioactivity of 4-*n*-NP [ring-U-^{14}C] is 25 µCi/ml), take appropriate amounts of the labelled compound to obtain a curve in the range from 0 to 150 nCi (e.g. 0; 12.5; 25; 50; 100; 150 nCi). For this purpose, the appropriate volumes of standard substance should be taken and placed directly in scintillation tubes. Complete all samples with the Ultima Gold scintillator to 2 ml. After the measurement is performed on the scintillation counter, a calibration curve should be plotted and calculations should be made.

10. After 72 h of incubation, take radioactivity measurements according to points 4−6 again and determine the amount of $^{14}CO_2$ in the individual $^{14}CO_2$ capturing containers. To do this, take 1 ml of Carbosorb E from each container and add 1 ml of Permafluor scintillator.

11. Take measurements on the scintillation counter.

Analysis of results

Based on the results obtained, prepare a circular graph of 4-*n*-NP [ring-U-^{14}C] radioactivity distribution before incubation and after 72 h of incubation. Draw conclusions about the ability of the examined microorganisms to mineralise the examined xenobiotic.

2.3. Chromatography

INTRODUCTION

Chromatographic techniques are currently used in many branches of science and industry as highly effective analytical and/or preparative methods, e.g. for cleaning or fractionation of a product. Chromatography is a method of physicochemical separation of compounds in a mixture due to their different partitioning rates into mobile and stationary phases. The mobile phase can be a gas, liquid or liquid in a supercritical state and the stationary (fixed) phase can be a solid or liquid. The dwell time of individual components in the chromatographic system is determined, among others, by the adsorption energy, the size of the distribution coefficient between the stationary and the mobile phase, the structure and physicochemical properties of the analytes and the stationary phase, the composition and physicochemical properties of the mobile phase, etc. For a complete separation of the components of a mixture to take place, they must differ from each other by the above parameters.

2.3.1. Basic parameters measured in chromatography

Chromatography as a technique for analysing the composition of a mixture of substances requires a mathematical description in the form of measurable parameters characterising a given chromatographic system. The basic parameters measured in chromatography include, among others:
- total retention time tR (OB segment) – this is the time an analyte (chemical constituent that is of interest in an analytical procedure) is retained in a column, from the time of dosage to recording the signal (peak) at the chromatogram. It is the sum of the time, in which the compound interacts with the chromatographic bed and the time taken to pass through a system;
- retention time of non-retained substance tM (OA segment) – this is the time taken by a substance which does not interact with the stationary phase, to pass through the chromatographic system;

- reduced retention time t'R (OB – OA segment) is the time when a given compound interacts with the chromatographic bed, and is characteristic for a given compound and a given filling.

If the values of individual retention times are multiplied by the volumetric flow rate, we obtain the appropriate values of retention volumes – the so-called chromatographic peak areas.

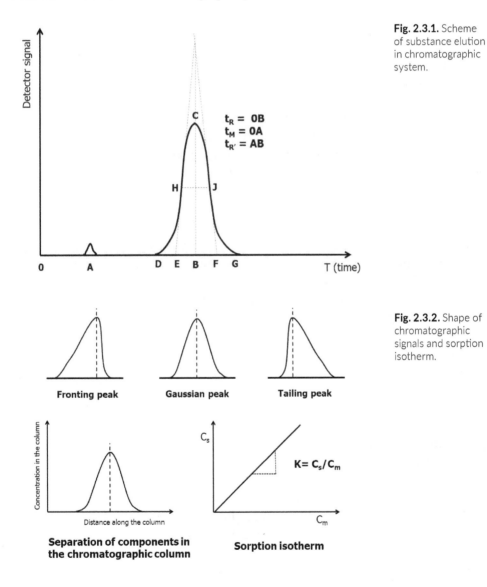

Fig. 2.3.1. Scheme of substance elution in chromatographic system.

$t_R = OB$
$t_M = OA$
$t_{R'} = AB$

Fig. 2.3.2. Shape of chromatographic signals and sorption isotherm.

Fronting peak   Gaussian peak   Tailing peak

Separation of components in the chromatographic column

Sorption isotherm

$K = C_s/C_m$

The retention factor (k) is another measure of retention. It is the ratio of the amount of time the dissolved substance spends in the stationary phase to the time it spends in the mobile phase. The retention factor is a measure of retention through the stationary phase. It is a relative measurement and has a linear character. For an unstable substance passing through a stationary phase, k = 0.

$$k = (t_R - t_M) / t_M = t'_R / t_M$$

where:
t_R – retention time,
t'_R – reduced retention time,
t_M – retention time of non-retained substance.

The amount of a particular substance absorbed by the stationary phase depends on the concentration in the mobile phase. The ratio between the amount absorbed and the concentration in the mobile phase at constant temperature is called the sorption isotherm. The course of the isotherm is linear when the ratio between the concentration of the test substance in the stationary phase (C_s) and the concentration of the dissolved substance in the mobile phase (C_m) is linear. The C_s/C_m ratio determines the slope of the curve and is described as the partition coefficient K. The chromatographic system, in which this relationship is linear, is described as linear chromatography, and the chromatographic shape of the separation is Gaussian.

The separation coefficient α (selectivity) is a measure of time or distance between the maximum values of two chromatographic peaks:

$$\alpha = k_2 / k_1$$

where:
k_1 – retention coefficient of the first peak,
k_2 – retention coefficient of the second peak.

Theoretical plates number (N) or column efficiency is the number of theoretical plates, which is an indirect measure of peak width for a peak with a specific retention time. It is assumed that columns with a higher number of plates are more efficient than columns with fewer plates. A column with a large number of plates will "produce" narrower peaks at a given retention time compared to a column with fewer plates:

$$N = 5{,}545 \, [(t_R) / w_h]^2$$
$$N = 16 \, [t_R / w_b]^2$$

Where:

N – number of theoretical plates,
t_R – retention time,
w_h – peak width at half height (in time units),
w_b – peak width at base (in time units).

Resolution (R) – the higher the resolution, the less the two peaks overlap. The resolution is not only the distance or time between the maximum values (a) of the two peaks. It also refers to both, selectivity (a) and peak width.

$$R = 1{,}18 \, [\, (t_{R2} - t_{R1}) / (w_{h1} + w_{h2})]$$
$$R = 2 \, (t_{R2} - t_{R1}) / (w_{b1} + w_{b2})$$

where:

t_{R1} – retention time of the first peak,
t_{R2} – retention time of the second peak,
w_{h1} – peak width at half height (in time units) – the first peak,
w_{h2} – peak width at half height (in time units) – the second peak,
w_{b1} – peak width at base (in time units) – the first peak,
w_{b2} – peak width at base (in time units) – the second peak.

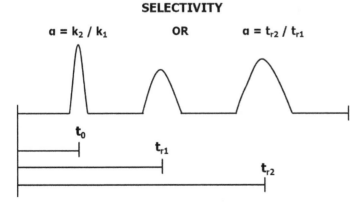

SELECTIVITY

$a = k_2 / k_1$ **OR** $a = t_{r2} / t_{r1}$

t_0 = dead time
t'_r = reduced retention time = $(t_r - t_0)$

Fig. 2.3.3. Selectivity of the chromatographic system.

The above-mentioned measurable chromatographic parameters are used in two basic applications of chromatographic techniques – qualitative and quantitative analysis. Qualitative analysis in chromatography can be performed in two ways.

Fig. 2.3.4. Resolution of a chromatographic system.

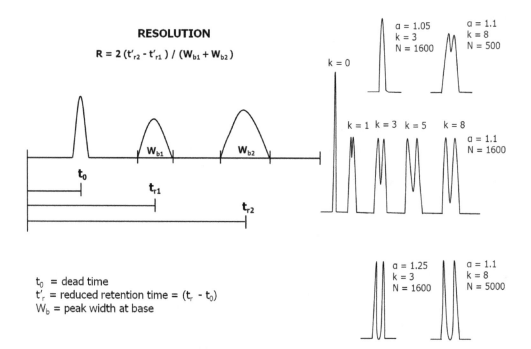

RESOLUTION

$$R = 2 \, (t'_{r2} - t'_{r1}) \, / \, (W_{b1} + W_{b2})$$

t_0 = dead time
t'_r = reduced retention time = $(t_r - t_0)$
W_b = peak width at base

The first, is to use the retention time or peak area, while the second is to use a chemical or physicochemical identification method (detector). In the case of quantitative analysis, the area/intensity of the peak/spot is proportional to amount of the analyte.

Fig. 2.3.5. Quantitative analysis by chromatographic techniques.

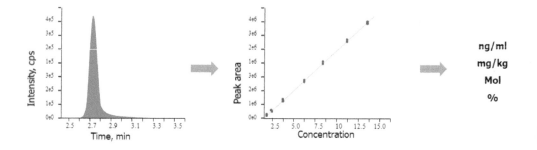

Any increase in the peak area must be proportional to the increase in the amount of the analyte. Quantitative analysis in chromatographic techniques is therefore based on the fact that the amount (mass or concentration) of the analyte in a mixture is proportional to the area or height of its peak or spot intensity/ area.

2.3.2. Liquid chromatography

In liquid chromatography (LC), the mobile phase is a liquid or liquid in a supercritical state, and the stationary (fixed) phase is a solid or liquid. Due to the nature of the phenomena, the basic chromatographic techniques used in liquid chromatography are divided into:
- adsorption chromatography – the separation takes place as a result of different affinity of the components of a mixture to the surface of the stationary phase, called adsorbent;
- partition chromatography – separation is related to different values of the coefficient of division of the mixture components between two unmixed phases, one of which is the stationary phase (liquid on the carrier or solid) and the other – the mobile phase (liquid or liquid in supercritical state);
- ion exchange chromatography – the basis for the separation are the differences in the strength of the intermolecular interactions between the ions from the solution and the ions associated with the stationary phase (called the ionite);
- affinity chromatography – involvees the use of specific chemical reactions between chemically reactive ligands associated with the surface of the adsorbent and unusual components of the separated solution;
- gel or exclusion chromatography – the separation of the mixture components is determined by particle sizes.

Mobile phase components of the chromatographic system can be divided into polar (soluble/mixing with water) and non-polar (not mixing with water). Polar solvents sorted by polarity/ water solubility are e.g.: water > methanol > acetonitrile > ethanol > propane-2-ol > oxydipropionitrile. Non-polar solvents sorted by their hydrophobicity are e.g.: n-decane > n-hexane > n-pentane > cyclohexane. Substances subjected to chromatographic analysis are dissolved in solvents that have similar properties. This in turn allows the separation of different compounds in different solvent combinations.

Due to the experimental techniques used, liquid chromatography can be divided into planar and columnar techniques.

Planar techniques

Planar chromatography is a technique in which chromatographic separation is carried out when the stationary phase is a thin layer. Planar techniques can be divided into: paper chromatography (PC) and thin layer chromatography (TLC).

Paper chromatography is a kind of partition chromatography in which substances from dried liquid samples are separated. The mobile phase is a solvent moving in a porous structure due to the forces of adhesion, cohesion and surface tension (capillary motion). The liquid moves upwards – these forces are stronger than gravity. The stationary phase is the water bound on the tissue paper surface.

Thin Layer Chromatography is a kind of adsorption chromatography. The mobile phase is a solvent of different composition. The stationary phase is a solid in the form of a thin layer placed on a glass, aluminium or plastic plate and is usually made of very polar silicate. Extract from a sample is applied in liquid form at the bottom of the plate in the form of spots (drying is necessary). In the case of TLC, visualisation methods such as UV and iodine staining as well as use of non-selective and selective dyes are necessary. After staining, the plate can be further examined (mainly using image analysis technique) in order to perform quantitative and/or qualitative determination (with respect to standards).

Planar techniques have found an application in determining: the quantity of an analyte, general assessment of product purity, preliminary identification of the substance, monitoring the composition of the fraction after preparative column chromatography or monitoring the course of reaction. However, in comparison to the column techniques, they are characterised by: poor resolution, poor or very poor sensitivity, inaccurate quantitative analysis – mainly in terms of repeatability and reproducibility of the process, thus only allowing a very general qualitative analysis. Taking these characteristics into account, these techniques are nowadays very rarely used and have been almost completely replaced by column techniques.

Column techniques

Apart from open column liquid chromatography (historical or of preparative type), modern liquid chromatography using columns is performed under high pressure (so called High-Performance/Pressure Liquid Chromatography – HPLC), under strictly controlled conditions. It requires specialised equipment including a liquid chromatograph where the operator is able to monitor and regulate the process. A modern liquid chromatograph is a modular device (allowing for expansion and/or modification), consisting of reservoirs for mobile phase components, pumps, autosampler (sample dosing), thermostat for chromatographic columns, detector, chromatographic waste tank and computer (Figure 2.3.6).

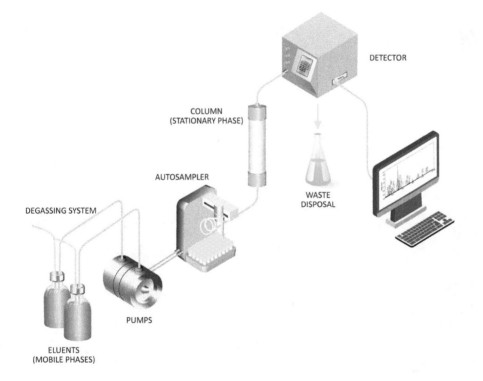

The heart of the chromatographic system is the chromatographic column. In HPLC, the columns must operate under high pressure and are usually made of steel. Typical sizes of LC columns are: 10–30 cm long and 4.6 mm in diameter. The columns are filled

Fig. 2.3.6. HPLC liquid chromatograph scheme.

with the chromatographic bed in the form of porous or non-porous granules coated with the stationary phase. Due to the mechanism of mixture separation in liquid chromatography, there are many types of fillings and countless stationary phases, that are composed mainly on the basis of variously modified silane derivatives and silanol. A typical chromatographic bed has a diameter of up to 5 μm. In many cases, so-called pre-columns are used to protect and extend the operation of the specific column.

Many years of development of the technology has led to further improvements and changes, which have provided further variants of this technique, i.e., uHPLC (ultra high-performance liquid chromatography), microLC (micro-liquid chromatography) and nanoLC (nano-liquid chromatography). The basic differences between these techniques are presented in Table 2.3.2.

Table 2.3.1. Examples of stationary phases used in column liquid chromatography

Stationary phase	Type of chromatography	-R		
HSi≡Si—R HSi≡Si—O 　　　＼R HSi≡Si—O 　　　＼SiR₃	RP-LC	$-C_8H_{17}$ $-CH_3$ $-C_{18}H_{37}$		⬡
	NP-LC	$-R-NO_2$; $-R-NH_2$; $-R-\equiv N$; $-R-CH-CH_2$ 　　　｜　　｜ 　　　OH　OH		

Parameter	High-performance chromatography			
	HPLC	uHPLC	microLC	nanoLC
Bed size [μm]	5	3–5	3–5	< 3
Column length [cm]	10–25	5–15	5–15	5–15
Internal diameter [mm]	4.6	1–4.6	0.3–0.5	0.01–0.1
Maximum flow of the mobile phase	2 ml/min	5 ml/min	60 μl/min	100 nl/min
Maximum pressure [Bar]	400	1400	700	800
Average separation cycle time [min]	15–25	5–10	3–6	3–30

Table 2.3.2. Types of high-performance liquid chromatography

Column techniques use all known methods of mixture separation. However, the most popular are separation methods, while adsorption, ion exchange, affinity or exclusion techniques are also used. The composition of the mobile phase can be constant during the chromatographic separation, which is called isocratic elution, or there can be defined changes of the mobile phase composition during the chromatographic separation and elution – which is called gradient elution. The separation can be carried out in the so-called normal phase system (NP-LC) – the least polar component of the mixture is washed out at the beginning, and the elution time decreases with the increase in polarity of the mobile phase, or in the reverse phase system (RP-LC) – the more polar the compound, the faster it is washed out, and the elution time extends with the increase in polarity of the mobile phase.

A separate branch of column liquid chromatography is preparative chromatography and although it is subject to exactly the same principles as HPLC, it meets other requirements – related to both laboratory and industrial scale processes. Thus, the size of the chromatographic columns (diameter and/or length), the rate of the mobile phase flow or the size/quantity of the sample itself subjected to the chromatographic process are much larger. Liquid chromatographs for preparative purposes are built in a very similar way to HPLC, the difference being the size and presence of fractional collector in the system.

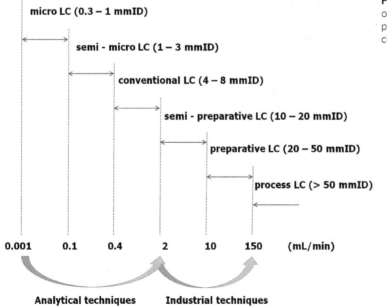

Fig. 2.3.7. Comparison of analytical and preparative liquid chromatography.

Analytical techniques must meet the following requirements: linearity, precision, accuracy, sensitivity, repeatability and resistance to variables that accumulate during serial analyses, which include: matrix effects (e.g. the effect of signal transfer, blanking or amplification), retention time fluctuations, reduction of the patency of the chromatographic system or decrease in chromatographic column performance. On the other hand, preparation techniques must meet the following criteria: recovery, product purity, low production cost with the maximum possible scale of the process, high throughput and the shortest possible process time.

Considering all the advantages and properties, column liquid chromatography is a very versatile method, applicable in the vast majority of modern chromatographic applications.

2.3.3. Gas chromatography

In gas chromatography (GC), the mobile phase is gas and the separation is based on the principle of partition chromatography. The mobile phase, the so-called carrier gas, must be chemically inert and stable over a wide temperature range. This criterion is perfectly fulfilled by noble gases such as helium, neon, argon and xenon. Few applications also use nitrogen. The carrier gas must not contain admixtures of oxygen, water vapour or other reactive gases, so additional filters/traps are often used. The stationary phase is a liquid bound to the substrate (column). The sample can be introduced into the GC in gaseous form using an injection valve or syringe and in liquid form with a micro syringe (1–10 μl).

The separation mechanism is different from liquid chromatography and the retention time of the analyte depends on: separation temperature, column length, chromatographic bed used, pressure and carrier gas flow. The standard GC analysis starts with the sample being applied to the inlet, where rapid evaporation of the sample takes place (the temperature is 30–50°C higher than the evaporation temperature of the tested compounds).

Then, the sample in gaseous form together with the carrier gas is transferred to a column where the analytes condense (temperature is held at a much lower level than the inlet) and bond with the stationary phase of the chromatographic system. Subsequently, the temperature is programmed to increase and when analytes reach their evaporation point they are released from the chromatographic bed into the detector.

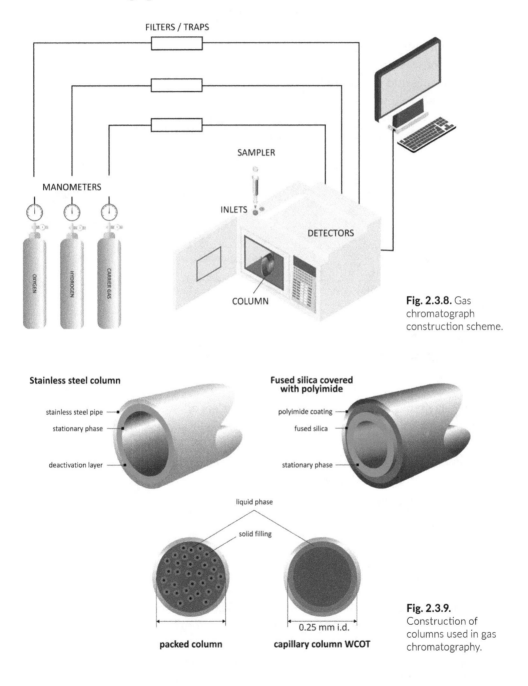

Fig. 2.3.8. Gas chromatograph construction scheme.

Fig. 2.3.9. Construction of columns used in gas chromatography.

Chromatographic columns used in GC are packed (granulate) or mostly capillary columns, made of metal, glass or fused silica with lengths from 30 to 120 m and internal diameter 0.1–1 µm. As the column length increases, its resolution and the theoretical number of chromatographic plates increases. The stationary phases used in gas chromatography are mainly based on polysiloxanes and glycols and their derivatives (Table 2.3.3). Carrier gas also affects the resolution and speed of analysis – with an increase in its flow the analysis time is shortened, but after exceeding certain limits at both small and large flows, the resolution decreases (selectivity and/or peak width).

Table 2.3.3. Stationary phases used in gas chromatography

Type of stationary phase	Type of chemical compounds analysed	Operating temperature
Poly(dimethylsiloxane)	Hydrocarbons, aldehydes, ketones, esters, volatile amines, glycols	–60°C–320°C
Poly(dimethylsiloxy)-Poly(1,4-bis(diethyl-siloxy)phenyl)siloxane	Pesticides, aromatic amines, petroleum fractions	–60°C–350°C
Poly(50%diphenyl-50%dimethyl-siloxane)	Phenols, phthalates, organochlorine pesticides	0–320°C
Poly(50% cyanopropylphenyl -50% methyl-siloxane	Isomers of fatty acid esters	up to 240°C
Polyethylene glycol Carbowax 20M	Alcohols, glycols, esters, polar solvents, fragrances	50°C–260°C

Due to its limitations, gas chromatography does not allow the analysis of a wide range of molecules and is successfully used in the separation of mixtures of compounds which are highly hydrophobic and volatile, e.g. aromatic substances, aromatic hydrocarbons or organochlorine compounds.

2.4. Mass spectrometry

INTRODUCTION

Mass spectrometry is an instrumental analytical technique is used to study, as well as confirm, the structure of organic compounds (qualitative analysis) and also to provide quantitative analysis of compounds. As a universal, sensitive and selective method, it has found very broad application in the analysis of various molecules in practically any material. Currently it is widely used as a tool for analysis by biologists, biochemists, chemists, physicists, criminologists, astronomers, in medicine, environmental protection and many other fields of science.

The beginning of mass spectrometry is considered to be the study of Sir Joseph John Thomson at the University of Cambridge, on the conduction of current through gases, which finally led to the discovery of the electron in 1897. Thomson received the Nobel Prize in Physics in 1906 for this research. The first mass spectrometer (parabola spectrograph) was developed between 1899 and 1911 as a result of further research by the scientist, during which he observed that the beam of electrons and other ions can be deflected by an electrostatic field. Later, Thomson's colleague Francis William Aston built a higher-resolution mass spectrometer to enable isotope observation (Nobel Prize in Chemistry in 1922). At the same time, Arthur Jeffrey Dempster, like Aston, improved the magnetic analyzer and also developed a method of ionisation of substances by means of an electron beam (Electron Ionisation (EI) ion source) which is still used today. Thomson, Aston and Dempster created the theoretical basis for the development of mass spectrometry. Today, more than a century after their first experiments, mass spectrometers are an irreplaceable research and routine tool in the work of numerous analytical laboratories.

2.4.1. Principle of mass spectrometer operation

The simplest type of mass spectrometer has four basic functions (at a vacuum of 10−6 mm Hg or less):

- To ionise the test substance (in the ion source), i.e., transforms evaporated molecules into ions (electron beam, electric potential of ionising gas, etc.);
- Accelerate the ions in an electric field;
- Separate the accelerated ions on the basis of the mass-to-charge ratio (m/z) in a magnetic or electric field (mass analysers);
- Determine ions that have a specific mass-to-charge ratio using a device (detector) that counts the number of ion collisions.

Mass spectrometer construction

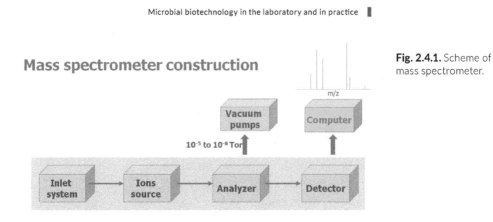

Fig. 2.4.1. Scheme of mass spectrometer.

2.4.2. Ion sources and ion types in mass spectrometry

The ionisation process in the mass spectrometer can be performed in several different ways within the so-called ion source. The selection of the ion source depends on the target purpose of the spectrometer – not every substance ionises well under all conditions. Ion sources can generally be divided into two main groups used in combination with gas or liquid chromatography. The first group includes sources such as EI, Chemical Ionisation (CI) and Field Ionisation (FI). For the second group, the choice is much wider and includes sources such as Field Desorption (FD), Electron Transfer Dissociation (ETD), Fast Atom Bombardment (FAB), Secondary Ion Mass Spectrometry (SIMS), Plasma Desorption (PD), Thermal Desorption (TD), Thermospray Ionisation (TS), Electrospray (ESI, APCI). A separate method, uses a MALDI (Matrix-Assisted Laser Desorption Ionisation) source to ionise substances from the crystalline phase. The most typical ion sources are described in Table 2.4.1.

The molecular weight of the compound or its molecular fragment can be read out based on the value of m/z. The ionisation process produces positive (+) and negative (–) ions. The molecular weight of a single charged ion is approximately equal to the molecular weight of the non-ionised substance only when ionisation is performed by attaching (–) or breaking the electron (+) (due to the very small electron weight). If a proton is attached (+) or disconnected (–) to the molecule, the weight of the ion is greater or correspondingly less than the weight of the non-ionised substance by the weight of the proton (1.00727646688 Da).

Table 2.4.1. Selected ionisation methods in mass spectrometry

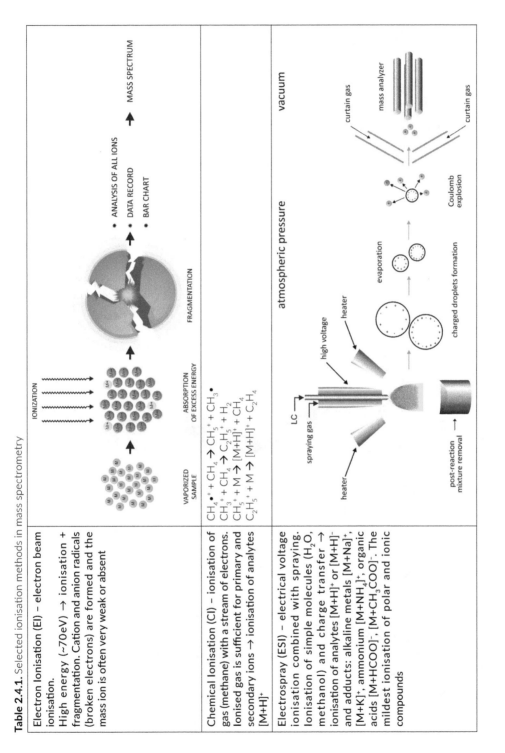

Electron Ionisation (EI) – electron beam ionisation. High energy (~70eV) → ionisation + fragmentation. Cation and anion radicals (broken electrons) are formed and the mass ion is often very weak or absent
Chemical Ionisation (CI) – ionisation of gas (methane) with a stream of electrons. Ionised gas is sufficient for primary and secondary ions → ionisation of analytes $[M+H]^+$
Electrospray (ESI) – electrical voltage ionisation combined with spraying. Ionisation of simple molecules (H_2O, methanol) and charge transfer → ionisation of analytes $[M+H]^+$ or $[M+H]^-$ and adducts: alkaline metals $[M+Na]^+$, $[M+K]^+$, ammonium $[M+NH_4]^+$, organic acids $[M+HCOO]^-$, $[M+CH_3COO]^-$. The mildest ionisation of polar and ionic compounds

IONIZATION

ABSORPTION OF EXCESS ENERGY

FRAGMENTATION

* ANALYSIS OF ALL IONS
* DATA RECORD
* BAR CHART

→ MASS SPECTRUM

VAPORIZED SAMPLE

$CH_4^{\bullet+} + CH_4 \rightarrow CH_5^+ + CH_3^{\bullet}$
$CH_3^+ + CH_4 \rightarrow C_2H_5^+ + H_2$
$CH_5^+ + M \rightarrow [M+H]^+ + CH_4$
$C_2H_5^+ + M \rightarrow [M+H]^+ + C_2H_4$

atmospheric pressure

vacuum

LC

spraying gas

heater

high voltage

heater

post-reaction mixture removal

charged droplets formation

evaporation

Coulomb explosion

curtain gas

curtain gas

mass analyzer

111

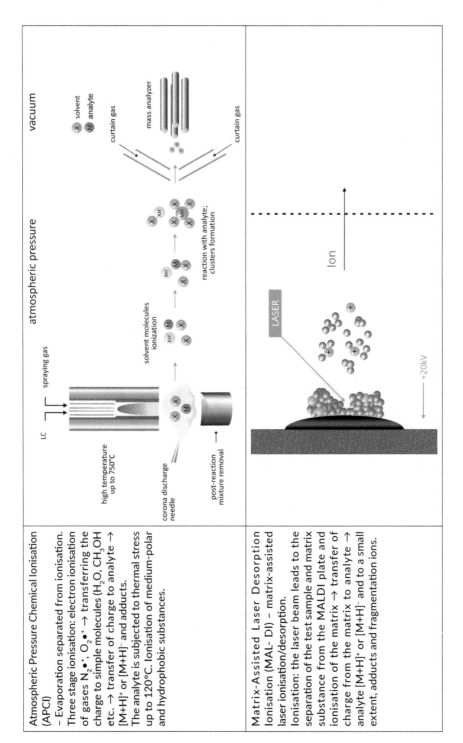

Atmospheric Pressure Chemical Ionisation (APCI)

– Evaporation separated from ionisation. Three stage ionisation: electron ionisation of gases $N_2^{\bullet+}$, $O_2^{\bullet+}$ → transferring the charge to simple molecules (H_2O, CH_3OH etc. → transfer of charge to analyte → $[M+H]^+$ or $[M+H]^-$ and adducts.

The analyte is subjected to thermal stress up to 120°C. Ionisation of medium-polar and hydrophobic substances.

Matrix-Assisted Laser Desorption Ionisation (MAL-DI) – matrix-assisted laser ionisation/desorption.

Ionisation: the laser beam leads to the separation of the test sample and matrix substance from the MALDI plate and ionisation of the matrix → transfer of charge from the matrix to analyte → analyte $[M+H]^+$ or $[M+H]^-$ and to a small extent, adducts and fragmentation ions.

If a given substance, e.g. peptide, is ionised multiple times (e.g. charged with 2 protons), according to the definition of m/z measurement, the ion observed in the mass spectrometer is 2 times smaller than the actual molecular weight of the peptide, e.g.: the mass of the peptide is 602 Da, after ionisation 2+ the mass ion is 301 m/z (because 602/2 = 301).

Three basic types of ions can be observed in mass spectrometry:

- pseudo-molecular ions (M) – anions or cations formed during the ionisation of the sample by the capture or loss of an electron, proton or larger molecule, by the molecule of the compound tested and composed of the lightest isotopes of the elements contained in the compound;
- isotopic ions (M+1, M+2, etc.) – pseudo-molecular ions are accompanied by peaks of m/z value greater by one, two or sometimes several units, corresponding to ions containing isotopes of larger atomic masses;
- fragment ions (F) – ions created by the fragmentation (disintegration) process of the substance during the ionisation process itself or obtained in collision cells in tandem mass spectrometry. On their basis, it is possible to obtain the so-called mass spectra (fragmentation spectra), which are the basis for analysing the structure of the substance (qualitative analysis) or creating selective quantitative methods.

2.4.3. Mass analysers

Ions created in the ion source are transferred to mass analysers for further "processing". There are several basic types of mass analysers, including: sector – magnetic or electric (S), ion cyclotron resonance (ICR), Fourier-transform ion cyclotron resonance (FT-ICR), quadrupole (Q), time-of-flight (TOF), ion trap (IT), linear ion trap (LIT) and orbitrap (O). The last five – Q, TOF, IT, LIT and O are currently the most common. Each of the listed mass analysers has a different mechanism of operation, which is reflected in partially or completely different functions (Table 2.4.2) and different applications.

Table 2.4.2. Comparison of functions and general principles of the most popular mass analysers

Analyser	Functions / operation modes	General principle of operation
Quadrupole (Q) ion transfer / ion removal / detector / voltage (DC&AC) / ion source	• ion transfer, • ion scanning (sorting by m/z value) • ion isolation (specified m/z values), • working as a collision cell • additional collision gas (compound fragmentation)	• two opposite rods of the quadrupole have positive polarisation, the other two, negative • polarisation of two pairs of rods is changed with high frequency (RF), which causes the circulation of ions between the four rods • potential difference shifts the circulating ions through the quadrupole
Ion trap (IT, LIT) 	• ion transfer • ion trapping (accumulation) and scanning • ion trapping, isolation, excitation (fragmentation) and scanning	• Trapping: the ions enter the trap and are trapped by the electrical potentials of the building blocks (they cannot leave the trap), • Scanning: new ions cannot enter the trap, changing/ramping the potentials gradually releases ions (specified m/z) from the trap, • fragmentation: trapping, excitation (by ion collisions), scanning
Time-of-flight analyzer (TOF) + kV	• ion transfer • ion scanning	• all ions are accelerated by the same potential at a certain point in time and drift through the overflow zone • ions are separated by mass-to-charge ratio. As lighter ions move faster than heavy ions, they hit the detector earlier than heavy ones • time-of-flight (TOF) is used to calculate the m/z value

Analyser	Functions / operation modes	General principle of operation
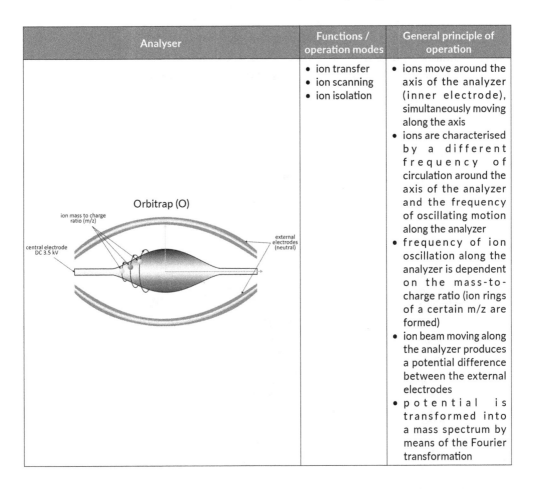 Orbitrap (O) ion mass to charge ratio (m/z) · central electrode DC 3.5 kV · external electrodes (neutral)	• ion transfer • ion scanning • ion isolation	• ions move around the axis of the analyzer (inner electrode), simultaneously moving along the axis • ions are characterised by a different frequency of circulation around the axis of the analyzer and the frequency of oscillating motion along the analyzer • frequency of ion oscillation along the analyzer is dependent on the mass-to-charge ratio (ion rings of a certain m/z are formed) • ion beam moving along the analyzer produces a potential difference between the external electrodes • potential is transformed into a mass spectrum by means of the Fourier transformation

Tandem mass spectrometry

Modern mass spectrometry primarily uses devices based on 2 or more mass analysers – the so-called tandem mass spectrometry. This solution significantly expands the analytical capabilities of both quantitative, qualitative and mixed analysis by using the properties of several types of mass analysers in one device. A general scheme of tandem mass spectrometer construction is presented in Figure 2.4.2.

Currently there are many tandem solutions available on the market. The most commonly used solutions include: triple quadrupole (QqQ), triple quadrupole with linear ion trap function (QTRAP), double quadrupole with time-of-flight analyzer (QTOF), double quadrupole with orbitrap (QO), double time-of-flight analyzer (TOF/TOF). Each of the above-mentioned devices has been used in many applications, but depending on the analysers used, each has some advantages in specific applications, as presented in Table 2.4.3.

Fig. 2.4.2. General scheme of the tandem mass spectrometer.

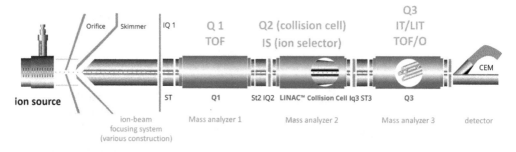

Mass analysers	Primary ion source	Ionization (+H+, −H+)	Mass range (Da)	Resolution	Quantitative analysis	Qualitative analysis
QqQ	ESI/APCI	± 1→n	5-2000/3000	0.1	+++	+
QTRAP	ESI/APCI	± 1→n	5 (50) – 2000/3000	0.1 /0.01	+++	+++
QqO	ESI/APCI	± 1→n	5 – 6000	0.1 /0.0001+	++	+++
QTOF	ESI/APCI	± 1→n	0 – 2000/40 000	0.1 /0.0001	++	+++
TOF/TOF	MALDI	± 1(2)	0 – ?	0.0001	+/–	+++

2.4.4. Basic scanning modes

Table 2.4.3. General comparison of the most popular tandem mass spectrometers

Considering the use of mass analysers in tandem mass spectrometers, scanning types can be divided into two main groups:
- MS scanning – transfer/selection/scanning of pseudo-molecular ions;
- MS/MS scanning – transfer/selection/scanning of pseudo-molecular ions and fragmentation (disintegration) of pseudo-molecular ions and transfer/selection/scanning of fragment ions.

The process of fragmentation (disintegration) of an ionised compound leads to the formation of charged fragments (fragment ions) and neutral fragments:

$$ABC^+ \rightarrow A^+ + B^{neutral} + C^+ + \text{etc.}$$
$$ABC^- \rightarrow A^- + B^{neutral} + C^- + \text{etc.}$$

116

It should be emphasised that secondary fragmentation may also take place, e.g.: $AB^- \rightarrow A^- \rightarrow A1^- + A2^- + A3^-$ etc. The fragmentation process in the tandem mass spectrometer mainly takes place in: collision cell (e.g. Q2), ion trap, ion source (EI/CI in GC-MS). If fragmentation occurs outside the above-mentioned zones, it is beyond the user's control and should be avoided. Basic scanning modes are presented in Table 2.4.4.

Table 2.4.4. Basic MS and MS/MS scanning modes

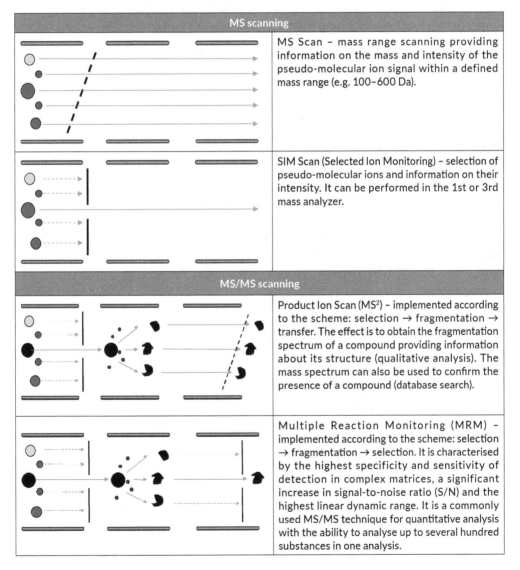

MS scanning	
	MS Scan – mass range scanning providing information on the mass and intensity of the pseudo-molecular ion signal within a defined mass range (e.g. 100–600 Da).
	SIM Scan (Selected Ion Monitoring) – selection of pseudo-molecular ions and information on their intensity. It can be performed in the 1st or 3rd mass analyzer.
MS/MS scanning	
	Product Ion Scan (MS^2) – implemented according to the scheme: selection → fragmentation → transfer. The effect is to obtain the fragmentation spectrum of a compound providing information about its structure (qualitative analysis). The mass spectrum can also be used to confirm the presence of a compound (database search).
	Multiple Reaction Monitoring (MRM) – implemented according to the scheme: selection → fragmentation → selection. It is characterised by the highest specificity and sensitivity of detection in complex matrices, a significant increase in signal-to-noise ratio (S/N) and the highest linear dynamic range. It is a commonly used MS/MS technique for quantitative analysis with the ability to analyse up to several hundred substances in one analysis.

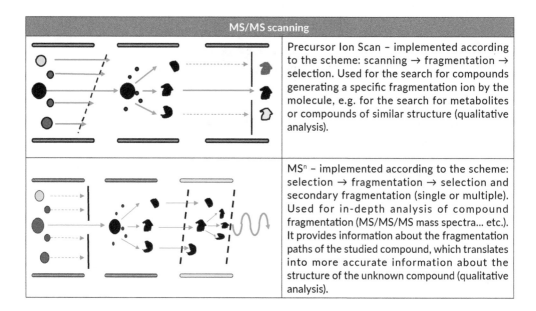

MS/MS scanning	
	Precursor Ion Scan – implemented according to the scheme: scanning → fragmentation → selection. Used for the search for compounds generating a specific fragmentation ion by the molecule, e.g. for the search for metabolites or compounds of similar structure (qualitative analysis).
	MS^n – implemented according to the scheme: selection → fragmentation → selection and secondary fragmentation (single or multiple). Used for in-depth analysis of compound fragmentation (MS/MS/MS mass spectra... etc.). It provides information about the fragmentation paths of the studied compound, which translates into more accurate information about the structure of the unknown compound (qualitative analysis).

2.4.5. Ion detection

The detector in the mass spectrometer is designed to record ions that have passed through the mass analyzer. The simplest and oldest ion detector is a photographic plate. Contemporary detectors transmit information in the form of electrical signals processed into a digital signal, which is then stored and analysed using IT tools.

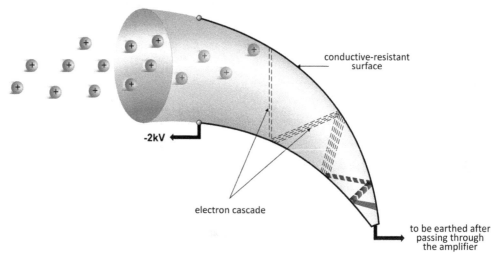

Fig. 2.4.3. Electron multiplier scheme.

conductive-resistant surface

-2kV

electron cascade

to be earthed after passing through the amplifier

The most frequently used detectors include: Faraday cup, electron duplicator, microchannel detector, photomultiplier detector or electrical potential measuring devices (e.g. ICR or Fourier-transform Orbitrap, which are concurrently mass analysers). The most popular solution is the electron multiplier due to its high signal and response in nanoseconds (Figure 2.4.3).

2.4.6. Applications of mass spectrometry

Mass spectrometry as a universal method, used both in quantitative and qualitative analysis, is very widely used in various applications and fields related to the analysis of organic and inorganic compounds. This method is used, among others, in research and routine analyses for:
- identification and quantification of individual components in organic mixtures;
- research on the kinetics and thermodynamics of chemical reactions;
- determination of isotopic composition of analysed substances;
- conducting of ultra-sensitive multi-element inorganic analyses;
- identification of compound structures;
- protein sequencing (including post-translation modifications);
- research on polysaccharide sequences and other biological and non-biological polymers;
- environmental pollution monitoring;
- detection of food adulteration and contamination;
- anti-doping tests;
- industrial process monitoring.

2.5. Atomic absorption spectroscopy

INTRODUCTION

Atomic absorption spectroscopy (AAS) is based on the phenomenon of absorption of electromagnetic radiation of a certain wavelength by free atoms.

The most important assumptions of this method are as follows:

- the source of the analysed absorption lines are free atoms;
- free atoms can absorb radiation of wavelengths that they themselves emit;
- the received absorption spectrum is characteristic for a given type of atoms.

This phenomenon was observed as early as the beginning of the 19th century, but its practical application in analytics was described by Alan Walsh in 1955. Absorption atomic spectrometry is an analytical spectroscopy technique, widely used for quantitative analysis of elements present in solutions (about 70 elements), mainly metals. The absorption of radiation depends on the number of free atoms, which is proportional to the concentration of the element in the sample, which makes it possible to quantify the concentration of the element. AAS is a comparative method. The concentration of the tested element is most often read on the basis of the standard curve (or adding a standard or internal standard). Reading the concentration of the tested element from the curve equation is possible only in the case of a straightforward dependence of absorbance on the concentration of the analysed element, which in practice is often connected with the necessity of sample dilution.

Construction of atomic absorption spectrometer

Atomic absorption spectrometer is an apparatus for quantitative analysis of elements by atomic absorption spectrometry and consists of the following elements:

- linear radiation sources;
- atomiser;
- monochromator;
- detector and amplifier;
- data collection and storage (recorder, computer).

A diagram of the structure of the atomic absorption spectrometer is shown in Figure 2.5.1.

There are single and double-beam devices. If the light from the radiation source after passing through the sample in concentrated form goes to the monochromator, it is a single-beam system.

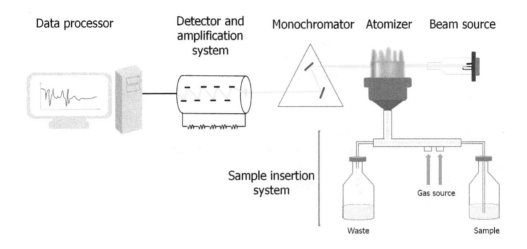

Data processor　　Detector and　Monochromator　Atomizer　Beam source
　　　　　　　　amplification
　　　　　　　　system

Sample insertion
system

Gas source

Waste　　　　　　　　　　Sample

It is then necessary to warm up the lamp before the measurement in order to stabilise it. However, if the light from the radiation source is divided into a beam passing through the sample and a reference beam that bypasses the sample insertion system, we are dealing with a two-beam system. The reading is expressed by the ratio of the beam passing through the sample to the reference beam. This modification serves to maintain the stability of the reading and counteracts changes in intensity of the radiation source. In double-beam systems, the analysis can be carried out without the need to warm up the lamp.

Fig. 2.5.1. Scheme of atomic absorption spectrometer construction.

Sources of radiation

The source of radiation are two types of lamps:
- hollow-cathode lamps (HCL);
- electrodeless discharge lamps (EDL).

The hollow-cathode lamp is a glass tube, filled with neon or argon. Inside, there is a tungsten anode and a hollow cathode made of metal whose resonance lines it emits (i.e., metal whose concentration is measured). The excited positively charged gas ions bombard the cathode and knock out metal atoms from it. The ejected gaseous atoms are excited and emit characteristic radiation. It is not possible to construct an effective lamp of such structure for certain elements, such as, As Sb, Se, Te. In this case, EDL lamps are used, where the excitation of metal atoms is caused by a high-frequency electromagnetic field. In practice, a separate lamp is usually needed for each element to be determined. Multi-element lamps can also be used to determine several elements, but the sensitivity of the analysis is then lower.

Since radiation of a certain wavelength, produced by the lamps, can only be absorbed by free atoms, the next stage of the analysis is atomisation, and the atomiser is an indispensable element of the apparatus. There are several types of atomisers and their selection depends on the element under analysis and the desired sensitivity of the analysis. The following types are known:

- flame atomisers;
- electrothermal atomisers;
- hydride atomisers;
- atomisers using mercury cold vapour.

Flame Atomisation (F-AAS) is the oldest and still most popular atomisation technique. In flame atomisers, the transition from solution to atomic gas takes place in two stages:

- nebulisation, i.e., spraying the analysed solution and introducing it as an aerosol into the flame;
- atomisation in the burner flame, where the oxidising gas is air, oxygen or nitrous oxide (N_2O) and the combustible gas is acetylene.

The most common oxidising gas is air. However, for elements that form stable oxides or hydroxides in the oxidising flame, it is necessary to use a reducing flame using nitrous oxide.

Thermal dissociation occurs in the burner flame, providing free atoms. Other reactions, such as ionisation, excitation or synthesis reactions, reduce the number of free atoms and thus negatively affect the sensitivity of the method.

Electrothermal atomisation (ET-AAS) enables the analysis of elements both in solution and in solid samples. Graphite cuvettes are used. During computer-controlled sample heating, the sample is dried, incinerated and atomised in the third phase. ET-AAS enables the analysis of trace amounts of elements, e.g. the detection limit of Zn in the electrothermal method is 0.001 µg/l and 1 µg/l in the flame atomisation method.

Hydride atomisation consists of the production of volatile metal hydrides, which undergo thermal decomposition to free atoms in an absorption cuvette. This technique is used for the analysis of elements which are difficult to vaporise, i.e., Sn, Se, As, Sb.

Cold vapour technique is an atomisation technique used only for mercury. This is due to the fact that Hg is the only metal that has a clear vapour pressure at room temperature and does not react practically with atmospheric oxygen. When reduced to metallic mercury, it can be marked as vapour in a flow cuvette.

Monochromators split electromagnetic radiation, allowing the separation of radiation of one wavelength from the spectrum (called resonance line) and pass it to the detector. The other wavelengths are eliminated. Monochromators are usually systems of slits and moving mirrors, thanks to which it is possible to select a certain wavelength from the spectrum.

Detectors are used to convert electromagnetic energy into electricity. The resulting electrical signal, proportional to the radiation intensity, is amplified and transmitted to the detector and recorder (computer).

Preparation of samples for analysis

Samples containing no organic matter, such as pure water, mineral water, only need acidification before analysis. Other samples, such as wastewater, soil, food, plant and animal tissues, must be mineralised before the measurement.

Mineralisation consists of decomposition and oxidation of organic compounds contained in the sample and transfer of the remaining components to solution. In practice, wet mineralisation in an open or closed (pressure) system is most often used. The decomposition of organic matter takes place using one or a mixture of concentrated mineral acids (HNO_3, H_2SO_4, $HClO_4$), sometimes with the addition of other strong oxidants, such as H_2O_2.

There are several ways of mineralisation:
- thermal heating, e.g. in an oven;
- use of UV radiation;
- use of microwave energy.

The most common are thermal or microwave mineralisation. The disadvantage in the case of thermal mineralisation is the longer duration of the process. The process usually takes place at temperatures from 100 to 150°C. Microwave mineralisation is faster, but requires more expensive apparatus. The sample is mineralised in an environment of concentrated mineral acids at elevated temperature, in a tightly closed vessel. As a result of the accumulation of gases, an increased pressure is created, which enables higher temperatures to be applied than the boiling point of the acids in an open system, and thus shortens the process time. Teflon or quartz vessels are used for microwave mineralisation, which are permeable to microwave radiation and simultaneously resistant to concentrated acids.

PRACTICAL PART

Quantitative analysis of metal content in mycelium using Spectra 240 FS (Agilent) atomic absorption spectrometer
— Sections 5.6.5 and 5.7.

2.6. Modern digital techniques used to record changes in the environment

Nowadays, digital techniques, remote sensing methods and GIS tools are more and more often used to monitor the urban environment and to assess the condition of green infrastructure elements like green areas, street wooded areas, waterbodies, agricultural and forest crops. Currently, there are different pitfalls of data acquisition: a. satellite imagery, aerial imagery, c. terrain measurement by using a drone. These techniques are constantly being improved and it can be expected that the range of their application will be extended for the needs of ecologists, landscape architects, microbiologists and environmental biotechnologists, as well as the increasingly frequent interdisciplinary research carried out with the participation of these specialists, as shown in Figure 2.6.1.1.

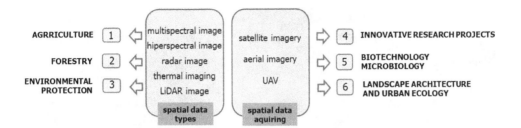

Explanations: 1 – estimation of damages in agriculture (drought, heavy rain, bacterial and fungal plant diseases); 2 – assessment of damages in the tree stand (windbreaks, pest invasion, phytopathogens); 3 – estimation of plant and animal populations; 4 – aerial inventory of vegetation; 5 – changes in land use (including results of chemical, microbiological and biotechnological analyses); 6 – current land configuration and anthropological changes

Fig. 2.6.1.1.
Development and application of modern digital techniques.

In example, during analysing aerial and satellite images of non-forest areas from different time periods it is possible to indicate the degree of secondary succession and the spread of invasive species. In the case of areas with dense tree cover (forest areas, city parks) it is possible to notice the places where the degradation of the natural environment took place (observation of trees infected by bacterial and fungal phytopathogens, reduction of tree crowns) or, on the contrary, areas where the proper development of vegetation takes place (disappearance of symptoms of plant diseases) or increase of tree density, especially in bigger scale (eg. green infrastructure of city/aglomeration or region). In turn, in greenery management, drones are mainly used to monitor forest tree stand in smaller scale (eg. selected object of given green area like park or investigated degraded area).

2.6.1. Satelitte imagery in analyzing historical and last landuse

Especially useful is the constantly expanding database of satelitte images which enables observation of the existing land cover or land use (eg green areas, built-up areas, degraded areas etc.) as well as picturesque lansapce with increasing spatial resolution of the images (Figure 2.6.1.2). and thus a better degree of accuracy.

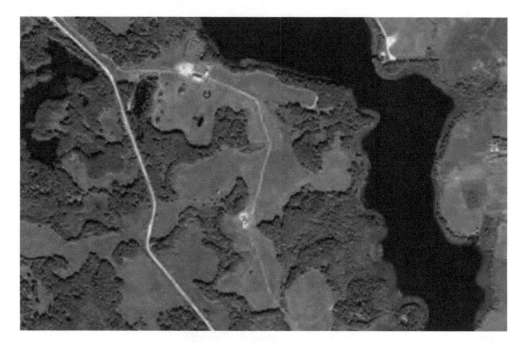

It is worth noting that it is possible to compare the latest satellite images with archival satellite images dating back to different years (for the indicated area both in the scale of the city, district and small and large green area object). The database of photographs spatial database is available on the websites of public administration offices (GUGiK) and the nationwide Geoportal platform in appropriate tabs with thematic layers. It is also possible to compare these maps by downloading maps to computer programmes, i.e., ArcGIS or QuantumGIS, which enable to place thematic layers ("one on top of the other") and to synthesise changes by noticing differences in the terrain cover. The exemplary dynamics of development changes for suburban areas as well as selected non-forest land is presented in Figures 2.6.1.3 and 2.6.1.4.

From the point of view of acquiring information about the area, the Landsat and Sentinel satellite imagery database is possible, as well as available basicly for free on the Copernicus website. However, these satellite data do not provide high resolution images[1], they are free and allow analysis of larger areas eg. non-urban open countryside with a large surface area measured in hectares.

Fig. 2.6.1.2. The dynamics of suburban area changes – construction of housing estates.

Photo taken by Techmex SA, photos from the satellite IKONOS, source: Okła 2010, p. 432 (Disclaimer: copying the illustration only with the author's consent).

---

1 The Sentinel-2 data resolution is 10m and the Landsat-8 data resolution is 30m.

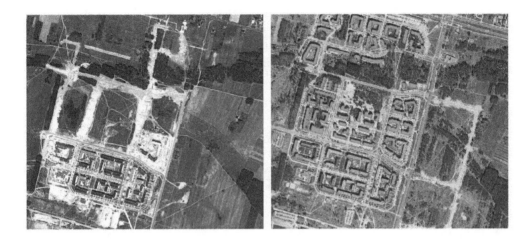

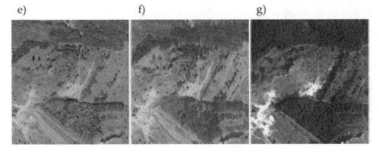

Fig. 2.6.1.3. The dynamics of suburban area changes – construction of housing estates

Fig. 2.6.1.4. The dynamics of changes in non-forest: a) year 1971, b) year 1982, c) year 1996, d) year 2003, e) year 2009, f) year 2012, g) year 2015
Source of illustration: Osińska-Skotak et al. 2019 (Disclaimer: copying the illustration only with the author's consent)

In the case of the Copernicus, apart from satellite images of a given area, it is also possible to download maps of urban agglomerations of Western and Central Europe, and from 2020 also major Polish agglomerations. In addition, specialized studies such as Urban Atlas, or a map of urban tree canopy, so-called Street Tree Layer (latest version from 2018) are also available based on these images. It is worth noting that these studies can serve as a source of knowledge about land cover and land use as well as showing the location of the area covered by the environmental analyses.Another source of available maps and geoinformations presents also in OpenStreetMaps organiation or ArcGIS online. After registration in these platforms, there is also a possiblility to downloaad given maps or make basic editing or creating spatial maps of given area (eg. city). After COVID-19 pandemic period, some spatial data are also available to download from national spatial services like ealier mentioned Geoportal or GUGIK websites.

2.6.2. Principles of detection selected remote sensing systems

In environmental research and urban ecology, constantly improved techniques from the group "remote sensing are increasingly being used to measure the variability of land use elements, allowing to determine the state of development and condition of vegetation. This issue is determined by various indexes. One such index is in example the Normalized Difference Vegetation Index (NDVI). The NDVI is one of the vegetation indexe[2] for collecting information about the photosynthetic rate and the biomass of a given area covered with vegetation. NDVI was first used by J. W. Rouse in 1973. NDVI is an indicator used very often nowadays in remote sensing measurements and is used practical in interdisciplinary research to show the development and condition of vegetation. NDVI is based on the contrast between the highest reflection in the near infrared band and absorption in the red band. NDVI is calculated using the formula:

$$NDVI = (NIR - VIS) / (NIR + VIS)$$

where:
VIS – reflection in the red band,
NIR – reflection in the infrared band.

---

2 A vegetation index is a combination of two or more selected reflectance values measured in the field, laboratory, or recorded from an aircraft or satellite. It is used to emphasize specific features of plants and to distinguish vegetation from other materials. Plant indicators were and are nowadays developed based on the spectral properties of plants.

Unmanned aerial vehicles, commonly called, are devices that are usually operated and under human control at the moment of work. The vehicles of this type do not have an on-board pilot, and the operator controls the vehicle from a remote terminal. This way of using the device without an on-board pilot is characterised by a smaller weight and size than a manned aircraft, which is an important economic issue. This method also has its limitations – the most important is the surface of given area. Countries such as China, Israel and the United States are widely recognised as leaders in UAV technology.

It should be noted that technologies based on these systems are currently being strongly developed in European countries and used in natural sciences (geography, ecology) and agricultural sciences (forestry, horticulture), e.g. in Germany such systems are used for research on urban ecology and green infrastructure, in the UK for cyclical measurements of damage and diseases of plants on farmland. In Poland, on the other hand, similar systems are used to monitor the urban environment, as well as to manage the State Forests. Drones with a standard test set enable monitoring of the level of emissions of particularly dangerous air pollutants, such as particulate matter PM 2.5 and PM 10, as well as carbon monoxide, sulphur oxide, hydrogen sulphide, methane, nitrogen, formaldehyde and others. An example of a drone with the apparatus described above is shown in Figure 2.6.1.5.

Fig. 2.6.1.5.
Unmanned aircraft with air pollution measurement equipment – model "SOWA", commonly called drone.

Source of illustration: USM Sp. z o.o. (Disclaimer: copying the illustration only with the author's consent).

In Polish conditions, drones are included in an airborne laser scanning system, most often using LiDAR technology, and are built as standard with a laser distance metre measuring the distance of the drone from the scanned area, software recording and scrolling spatial information, GPS transmitter and imaging devices (digital cameras). Thanks to legal regulations on application of drones, in Poland since 2018, there is already a considerable number of companies (including Polish ones) dealing with taking pictures of unmanned craft cooperating with universities and research centres (training, joint scientific research. In Poland, drone research is becoming more and more common.

2.6.3. Airplane scanning in terrain measurement

However, at larger scales of terrain measurement, terrain imaging methods using airborne plane laser scanning (ALS) are more applicable. The first ALS photos (so-called point cloud) were taken as early as in the 1970s in the USSR for the purposes of forestry and landscape research of non-urban. The dynamic development of this technology took place only in the 1990s, mainly in the USA and Canada. However, there is still a lack of current legal regulations that would consider the use and legal application of drones for educational and scientific purposes. Despite the lack of these regulations, however.

In turn, last interesting study from Poland using this data acquisition technique is shown in the crown tree map (Mapa koron drzew). The map contains arrangement with geoinformation of existing trees in the city of Warsaw and is the first interactive study available in Poland by using the Internet. Additionally, it is possible to include layers that provide information about the tree species and their health based on information about leaf discoloration or defoliation. An example photo illustrating variety of tree species 6 by using airplane scanning is shown in Figure 2.6.1. (letter abbreviations in the legend correspond to tree species names, e. g. MT_FaSylva is Fagus sylvatica, MT-Pin is Pinus sp. etc. according to the abbreviation list for the tree stand database).

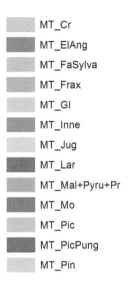

	MT_Cr
	MT_ElAng
	MT_FaSylva
	MT_Frax
	MT_Gl
	MT_Inne
	MT_Jug
	MT_Lar
	MT_Mal+Pyru+Pr
	MT_Mo
	MT_Pic
	MT_PicPung
	MT_Pin

Spatial data Point cloud data acquired drones airborne laser scanning may be used to measures area along tree crowns or lower plants (shrubs, herbaceous plants), development elements (e.g. outbuildings) or on the ground (Figure 2.6.1.7). The accuracy of the image depends on, among others, the density of plant cover, specific terrain conditions, weather, time of day, flight parameters.

Thanks to drones and other mobile airplane scanning and stationary devices recorded as image data from the LIDAR group (Light Detection and Ranging – a device operating on a similar principle to radar, but using visible laser light instead of microwaves), with appropriately set parameters of the equipment it is possible to measure the spatial structure of forests and tree density with high accuracy, as well as to indicate degraded sites, e.g. drought of the stand caused by insect invasion or plant pathogenic microorganisms. Figure 2.6.1.8 shows an infrared photo of a fragment of the damaged forest.

It is worth noting that, as it is currently the case with the discussed research on detection systems for the State Forests, also green areas can be monitored, including areas degraded as a result of xenobiotic contamination and/or development of phytopathogens, which will require revitalisation (see Section 5.1).

Fig. 2.6.1.6.
Photograph taken by a drone airplane scanning illustrating the reach of deciduous tree species classification results using aerial data fusion in fragment of Bielanski Forest in Warsaw (Tree Crown Map).

Source of illustration: Mapa Koron Drzew [Tree Crown map], Map and legend elaborared by MSc Thilo Wellmann, HU Berlin based on spatial data resulted from Geodesy Faculty of Warsaw City Council (Disclaimer: copying the illustration only with the author's and Warsaw City Council Department consent)

Fig. 2.6.1.7. Image obtained from the so-called point cloud when recording forest area with a drone with LiDAR technology airplane scanning

Source of illustration: ProGea consulting, Zakopane 2007, from: Będkowski et al. 2010 (Disclaimer: copying of illustration only with author's consent)

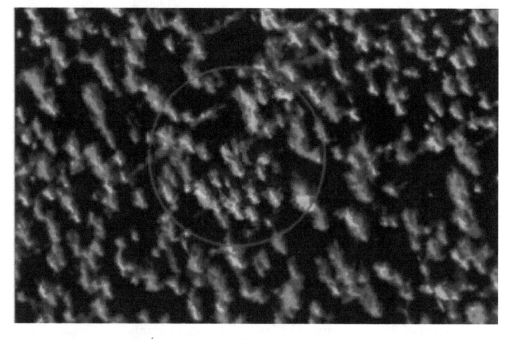

A photograph taken by J. Irlik, Świerklaniec Forest District, source of illustration: Okła 2010, p. 429 (disclaimer: copying the illustration only with author's permission)

Fig. 2.6.1.8. Drought of the stand of a forest fragment caused by the invasion of the *Heterobasidion annosum* fungus causing root rot

Modern imaging and detection techniques are of great interest to scientists due to the speed of spatial data acquisition and the ability to analyse large urban and extra-urban areas often without the need to cheque the parameters of plants and other elements of land use in the field.

Literature

Books

Będkowski, K., Wężyk, P., 2010. *Skaning laserowy*, w: Okła, K. (red.), *Możliwości wykorzystania teledetekcji i fotogrametrii w Lasach Państwowych*. Centrum Informacyjne Lasów Państwowych, Bedoń, s. 327–343.

Borkowska-Wykręt, D., Kurczyńska, E., 2017. *Mikroskopia świetlna w badaniach komórki roślinnej*. Wydawnictwo Naukowe PWN, Warszawa.

Długoński, J. (red.), 1997. *Biotechnologia mikrobiologiczna – ćwiczenia i pracownie specjalistyczne*. Wydawnictwo Uniwersytetu Łódzkiego, Łódź.

Długoński, J. (red.), 2016. *Microbial Biodegradation From Omics to Function and Application*. Caister Academic Press, Norfolk.

Dong, M.W., 2006. *Modern HPLC for Practicing Scientists*. John Wiley and Sons, Hoboken, NJ.

Greaves, J., Roboz, J., 2013. *Mass spectrometry for the novice*. CRC Press, Taylor and Francis Group, Boca Raton, FL.

McNair, H.M., Miller, J.M., Snow, N.H., 2019. *Basic Gas Chromatography*. John Wiley and Sons, Hoboken, NJ.

Okła, K., 2010. *Możliwości wykorzystania teledetekcji i fotogrametrii w Lasach Państwowych*, w: Okła, K., (red.), *Geomatyka w Lasach Państwowych. Część I: Podstawy*, Centrum Informacyjne Lasów Państwowych, Bedoń.

Smoluch, M., Grasso, G., Suder, P., Silberring, J., 2019. *Mass Spectrometry: An Applied Approach*. John Wiley and Sons, Hoboken, NJ.

Snyder, L.R., Kirkland, J.J., Dolan, J.W., 2011. *Introduction to Modern Liquid Chromatography*. John Wiley and Sons, Hoboken, NJ.

Szczepaniak, W., 2012. *Metody instrumentalne w analizie chemicznej*. Wydawnictwo Naukowe PWN, Warszawa.

Wilde, K.D., Engewald, W., 2014. *Practical Gas Chromatography: A Comprehensive Reference*. Springer-Verlag, Berlin–Heidelberg.

Original scientific papers

Costantini, L.M., Baloban, M., Markwardt, M.L., Rizzo, M., Guo, F., Verkhusha, V.V., Snapp, E.L., 2015. *A palette of fluorescent proteins optimized for diverse cellular environments.* Nat. Commun. 6, 7670.

Haase, D., Clemens, J., Wellmann, T., 2019. *Front and back yard green analysis with subpixel vegetation fractions from earth observation data in a city.* Landscape Urban Plan. 182, 44–54.

Janicki, T., Długoński, J., Krupiński, M., 2018. *Detoxification and simultaneous removal of phenolic xenobiotics and heavy metals with endocrine-disrupting activity by the non-ligninolytic fungus Umbelopsis isabellina.* J. Hazard. Mater. 360, 661–669.

Kopeć, D., Woziwoda, B., Fortysiak, J., Sławik, J., Ptak, A., Charąża, E., 2016. *The use of ALS, botanical, and soil data to monitor the environment hazards and regeneration capacity of areas devastated by highway construction.* Environ. Sci. Pollut. Res. 23, 13718–13731.

Krupiński, M., Janicki, T., Pałecz, B., Długoński, J., 2014. *Biodegradation and utilization of 4-n-nonylphenol by Aspergillus versicolor as a sole carbon and energy source.* J. Hazard. Mater. 280, 678–684.

Osińska-Skotak, K., Jełowicki, Ł., Bakuła, K., Michalska-Hejduk, D., Wylazłowska, J., Kopeć, D., 2019. *Analysis of Using Dense Image Matching Techniques to Study the Process of Secondary Succession in Non-Forest Natura 2000 Habitats.* Remote Sens. 11, 893.

Różalska, S., Soboń, A., Pawłowska, J., Wrzosek, M., Długoński, J., 2015. *Biodegradation of nonylphenol by a novel entomopathogenic Metarhizium robertsii strain.* Bioresour. Technol. 191, 166–172.

Wellmann, T., Haase, D., Knapp, S., Salbach, Ch., Selsamd, P., Lausch, A., 2018. *Urban land use intensity assessment: the potential of spatio-temporal spectral traits with remote sensing.* Ecol. Indic. 85, 190–203.

Websites

Dennis Ch., Beales P., 2019. *How we could use drones to detect plant health threats – APHA Science Blog*, https://aphascience.blog.gov.uk/2019/09/20/drones (access: 20.09.2019).

Geoportal – Geoportal Infrastruktury Informacji Przestrzennej, https://www.geoportal.gov.pl (access: 20.09.2019).

SmallGIS, https://teledetekcja.smallgis.pl/index.php?a=usluga&b=zmiany_zabudowy (access: 2.12.2019).

USM Sp. z o.o., Mobilny System Obserwacji i Wspomagania Analizy Powietrza SOWA, http://usm.net.pl/sowa-2 (access: 2.12.2019).

Legal acts

Ustawa z dnia 29 listopada 2000 r. Prawo atomowe, Dz. U. z 2018 r., poz. 792, nr 1669.

Rozporządzenie Rady Ministrów z dnia 12 lipca 2006 r. w sprawie szczegółowych warunków bezpiecznej pracy ze źródłami promieniowania jonizującego, Dz. U. z 2006 r., nr 140, poz. 994.

Rozporządzenie Rady Ministrów z dnia 2 września 2016 r. w sprawie stanowiska mającego istotne znaczenie dla zapewnienia bezpieczeństwa jądrowego i ochrony radiologicznej oraz inspektorów ochrony radiologicznej, Dz. U. z 2016 r., poz. 1513.

3
**Determining the taxonomic affiliation
of microorganisms**

3.1. Genotypic techniques for differentiation and identification of bacteria

INTRODUCTION

The determination of the affiliation of a strain of a microorganism to a relevant taxonomic unit is important to determine its biological diversity, functional aspects and relationships with other organisms. The isolation and characterisation of bacteria is based on classical methods (phenotypic techniques), based mainly on culturing and methods using genotypic characteristics. Classical techniques are usually time-consuming, depend on many environmental factors and often do not lead to accurate classification of the tested strain (e.g. determination of species affiliation). Techniques based on the analysis and properties of nucleic acids (genotyping techniques), due to their speed of analysis, high specificity and repeatability, lower probability of error than in classical methods and the possibility of differentiation below the level of species or subspecies, have become a beneficial tool in taxonomic research. The most important concepts and methods for the detection and identification of bacteria based on genetic analysis are presented in Figure 3.1.1.

Biotyping techniques based on DNA analysis usually use the DNA fingerprinting method, which results in the so-called genetic fingerprint, i.e., a unique, electrophoretic system of replicate DNA fragments corresponding to individual bacterial genotypes being the basis for their differentiation. Many identification methods are also based on hybridisation techniques, i.e., obtaining a hybrid between the examined DNA fragment and a molecular probe. Precise sequencing techniques and subsequent analysis of length and/or sequence of single genes, operons, selected polymorphic DNA fragments or even entire genomes are also used in common practice. Genetic methods for selecting and identifying bacteria can generally be divided into systems based on the analyses of plasmid DNA or genomic DNA.

The two principal methods of genotyping based on plasmid DNA are Plasmid Profile Analysis (PPA) and Restriction Enzyme Analysis of Plasmid DNA (REAP). It should be emphasised that both techniques are limited to the testing of strains with plasmids and may give erroneous results because many species and even strains of bacteria acquire and lose plasmids spontaneously, so that these microorganisms may have a similar plasmid profile.

Widely used methods of identification of bacteria, based on chromosomal DNA analysis, include the study of fragments of rrn operon sequences, including genes encoding ribosomal RNA (ribotyping). The most common sequences used in ribosomal analyses, based mostly on PCR and sequencing, are: (1) hyper variable regions of the 16S rRNA encoding gene (mainly the V1-V3 region); (2) polymorphic regions between the 16S rRNA and 23S rRNA (ribosomal RNA) encoding genes (Internal Transcribed Spacer PCR – ITS PCR); (3) 5S, 16S and 23S rRNA encoding sequences; (4) tRNA encoding region and (5) intergenic region (16S-23S rRNA), which includes the tRNA encoding sequences (tRNA Intergenic Spacer PCR – tDNA-PCR).

Popular ribotyping techniques also include Amplified Ribosomal DNA Restriction Analysis (ARDRA), which consists of obtaining restriction profiles of specific strains based on specific restriction enzyme digestion of an amplified rDNA fragment that usually encodes a 16S rRNA subunit.

Examples of many commonly used molecular bacteria identification and differentiation methods based on DNA analysis include:

- Restriction Enzyme Analysis Pulsed Field Gel Electrophoresis (REA-PFGE) a technique that allows electrophoretic separation of a small number of large DNA restriction fragments in order to obtain a specific profile ("genetic fingerprint") characteristic of the tested microorganism;

- Polymerase Chain Reaction-Denaturing Gradient Gel Electrophoresis (PCR-DGGE), which enables the differentiation of DNA regions containing several nucleotide polymorphisms;
- Randomly Amplified Polymorphic DNA (RAPD), which, by hybridising of a primer or primers to a matrix with any or all or partially arbitrary sequence, allows a profile of largely non-allelical PCR products to be obtained, reflecting the differences in amplified nucleotide sequences between primer-binding sites;
- Multilocus Sequence Typing (MLST) method, which is a comparative analysis of a sequence of internal fragments of (usually) seven housekeeping genes. This method usually analyses gene fragments of about 450–500 base pairs, which have sufficient variability to differentiate billions of distinct allelic profiles;
- Multiple Locus Variable Number Tandem Repeats Analysis (MLVA), based on PCR amplification and sequencing of rapidly mutating, repeating DNA fragments, called tandem repetitions. In this method, both minisatellite regions (2–6 nb per section) and macrosatellite regions (10–60 nb per section) are duplicated;
- Rep-PCR (Repetitive Sequence – Based PCR) technique, based on the amplification of repetitive REP (Repetitive Extragenic Palindromic) sequences of about 35 nb, separating longer sequences of single DNA copies, which in the electrophoretic chapter present a specific profile of amplicons of different lengths;
- DNA microarrays, i.e., a set of thousands of fragments of DNA probes with known sequences, connected on a solid base in an orderly pattern, whose principle of operation is based on the hybridisation of complementary nucleic acid molecules. There are two basic types of DNA microarrays: oligonucleotide microarrays (so-called DNA chips) and cDNA microarrays, which differ in the way they are obtained, their structure and type of probes used;

Whole Genome Sequencing (WGS) with a special focus on Next Generation Sequencing (NGS) methods, based on parallel mass sequencing of up to several hundred million matrices.

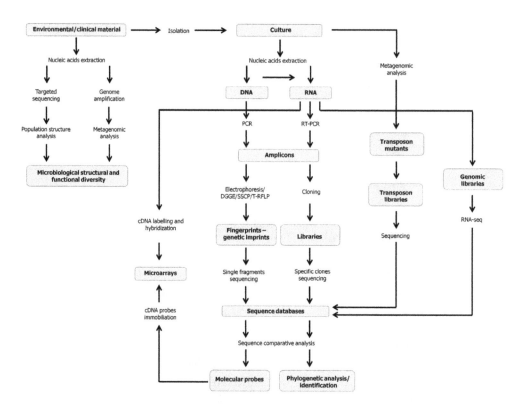

Fig. 3.1.1. Main pathways for differentiation and identification of bacteria by genetic methods

3.1.1. Isolation from the soil and identification of anaerobes by multiplex PCR

INTRODUCTION

The multiplex PCR (M-PCR) method is a modification of the polymerase chain reaction (PCR) that enables amplification of two or more genome sequences by using several primers in a single reaction mixture. As a result, amplicons of different sizes, specific to the respective DNA fragments are produced. The M-PCR technique therefore not only reduces the work steps and thus the time, but also saves genetic material (one reaction is enough to amplify several target sequences). Although M-PCR reduces testing costs and allows for fast, simultaneous and precise detection of multiple sequences, optimisation of the reaction can pose a number of difficulties, including low sensitivity and specificity. An appropriate selection of primers is the key issue for the effectiveness of the multiplex PCR reaction. Proper design of primer parameters, such as length, GC vapour content, attaching (hybridisation) temperature, homology with the target sequence and concentration are essential to achieve high specificity and efficiency of the reaction. The primers used in M-PCR should have an appropriate length, about 18–22 pz., GC content in the range of 45–60% and comparable melting point (Tm) in the range of 55–60°C (75–80°C for primers with higher amounts of guanine and cytosine in the chain). Inadequate optimisation of the above-mentioned factors results in lower amplification efficiency as well as promotes non-specific binding of the primers to the matrix and consequently the formation of inappropriate PCR reaction products. The presence of more than one primer pair in the reaction environment also increases the likelihood of obtaining undesirable amplification products, due to the possibility of their cross-interference and the formation of primer pairs. Therefore, an important step in reaction planning should be to control the primers against the possibility of creating their dimers, with all primers present in the reaction mixture. The effectiveness of the M-PCR reaction may also be affected by the phenomenon of preferential amplification of one target sequence over another. One of the processes determining this phenomenon is the so-called PCR drift, i.e., the fluctuation of interactions between individual components of the reaction mixture, taking place especially in the early phases of the PCR reaction cycle and in the presence of very low matrix concentration.

The multiplex PCR can basically be divided into two groups: a technique using a single matrix or a reaction with multiple matrices, which aims to amplify specific regions within all the matrices used.

The multiplex PCR technique has found application in many areas of DNA testing, including: (1) in the detection of deletions in target sequences, (2) in the analysis of mutations and polymorphisms, (3) in reverse transcription PCR reactions (RT-PCR) or (4) in quantitative analyses. Given the practical aspect of M-PCR, it is highly desirable for research aimed at achieving simultaneous detection of multiple agents that cause similar or identical clinical syndromes and/or have similar epidemiological characteristics. Therefore, this technique has proven to be a valuable tool for identifying viruses, bacteria and parasites in the aetiology of infectious diseases. An example of using

M-PCR in diagnostics is the use of mRT-PCR (Multiplex Reverse Transcription Polymerase Chain Reaction) to simultaneously detect certain human enteroviruses.

Another example of using the M-PCR reaction is the identification of *Clostridium perfringens* strains and the determination of toxigenicity. Classic identification of *C. perfringens* bacteria is based on microscopic observations, examination of biochemical characteristics and antibiotic susceptibility.

C. perfringens (gas gangrene bacteria) is a Gram-positive, mostly spore-forming, anaerobic rod-shaped bacteria. Unlike all other pathogenic bacteria of the genus *Clostridium*, the cells have no flagellum and therefore no ability to move. *C. perfringens* bacteria produce four major lethal protein toxins called alpha (CPA), beta (CPB), epsilon (ETX), and jota (ITX), under which strains of this species are classified into five toxinotypes (biotypes) from A to E, each of which is responsible for a different disease in humans and animals.

In addition, each *C. perfringens* toxinotype can (but does not always), synthesise other toxins and virulence-enhancing enzymes such as enterotoxin (CPE) or perfringolysin O (PFO). Type A strains are considered to be ubiquitous and most common in the environment. They are found primarily in soil and sewage, but can also constitute some of the natural flora in the digestive tract of humans and other animals. *C. perfringens* type A is a well recognised cause of food poisoning caused mainly by consumption of food (especially meat) contaminated with strains capable of producing enterotoxin. However, the bacteria of this toxinotype primarily produce significant amounts of alpha toxin, which shows hemolytic properties and degrades phosphatidylcholine and sphingomyelin, causing damage to cell membranes of eukaryotes. As a result, the infection leads to the destruction of tissues, their ischaemia and hypoxia. Alpha toxin also activates the arachidonic acid cascade, resulting in formation of prostaglandins, leukotrienes and thromboxanes, which generate an inflammatory reaction and cause vasoconstriction. *C. perfringens* type A is primarily the cause of infections of the skin and subcutaneous tissue, but the infection can also spread to other organs and systems. Alpha toxin can affect, among others, myocardial function, causing pressure reduction and bradycardia.

PRACTICAL PART

Materials and media

1. Media: Blood agar with sheep blood, McClungToabe (with egg yolk), enriched broth, broth enriched with tryptophan, medium with urea according to Christensen.
2. Reagents: Ehrlich reagent, Wirtz and Gram staining reagents, agarose, DNA size standard 100 bp, TAE buffer, load buffer for applying DNA to the gel wells (e.g. 6 x load buffer by A&A), dye for DNA visualisation in the gel.
3. Apparatus: ANOXOMAT device for preparation and culture of bacteria in an anaerobic atmosphere, jars for culture of bacteria in anaerobic conditions, incubator, rotary shaker, optical microscope with 100x lens magnification, vortex shaker, thermoblock, thermocycler, horizontal electrophoresis kit with power supply, laboratory mini centrifuge, spectrophotometer for measuring DNA microvolume (e.g. NanoDropTM from Thermo Scientific), transilluminator.
4. Other materials: soil (10 g), serological pipettes, plastic bacteriological tests, sterile NaCl solution (0.85%), glass beads 0.5 – 1 cm in diameter, bacteriological tubes.

Aim of the exercise
Isolation of anaerobic microorganisms from the soil and analysis of isolated bacteria for the presence of toxicogenic C. perfringens strains.

Procedure

1. Isolation of anaerobic bacteria and separation of the *C. perfringens* strains among them:
   a) introduce the material to be tested (10 g of soil) into Erlenmeyer flasks (300 ml) containing 100 ml 0.85% NaCl with sterile glass beads. Heat the flasks in an incubator for 20 min at 80°C and then incubate on a rotary shaker (180 rpm) for 15 min at room temperature;

145

b) prepare a series of dilutions of the soil samples to be tested, then inoculate 0.1 ml of each of them on agar plates with the addition of sheep blood and incubate in anaerobic jars for 24 h at 37°C under anaerobic conditions;

c) from single colonies indicating beta-hemolytic activity (complete destruction of the blood cells manifesting itself by a clear brightening around the colony, most often there is a double hemolysis zone – total hemolysis closer to the colony, partial hemolysis farther away from the colony), inoculate the plates with addition of sheep blood (confirmation of beta hemolysis and multiplication of pure cultures); incubate the plates in anaerobic jars for 24 h at 37°C in an anaerobic atmosphere;

d) perform microscopic preparations from pure bacteria cultures obtained on blood agar plates (confirming beta-hemolytic activity) stained with the Gram method. For selected Gram-positive bacteria, perform staining with Wirtz method (bacteria of genus *C. perfringens* produce endospores in the form of a centrally placed "barrel", but under in vitro conditions cultured on agar media they very rarely form spore forms; there are also non-sporulating *C. perfringens* strains);

e) from the same 24 h bacterial cultures on blood agar plates (point c), perform a growth culture on McClunge's agar plates and a liquid broth medium (2 times 5 ml of growth medium in bacteriological tubes) and a medium according to Christensen. With the addition of tryptophan; incubate all cultures for 24 h at 37°C under anaerobic conditions. In case of the culture in liquid broth, incubate the bacteria also under aerobic conditions in order to control growth in an atmosphere with oxygen;

f) after 24 h of incubation, colonies suspected of belonging to *C. perfringens* (no growth under aerobic conditions, Gram-positive, beta-hemolysis on agar with sheep blood, decomposition of lecithin on McClunge's medium to phospholine and bicyrol (characteristic ring of precipitate around the colony), negative biochemical tests for the ability to decompose urea and tryptophan to indole) should be inoculated into enriched broth and incubated for 24 h at 37°C under anaerobic conditions in order to obtain cultures for the detection of genes encoding major lethal toxins (alpha, beta, epsilon and jota) produced by *C. perfringens* by multiplex PCR.

2. Genetic identification of toxinotypes A-E of *C. perfringens* strains.

Isolate the bacterial DNA using a commercial DNA isolation kit (e.g. Easy-DNATM Kit from Thermo Fisher Scientific) according to the manufacturer's instructions. Determine the concentration and purity of the isolated DNA using the NanoDrop device (determine values at wavelengths A260, A280, A260/230 and A260/280):

a) prepare the reaction mixture for the PCR reaction:

- 25 μl of ready-made commercial amplification mixture, containing DNA polymerase, $MgCl_2$ and
- dNTPs (e.g. ACCUZYMETM Mix of Bioline);
- 1 μl of each primer (according to Table 3.1.1) at a concentration of 10 μM (final primer concentration in a reaction mixture 0.2 μM);
- 1 μl DNA matrix;
- 16 μl double distilled water (ddH_2O).

Toxin	Sequence (from 5' to 3')	Amplicon size (pz.)
CPA	GCTAATGTTACTGCCGTTGACC	324
	TCTGATACATCGTGTAAG	
CPB	GCGAATATGCTGAATCATCTA	196
	GCAGGAACATTAGTATATCTTC	
ETX	GCGGTGATATCCATCTATTC	655
	CCACTTACTTGTCCTACTAAC	
ITX	ACTACTCTCAGACAAGACAG	446
	CTTTCCTTCTATACTATACG	

Table 3.1.1.
Amplification primer sequences of genes encoding the four main toxins produced by *C. perfringens* strains

b) carry out the amplification reaction in a thermocycler under the following thermal and time conditions: pre-denaturation at 95°C for 5 min, then 35 cycles consisting of: denaturation at 95°C for 1 min, connexion (hybridisation) of primers at 55°C for 1 min, elongation at 72°C for 1 min and final elongation at 72°C for 5 min after the last cycle;

c) separate the PCR products electrophoretically in TAE buffer on 1% agarose gel, applying 10 μl of the matrix previously mixed with a load dye at a ration 10:1 to the wells; also apply a DNA mass marker (100 bp) on the gel; after the electrophoretic separation, stain the gel with a reagent that allows visualisation of the DNA under UV light (e.g. SYBR Green).

Analysis of results

On the basis of the electrophoretically separated stripe pattern of DNA amplification fragments and on the basis of Table 3.1.2 presenting the classification of toxinotypes of *C. perfringens* strains, determine the presence of individual bacterial biotypes in the tested soil samples.

Table 3.1.2.
Classification of
C. perfringens strains
on the basis of their
ability to produce four
major lethal toxins

Toxin	Location*	Biological activity	Toxinotype				
			A	B	C	D	E
CPA	C	necrotic diseases, hemolytic activity, smooth muscle contraction	+	+	+	+	+
CPB	P	Necrosis, oedema, enterotoxicity	-	+	+	-	-
ETX	P	Skin necrosis, swelling, smooth muscle contraction	-	+	-	+	-
ITX	P	Necrosis	-	-	-	-	+

Explanations: * C – chromosomal DNA, P – plasmid DNA.

3.1.2. Taxonomic classification of the genus *Streptomyces* based on PCR 16S rRNA method

INTRODUCTION

Actinomycete (*Actinobacteria* phylum) are Gram-positive, mostly acid-resistant bacteria, capable of using various nutrients as a primary source of carbon and energy (chemoorganotrophs) and with a high GC pairs vapour content in the DNA molecule (69–78%).

Most of the species are obligate aerobes, but there are also microaerophilic and relatively anaerobic actinomycetes (e.g. from the genus *Actinomyces* sp.). Many types of bacteria belonging to this taxonomic type occur in different pleomorphic forms (*Corynebacterium* sp., *Gardnerella* sp.). Some of actinomycete genera, including most of biotechnological significance, are classified into an informal group called actinomycetes. These include the following genera: *Streptomyces, Frankia, Nocardia and Actinomyces*. Colony morphology, nutritional requirements, life cycle and form of reproduction make this group similar to microfungi. This similarity can be primarily seen in the morphology of actinomycete surface culture, in which they grow in the form of filamentous, often branching cells, macroscopically resembling the structure of the mycelium. In the first phase of growth, most of the elongated actinomycete cells grow deep into the medium, creating a basic (substrate) pseudomycelium, smooth in terms of surface and loose in the spaces between the hyphae, which usually forms a radial arrangement from the place where the colonies begin to grow. In the next stage of its life cycle, it differentiates into a secondary (aerial) pseudomycelium, growing above the basic pseudomycelium, the hyphae have a more compact structure and its surface takes on a granular, flocculent or powdered appearance. The division into primary and secondary pseudomycelium is a functional classification, as there are usually no significant differences in morphological structure between the hyphae in both forms of bacterial colonies. It happens, however, that filamentous cells of the substrate pseudomycelium may be thinner and contain less vacuoles than the hyphae forming the aerial pseudomycelium.

Actinomycetes proliferate vegetatively (by fragmentation of the pseudomycelium) or by spore production. Under favourable conditions, the hyphae of the aerial pseudomycelium increase in thickness, transforming into conidiophores, at the top of which there is a process of sporulation, i.e., production of conidial spores. Mostly spherical and immobile spores are formed as a result of conidiophores segmentation, which determines their considerable morphological differentiation.

The structure of pore-producing hyphae then takes on the characteristic arrangement of straight, wavy, spirally twisted or pastoral-like spore chains and is one of the most important diagnostic features of actinomycetes. In some species, spores are formed in special sporangia at the top of conidiophores (e.g. *Streptosporangium* sp.). Others, on the other hand, may produce ciliated spores capable of movement (e.g. *Actinoplanes* sp.). A characteristic feature of the majority of actinomycetes is the ability to produce numerous pigments that can give colour to the colonies themselves (endopigments), as well as to colour the growth environment by diffusing pigments through the substrate (exopigments).

The high adaptability of the actinomycetes to the prevailing environmental conditions and the wide spectrum of organic substances they decompose makes them common bacteria found in different environments. The most common source is the soil ecosystems. These bacteria also occur in marine and fresh water, bottom sediments, organic waste, decomposing plant matter, but have also been isolated from caves or glaciers from the Antarctic. Actinomycetes constitute an important population of soil microorganisms as they play a key role in the cycling of elements in nature and contribute to the formation of humic compounds. They fertilise and mineralise the soil, decomposing various, complex chemical compounds, including those that are difficult to degrade by other microorganisms, such as chitin, lignins, cellulose, starch, pectin. Many *Actinobacteria* species are also characterised by the ability to degrade substances of anthropogenic origin with different physicochemical properties, including: pesticides, aliphatic hydrocarbons or compounds containing aromatic rings, thus helping in bioremediation of contaminated environments. In addition, Actinomycetes can also produce enzymes, such as lysozyme or glucanases, which cause lysis of cell walls of bacteria and filamentous fungi. Some of the species are in symbiosis with insects (*Gordonibacter* sp.) or vascular plants (*Frankia* sp.), while others have acquired pathogenic characteristics such as *Mycobacterium* sp.

One of the main aspects of industrial interest in actinomycetes is their ability to produce many biologically active compounds, including antibiotics. Most species synthesising antibiotic substances used in medicine, veterinary practice and agriculture belong to the genus *Streptomyces* sp. About two thirds of antibiotics of natural origin are produced by this type of bacteria. Examples of known antibiotic metabolites produced by *Streptomyces* sp. include streptomycin (*S. griseus*), tetracycline (*S. aureofaciens*), amphotericin B (*S. nodosus*), nystatin (*S. nursei*), chloramphenicol (*S. venezuelae*) and neomycin (*S. fradiae*). These organisms are therefore subject to intensive research in the expectation that they will contribute significantly to the provision of new drugs to combat the global emergence of antibiotic resistance among pathogenic microorganisms.

In terms of the number and diversity of the known species, the genus *Streptomyces* sp. is one of the largest taxonomic positions of actinomycetes. The identification of this group of bacteria is possible, among others, through the assessment of physiological and morphological features (types of conidiophores, colour of aerial mycelium, type of spore surface), biochemical tests (e.g. determination of sugar fermentation capacity), assessment of chemical composition of the cell wall, including determination of peptidoglycan type, characterisation of fatty acid and phospholipid chains in the cell membrane, determination of GC vapour content, 16S rDNA sequence analysis and DNA-DNA hybridisation.

There is a great variety of bacteria in the environment, so the nature and complexity of the molecular techniques used to identify them are very diverse. The analysis of the nucleotide sequence of the gene encoding the 16S rRNA ribosomal subunit is one of the most commonly used methods to determine the taxonomic position of bacteria, as well as to establish the phylogenetic affinity of the examined organisms. The main advantages of using the 16S rRNA gene as a molecular marker in identification include: (1) affiliation to a group of housekeeping genes, i.e., coding sequences that occur in all bacteria and are responsible for the most important metabolic processes in the cell, and their expression is relatively constant regardless of physiological state; (2) having strongly conservative regions preceded by fragments of high polymorphism (variable and hypervariable sequences, unique for the genus or species of bacteria), (3) a small length, about 1500 pz, which allows its relatively easy amplification and sequencing, and at the same time contains enough information to allow taxonomic classification and phylogenetic analysis.

The use of 16S rRNA analysis for bacterial differentiation consists of amplification and sequencing of the whole or smaller fragments (characteristic for a given species) of the gene encoding the 16S rRNA subunit, which are then merged into the whole, and homology is compared with gene sequence patterns found in specialised databases (e.g. RDP-II – Ribosomal Database Project, RIDOM – Ribosomal Differentiation of Medical Microorganisms or MicroSeq). It is generally accepted that a higher than 99% similarity of the tested sequence to the reference sequence, allows the test organism to be assigned to the species, while a similarity of at least 97% allows the bacteria to be assigned to the genus.

By analysing the 16S rDNA sequence, many genera and species of bacteria have been taxonomically reclassified, the systematisation of numerous slow-growing microorganisms and so-called unculturable microorganisms (i.e., those that cannot be isolated by culturing methods, for which no culture media and conditions have been developed) has been carried out, and this method has allowed the discovery of a number of new species and strains of prokaryotes. The use of 16S rDNA sequencing is not only limited to research purposes, but is also successfully used in clinical microbiological laboratories to identify bacterial isolates from different diseases. In many cases, the performance of such molecular analyses provides the basis for microbiological diagnosis or is an important complement to conventional bacteriological tests based on phenotypic and biochemical tests.

It should be noted that the differentiation of some bacteria (e.g. closely related Staphylococcus or Mycobacterium species) through 16S rDNA sequence analysis is not possible due to the highly conservative nature of this region. Although this method has some limitations of use, 16S rDNA sequencing gives a higher percentage of species identification than routine phenotypic methods, and the success rate of identification with this method varies from 62 to 92% depending on the bacteria group and classification criteria. The high sensitivity and repeatability of the 16S rRNA analysis method, it's relatively low cost and the constantly expanding databases with new 16S rDNA sequences make this technique a valuable tool for both scientific research and laboratory diagnostics.

PRACTICAL PART

Materials and media

1. Organisms: soil actinomycete strain isolate, electrocompetent E. coli DH5α cells.
2. Reagents: agarose, ampicillin, X-gal, ITPG, DNA size standard 100–3000 bp, TBE buffer, load buffer for applying DNA to the gel wells (e.g. 6 times load buffer from A&A), dye for DNA visualisation in the gel (e.g. SYBR Green), EcoRI restriction enzyme with DNA digestion reagents.
3. Media: Luria-Bertani (LB), SOC, LB selective medium with ampicillin, X-gal and IPTG.
4. Ready-made commercial sets for: plasmid DNA isolation, purification of PCR products, ligation of PCR product to the DNA of the plasmid vector, plasmid DNA purification.
5. Apparatus: incubator, thermoblock, thermocycler, laboratory mini-centrifuge, electroporator (e.g. Gene Pulser XcellTM Electroporation System from Bio-Rad), horizontal electrophoresis kit with power supply, spectrophotometer e.g. NanoDropTM from Thermo Scientific, transilluminator.
6. Other materials: Petri dishes, serological pipettes, automatic pipettes, sterile tips for automatic pipettes, bacteriological loops.

Aim of the exercise

Taxonomic classification of the genus Streptomyces based on PCR 16S rRNA method.

Procedure

1. Isolation of chromosomal DNA.

   Perform DNA isolation using a commercial DNA isolation kit according to the procedure provided by the manufacturer. After isolation, perform spectrophotometric measurement (A230, A260 and A280) with the NanoDrop device to determine the concentration and purity of DNA.

2. Amplification of the 16S rDNA fragment by PCR:

   a) prepare the reaction mixture for the PCR reaction:
   - 25 µl of ready-made, commercial amplification mixture containing DNA polymerase, $MgCl_2$ and dNTPs (e.g. ACCUZYMETM Mix from Bioline);
   - 1 µl of each primer (final concentration 5 pM µl^{-1}):
   - primer 1 (5' AAGGAGGTGATCCAGCCGCA),
     primer 2 (5' AGAGTTTGATCCTGGCTGAG);
   - 2 µl DNA matrix;
   - 2,5 µl DMSO;
   - 21 µl double distilled water (ddH_2O);

   b) carry out the amplification reaction in the thermocycler under the following thermal and time conditions: pre-denaturation at 94°C for 4 min, then 30 cycles, consisting of: denaturation at 94°C for 0.45 min, connexion (hybridisation) of primers at 55°C for 0.45 min, elongation at 72°C for 1.30 min and after the last cycle final elongation at 72°C for 5 min.

3. Separation of DNA fragments on agarose gel. Prepare a 1% agarose solution in 1 × TBE in a volume adapted to the apparatus used, then apply 10 µl of isolated DNA from the previously mixed 1 µl load buffer to the wells of the gel; apply a DNA size marker to one of the wells; carry out electrophoresis in 1 × TBE buffer at 100 V for 30 min; after electrophoresis, stain the gel with a reagent allowing

for visualisation of the DNA under UV light (e.g., SYBR Green); assess the presence of amplicon (~1500 pz).

4. PCR product purification.
Purify PCR products with a ready-made commercial kit (e.g. GeneJetTM PCR Purification Kit from Thermo Scientific) according to the procedure provided by the manufacturer.

5. Ligation of PCR product to pGEM T-Easy vector.
Carry out the ligation with a ready-made, commercial pGEM – T-Easy Vector System I from Promega according to the procedure provided by the manufacturer.

6. Electrotransformation of *E. coli* DH5α competent cells.
Place the suspension of electrocompetent DH5α cells (50 μl) in an electroporation cuvette together with 5 μl of DNA solution (after ligation); place the cuvette in the electroporation chamber and trigger an electrical impulse; immediately after the cells are treated with an impulse, suspend the material in 2 ml of SOC medium and incubate for 1 h at 37°C; electroporation parameters: voltage: 1500 V, capacitor capacity: 50 μF, external resistance: 100 ohms.

7. Culture on LB ampicillin, X-gal and IPTG selective media.
In order to select recombinants, inoculate a bacterial suspension on a selective medium (ampicillin, X-gal and ITPG); white colonies – cells with a vector with the desired insert

8. Multiplication of recombinant clones.
Inoculate selected colonies on 2 ml of LB medium with ampicillin (100 μg ml^{-1}) and incubate overnight (16–18 h) at 37°C.

9. Isolation and purification of plasmid DNA.
Isolate and purify the plasmid DNA using a ready-made, commercial kit (e.g. QIAprep Spin Miniprep by QIAGEN) according to the procedure provided by the manufacturer.

10. DNA digestion with EcoRI restriction enzyme to check the presence and size of the insert.
The obtained plasmid DNA should be digested with EcoRI restriction enzyme according to the procedure provided by the manufacturer of the enzyme; the size of the cut insert

should be checked electrophoretically on 1% agarose gel; the electrophoresis should be carried out in 1 × TBE buffer at 100 V for 30 min; after the electrophoretic separation, the gel should be stained with a reagent that allows visualisation of the DNA in UV light (e.g. SYBR Green); a DNA size marker should be applied to one of the wells. ATTENTION! The insert has an internal digestion site for EcoRI, so two additional fragments of about 850 and 650 pz should be visible on the gel.

Analysis of results

Plasmid DNA with a cloned insert should be forwarded for sequencing and sequence analysis in genomic databases. Once the results are obtained, the microorganism should be assigned to a specific taxonomic unit and the affinity analysis and phylogenetic relationship should be determined.

3.2. Fungi of the phyla Mucoromycota, Ascomycota and Basidiomycota – morphological, biochemical features and genetic analysis

INTRODUCTION

Fungi are heterotrophic organisms, which can decompose complex polymers (with the participation of extracellular enzymes) and use them as a source of carbon, nitrogen and energy. They have a cell wall and a clearly separated cell nucleus, so they belong to the domain Eukaryote. In older taxonomic systems, fungi belonged to the plant kingdom (mainly because of the presence of the wall). Since 1969 they are classified to the fungi kingdom (Fungi or Mycota). The most characteristic morphological form of the fungus is a hypha. It is not observed in any other group of organisms. However, this is not a criterion for belonging to this kingdom, since fungi also include yeast, which occur in single cell form (described in Section 3.2.3). The growth of hyphae allows the penetration of solid media and the acquisition of new nutrients. Fungal organisms are made up of hyphae, organised into mycelia. The mycelia of Ascomycota and Basidiomycota contain septa. In turn, Mucoromycota are multinucleated. There are many nuclei in their hyphae, because there are no septa. The cell membranes of fungi contain ergosterol, a specific sterol, absent in the cells of other organisms. This difference in structure is of great practical importance in the fight against fungi. It enables selective action of antifungal agents, which interfere with ergosterol synthesis. Glycogen, volutin and fats are the spare materials found in fungal cells. The structure of the fungal hypha is presented in Figure 3.2.1.

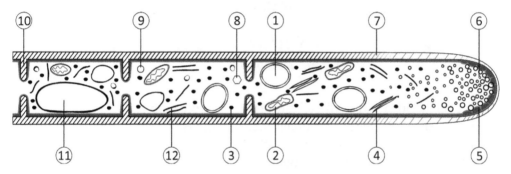

Explanations: 1 – cell nucleus, 2 – mitochondrion, 3 – ribosome, 4 – Golgi apparatus, 5 – cell membrane, 6 – vesicles, 7 – cell wall, 8 – crystals, 9 – spare materials (lipid bodies), 10 – septum, 11 – vacuole, 12 – endoplasmic reticulum.

Fig. 3.2.1. Scheme of hypha structure.

An organism's affiliation to the kingdom of fungi is determined primarily by the presence of chitin in the cell wall. This polymer, composed of N-acetylglucosamine, is also present (apart from fungi) in a number of invertebrates, playing an important role in their life processes. The cell wall gives shape to cells and hyphae, protects against harmful environmental factors and prevents osmotic lysis. Fungi can be deprived of the cell wall by digesting hyphae with a complex of lytic enzymes (protoplasts – Subsection 1.5). The cell wall thickness is usually between 150 and 400 nm. It is mainly composed of polysaccharides, which constitute about 80% of the components. The cell wall consists of four layers (Figure 3.2.2). The outer layer consists of amorphous β-1,3 and β-1,6-glucans (1). Then there is a network of glycoprotein with glucans, embedded in proteins (2). Below, there is a layer of proteins (3). The innermost layer is composed of chitin microfibrilles immersed in proteins (4). Chitin is a structural component of the wall (scaffolding), responsible for its elasticity and durability.

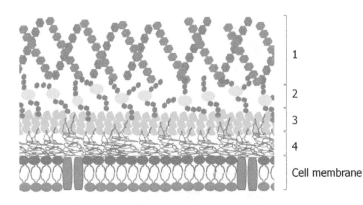

1

2

3

4

Cell membrane

Fig. 3.2.2. Scheme of fungal cell wall structure (explanations in the text).

The composition of the cell wall differs in the representatives of individual taxonomic groups (Table 3.2.1). Chitin is present in the cell wall in all representatives of the kingdom fungi, but is not always observed in all morphological forms or growth phases, e.g. in yeast belonging to the Ascomycota, chitin occurs in the wall only in the bud scar, and in Microsporidia only in the endospore. The presence of chitosan, i.e., deacetylated chitin derivative, is characteristic of Mucoromycota representatives. Differences in the composition of the cell wall occur even in representatives of one species and are related to the morphological form and the life cycle phase.

Table 3.2.1. The most important polysaccharides of the fungal cell wall.

Phylum	Fibrillar components	Amorphous components (matrix)
Mucoromycota	Chitin and chitosan	Polyglucuronic acid, glucuronomannoproteins
Ascomycota	Chitin, β-(1,3) and β-(1,6) glucans	α-(1,3)-glucan, galactomannoproteins
Basidiomycota	Chitin, β-(1,3) and β-(1,6) glucans	α-(1,3)-glucan, xylomannoproteins

It is difficult to determine the number of fungal species found on Earth. So far, a hundred thousand species of fungi have been described, but it is estimated that there may be up to one million. Some of the fungi are non-culturable, i.e., they are not capable of growing on laboratory media. Fungi often live inside other organisms as endophytes or parasites, as well as in extreme environments where they are difficult to isolate. In addition, modern molecular methods make it possible to isolate new species on the basis of genetic differences that are impossible to be observed at the level of morphological studies (e.g. a new species of *Aspergillus brasiliensis* was isolated from the *A. niger* species on the basis of β-tubulin and calmodulin gene sequence analysis). The key criterion in the natural taxonomy of fungi is the method of sexual reproduction. On this basis, fungi are divided into several types (Table 3.2.2).

Table 3.2.2. The most important features of various fungi phyla.

Phylum	Estimated number of identified species	Representatives	Reproduction	Significance
Microsporidia	1,250	*Nosema bombycis* *Areospora* sp. *Encephalitozoon* sp.	Asexual: spores with infectious thread Sexual: plasmogamy, karyogamy and meiosis	Obligate intracellular parasites of invertebrates, fish and humans with reduced immunity
Chytridiomycota	980	*Neocallimastix* sp. *Anaeromyces* sp. *Batrachochytrium dendrobatidis*	Sexual and asexual: zoospores	Most often, unicellular, anaerobically decompose glucose in ruminant stomachs (symbiosis)
Previously: Zygomycota; Mucoromycota, Zoopagomycota have been separated in 2016	1,600 760 900	*Rhizopus stolonifer* *Cunninghamella elegans* *Thamnidium elegans* *Erynia* *Entomophthora* *Furia*	Sexual: fusion of gametangia, zygospore is created. Asexual: sporangiospores, Entomophthorales: conidiophores	Decomposition of organic matter, food spoilage. Production of enzymes, chitosan, unsaturated fatty acids, β-carotene. Mucoraceae yeast – alcohol production. Entomophthorales – biocontrol
Ascomycota	90,000	*Chaetomium,* *Aspergillus,* *Penicillium,* *Trichoderma,* *Fusarium*	Sexual: ascospores in bags, there are fruiting bodies. Asexual: conidiospores	The most numerous group, there are saprophytes, symbionts and numerous parasites of plants, animals and humans. Production of important secondary metabolites such as antibiotics and mycotoxins. Yeast – production of alcohol and biomass
Basidiomycota	50,000	*Trametes versicolor* *Phanerodontia chrysosporium* *Agaricus bisporus*	Sexual: basidiospores on the bases, there are fruiting bodies. Asexual, rare: conidiospores	Macroscopic edible and poisonous mushrooms. Numerous plant parasites. Fungi of wood rot

The most represented and important for the environment and man belong to the phyla with the most numerous species: Mucoromycota, Ascomycota and Basidiomycota. Fungi in which no sexual reproduction was observed were collected for practical reasons in an artificial taxon of Deuteromycota. Taxonomic differentiation of these fungi considers macroscopic and microscopic morphology, including the way in which conidial spores are produced, their shape, colour and location on the conidiophore. In natural taxonomy, however, they are included in Ascomycota, without affiliation to particular classes, as mitosporic fungi (forming conidia, asexual spores formed as a result of mitosis). The classification of fungi is still subject to changes, e.g. in 2016 the polyphyletic phylum Zygomycota was replaced by Mucoromycota and Zoopagomycota.

3.2.1. Mucoromycota

These are terrestrial fungi, mainly saprotrophic ones. There are no septa in their mycelium (multinucleated). Sexual reproduction in this taxonomic group is of a nature of zygogamy and gametangiogamy. Two morphologically identical (isogamy), but sexually distinct (male and female) gametangia are fused. A thick-walled zygospore is formed, which must undergo a period of dormancy. Before germination, the diploid nucleus of the zygospore passes through meiosis and germinates into haploid vegetative mycelium or directly into sporangiophore, i.e., spore-bearing stalk. The haploid phase predominates in the life cycle. The life cycle of a Mucoromycota representative is shown in Figure 3.2.3.

Asexual reproduction: this group includes the formation of numerous sporangiospores, which are formed in sporangia, e.g. in *Rhizopus, Mucor* (Figure 8.1.2.C-D). In the more evolutionarily advanced representatives of Mucoromycota, there has been a reduction in the number of asexual spores that are produced in the sporangioles (from several to one), e.g. *Blakeslea* or *Cunninghamella* (Figure 8.1.6.I). In some representatives of the Mucorales order, there are additional elements related to the sporangium and sporangiophore: columella, collar or apophysis, which are helpful in assessing systematic adherence within the Mucorales (Figure 8.1.12.D).

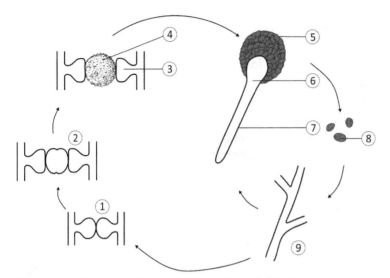

Fig. 3.2.3. Life cycle of Mucoromycota.

Explanations: 1 – progametangia, 2 – fusion of gametangia, 3 – suspensor, 4 – zygospore, 5 – sporangium, 6 – columella, 7 – sporangiophore, 8 – sporangiospores (asexual spores), 9 – somatic hyphae.

The columella is the swollen, top part of sporangiophore, directed inside the sporangium. The collar is the lower part of the sporangium wall, at the top of sporangiophore. Sometimes the sporangiophore is extended, directly at its peak below the sporangium, forming an apophysis.

3.2.2. Ascomycota

It is the most numerous and diverse type of fungi (about 90 thousand species) of great practical importance. Ascomycota are mostly terrestrial fungi. They are dominated by saprotrophs, but there are also numerous plant and animal parasites, as well as strict symbionts (lichens). These fungi have septa in the hyphae. The life cycle is dominated by haplophase, but also dicariophase, i.e., the phase of conjugated nuclei, is observed. The dicariophase is specific only for Ascomycota and Basidiomycota.

A characteristic feature of Ascomycota is the production of sac-like structures (asci) in the sexual process, in which meiotic sexual spores – ascospores – are formed endogenously, usually in the amount of 8 pieces. At the same time many asci are formed (creating a hymenium, i.e., spore-bearing layer) and they are

161

surrounded or supported by a wall, thus creating a fruiting body, which is called an ascocarp.

There are usually 3 types of ascocarps in Ascomycota:
- bowl-shaped apothecium;
- bottle-shaped perithecium;
- closed spherical cleistothecium (Figure 3.2.4).

Some phytopathogenic Ascomycota have an ascostroma – a specific fruiting body when chambers containing asci are formed within the stroma.

The role of the fruiting bodies is to protect the asci and contribute to the spread of ascospores.

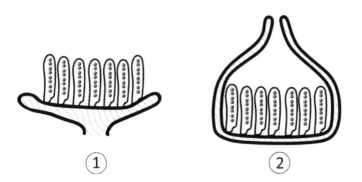 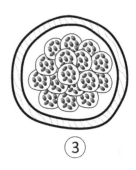

Explanations: 1 – apothecium, 2 – perithecium, 3 – cleistothecium.

Fig. 3.2.4. Structure of the ascocarps in Ascomycota.

Sexual reproduction of Ascomycota is presented in Figure 3.2.5. The process begins with a combination of two morphologically different gametangia or a somatic fusion of two vegetative, sexually differentiated hyphae. The nuclei of the different sexual types of hyphae are placed close to each other to form dicarions. In the case of gametangia fusion, the dikaryophase begins when the nuclei from the antheridium pass to the oogonium and locate close to the nuclei of the oogonium. Then, dikaryotic asci-forming hyphae are produced. The apex part of the hypha is bent and creates a crozier. At the same time the pair of coupled nuclei is divided. The upper pair of nuclei is separated from the basal cell and from the crozier. The basal cell fuses with the crozier, again creating a two-nuclear cell. The apical cell becomes a young ascus in which karyogamy and then meiosis and mitosis occur. The result is 8 haploid ascospores.

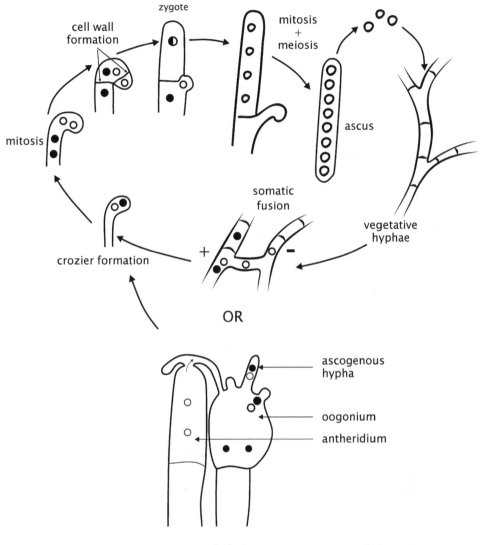

zygote

cell wall
formation

mitosis
+
meiosis

ascus

mitosis

somatic
fusion

vegetative
hyphae

crozier formation

+

−

OR

ascogenous
hypha

oogonium

antheridium

gametangia fusion

The asci stage, i.e., the production of ascospores in asci, is the final phase of sexual reproduction and is called teleomorphic stage (teleomorph).

A representative of Ascomycota, the *Chaetomium globosum*, produces numerous olive perithecia containing asci with ascospores on solid media containing cellulose as a carbon source (Figure 8.1.5. C-D).

Fig. 3.2.5. Sexual reproduction of Ascomycota (explanations in the text).

Numerous Ascomycota reproduce asexually with conidial spores (anamorphic stage, anamorph). Asci and conidia can develop on the same mycelium. Conidia (conidiospores) are asexual spores, formed by mitosis and are widespread in fungi.

Conidiospores differ in the way of formation, shape, size and number of cells, e.g. fungi of the genus *Fusarium* have macroconidia characteristic for this genus, consisting of several sickle-shaped or boomerang-shaped cells. Conidia are most often formed on conidiophors in the conidia-forming cells, i.e., the phialides from which they are released, often forming short chains. In some fungi, e.g. *Aspergillus*, *Penicillium*, characteristic conidia-forming systems are formed (Figure 8.1.4. D-F), composed of conidophore, phialides and numerous conidia (sometimes additional auxiliary cells, e.g. metules, may be present between the conidophore and phialides). Some of the conidia are formed from vegetative hyphae (arthrospores, oidia, chlamydospores). Chlamydospores (resting spores) are characterised by a thick cell wall and often contain melanin pigments. They are produced by fungi belonging to different phyla of Ascomycota, Mucoromycota, Basidiomycota, in old mycelia or in unfavourable environmental conditions.

3.2.3. Basidiomycota

Basidiomycota are a numerous and important phylum of fungi. The hyphae of Basidiomycota contain septa and they have a specific structure (dolipores). Basidiomycota usually reproduce sexually, producing numerous sex spores – basidiospores. Sexual reproduction is simplified, i.e., no gametangia is produced, but a fusion of 2 haploid vegetative hyphae (sexually distinct) occurs. Dikaryotic hyphae are formed from them to produce dikaryotic fruiting bodies (basidiocarps). Basidia are formed in the hymenial layer of the fruiting body. The fusion of nuclei, i.e., karyogamy and meiosis, takes place in basidia. Next, the whole basidium is filled with a vacuole, which pushes the nuclei outwards through the sterigmata. As a result of the sexual process, 4 haploid basidiospores are formed on the basidia (Figure 3.2.6). Dikaryophase dominates the life cycle (n + n).

Basidiomycota include:
- macroscopic edible and poisonous fungi, most often forming fruiting bodies consisting of a cap and a stem;
- wood rot fungi (bracket fungi or resupinate type fruiting bodies);
- dangerous plant parasites, such as rusts, roussels and smuts (they have a complex life cycle, produce several types of specific spores and may require a change of host (rusts)).

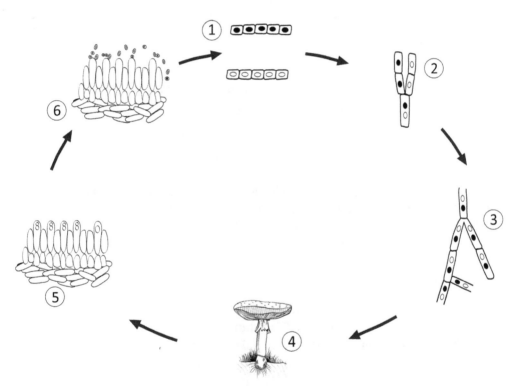

Explanations: 1 – haploid primary mycelium of different sex types, 2 – somatic fusion, 3 – dikaryotic mycelium, 4 – fruiting body, 5 – basidia, 6 – release of basidial spores.

Fig. 3.2.6. Sexual reproduction of Basidiomycota.

165

PRACTICAL PART

Materials and media

1. Species of filamentous fungi, representatives of particular phyla:

   Phylum Mucoromycota: *Rhizopus stolonifer, Cunninghamella elegans, Thamnidium elegans.*

   Phylum Ascomycota: *Chaetomium globosum, Talaromyces ruber (Penicillium rubrum), Fusarium graminearum, Cladosporium herbarum, Aspergillus nidulans.*

   Phylum Basidiomycota: *Trametes versicolor, Phanerodontia chrysosporium (Phanerochaete chrysosporium).*

2. Media: ZT plates, plates with synthetic medium according to Fred with cellulose strips to stimulate the production of fruiting bodies by *Chaetomium globosum.*

3. Apparatus: light microscope.

Aim of the exercise
Comparison of macroscopic and microscopic morphology of selected representatives of particular fungi phyla. Observation of characteristic structures and spores associated with sexual (Chaetomium) and asexual reproduction (other fungi).

Procedure

1. Direct observation and description of macroscopic morphology of fungi on solid ZT medium.
2. Microscopic observation of the material taken up in a drop of water on a microscope slide, covered with a cover slide (live specimens), magnification 400 times.
3. Draw and describe images from the live specimens, microscope magnification 400 times, for the fungal species listed below.

In macroscopic observations, attention should be paid to:

- growth intensity;
- nature of mycelium growth (superficial, growing into medium, aerial);

- mycelium consistency (loose, compact, pressed, smooth, corrugated, shaped radially, concentrically on the surface);
- mycelium colour from the top and bottom (reverse), possibly the secretion of dyes into the medium, presence of coloured clusters of spores on the surface.

In microscopic observations, attention should be paid to:

- presence of septa in mycelium hyphae;
- type of spores, possible presence of resting spores (chlamydospores);
- and the presence of the following elements:
  - Mucoromycota: sporangiospores, sporangia, sporangiophores, columella, collar;
  - Ascomycota: perithecium, asci, ascospores, conidiophores, phialids, metules, conidiospores (conidia).

Analysis of results

On the basis of microscopic observations, indicate which structures associated with reproduction are characteristic of Mucoromycota and which of Ascomycota.

3.2.4. Molecular identification of filamentous fungi

In the previous section, the methods of identification of fungi using micro and macroscopic techniques are presented. However, for a dozen or so years now, taxonomic studies of fungi also use molecular techniques (DNA barcoding), which compare sequences of a selected DNA fragment (e.g. ITS) with sequences available in databases. In contrast to the methods used for identification of bacteria and Actinomycetes, the taxonomic affiliation of fungi should include both molecular and micro and macroscopic analysis.

Molecular identification of fungi includes the following steps:
- DNA isolation from a pure fungal culture;
- DNA fragment duplication by PCR technique based on known primers;
- sequencing the obtained DNA fragment;

- searching for homologous sequences in databases (BLAST search);
- construction of a phylogenetic tree, on the basis of which it is possible to determine the species affiliation and affinity of the examined fungus.

The most important marker, enabling the identification of fungi up to species level, is the ITS region (the fastest evolving part of the rRNA cistron). Figure 3.2.7 presents graphically a fragment of this region with marked places where the primers are attached.

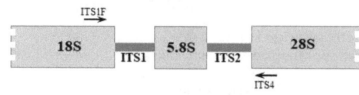

Fig. 3.2.7. Amplification of ITS region fragment.

Explanations: the figure shows the connexion points for ITS1F (forward) and ITS4 (reverse) primers; the primer sequences are given in the practical part.

For some species of fungi belonging to the genera *Aspergillus, Fusarium, Penicilium, Trichoderma or Metarhizium*, the use of only the ITS region for phylogenetic analyses is insufficient, as this fragment does not show sufficient resolution. In this case, additional (secondary) marker genes encoding specific proteins are used (Table 3.2.3).

Marker	Name	Primers
Tef1	Translation elongation factor 1-α gene	EF1-983F oraz EF1-2218R EF1-1018F oraz EF1-1620R ef1 oraz ef2
RPB1	RNA polymerase II subunit 1	RPB1af RPB1cr
RPB2	RNA polymerase II subunit 2	RPB2-5f RPB2-7cR
Tub2	Beta-tubulin	Bt2a Bt2b
CaM	calmodulin	CMD5 CMD6
MCM7	minichromosome maintenance proteins	Mcm7-709for Mcm7-1348rev

Table 3.2.3. Additional marker genes used to identify fungi.

As mentioned before, the obtained sequences (both the ITS fragment and other markers) should be compared with homologous sequences deposited in databases (e.g. GenBank, CBS-KNAW). It is extremely important to compare the obtained sequences with those from reliable sources.

PRACTICAL PART

Materials and reagents

1. Microorganisms: slants with selected strains of filamentous fungi.
2. Reagents: sterile water, mixture of chloroform: isoamyl alcohol 24:1, CTAB/NaCl mixture, isopropanol, ethyl alcohol 100%, TE buffer (pH 8.0), RNAse A (solution 10 mg/ml), proteinase K (20 mg/l), 5 M NaCl, 4 M potassium acetate.
3. Materials: glass beads (1 mm), sterile Eppendorf tubes, tips, automatic pipettes.

Aim of the exercise
Molecular identification of filamentous fungi.

Procedure

NOTE: DNA isolation can also be performed with commercially available sets.

1. DNA isolation:
   a) rinse the slants with sterile deionised water, centrifuge spores and mycelium fragments and suspend in 300 µl TE buffer. Add glass beads, 3 µl proteinase K (20 mg/l) and 30 µl 10% SDS;
   b) homogenize the cells mechanically for 3 min using FastPrep-24 (MP Biomedicals);
   c) incubate for 30 min in a water bath at 55°C,

d) add 130 μl 5 M NaCl and 46 μl 10% CTAB/NaCl (previously heated to 65°C). Mix the sample thoroughly using a vortex mixer and then incubate for 10 min at 65°C;

e) after cooling, add 0.5 ml of chloroform: isoamyl alcohol mixture (24:1) and mix thoroughly by inversion, then centrifuge for 15 min, 13,000 × g at 20°C.;

f) transfer the top layer to a new Eppendorf tube;

g) add 0.1 volume of ACK and 1 volume of isopropanol. Incubate 15 min at −20°C.;

h) centrifuge samples at 4°C, 15 min, 13,000 × g, pour off the supernatant and rinse the precipitate with 1 ml of 70% ethanol, centrifuge again and dry the resulting precipitate.;

i) add 20 μl TE buffer and 2 μl RNAse, incubate at 37°C for 30 min. After incubation, freeze samples at −20°C.;

j) determine the purity of the isolated DNA on a UV-VIS microspectrophotometer.

2. PCR reaction:

a) prepare the reaction mixture for the PCR reaction:
   - 10 μl ready-to-use, commercial amplification mixture containing DNA polymerase, $MgCl_2$ and dNTPs (e.g. ACCUZYME™ Mix from Bioline);
   - 1.5 μl ITS1 (5'-TCCGTAGGTGAACCTGCGG-3') and ITS4 (5'-TCCTCCGCTTATTGATATGC-3') 10 μM primers (final primers concentration in the reaction mixture 0.2 μM);
   - 3 μl DNA matrix;
   - 4 μl double distilled water (ddH_2O). Carry out the amplification reaction in a thermocycler under the following thermal and time conditions: initial denaturation at 95°C for 5 min, followed by 35 cycles, consisting of: denaturation at 95°C for 1 min, connexion (hybridisation) of primers at 55°C for 1 min, elongation at 72°C for 1 min and final elongation at 72°C for 5 min after the last cycle;

b) after completing the PCR reaction, perform electrophoretic separation in TAE buffer on 1% agarose gel, applying 10 µl of gel matrix mixed previously with a 10:1 load dye to the wells; apply a DNA mass marker to the gel; after electrophoretic separation, stain the gel with a reagent that allows visualisation of the DNA under UV light (e.g. SYBR Green).

3. Isolate the amplicons from agarose gel with commercially available sets.

4. Sanger sequencing. Prepared samples are sent for sequencing to a specialised laboratory.

Analysis of results

The obtained sequences should be compared with those available in databases (BLAST search). On this basis, determine the probability of the analysed strain belonging to a particular genus or species.

3.2.5. Yeast

Yeast are microfungi, but not filamentous ones, because they occur as a single cell. Most of the fungi are present in one of the two forms of growth: a single rounded yeast cell or a hypha. Some fungi, so-called dimorphic fungi (yeast), can take both morphological forms, depending on environmental conditions. A yeast cell has a cell wall, the composition of which varies according to the age of the cell, the morphological form, the culture conditions and, above all, the systematic affiliation. The wall of the yeast belonging to the Ascomycota contains mainly mannan and glucan (about 30% each) and only traces of chitin (1–2%). In comparison, the cell wall of the Basidiomycota is mainly composed of chitin and mannan. The protein-lipid cell membrane of the yeast contains, as in all fungi, ergosterol. The cell nucleus is relatively small compared to the size of other organelles. The vacuole is large, occupies a central place in the cell and is the only organellum visible in the yeast under a light microscope. Moreover, the cell contains a cytoplasm, endoplasmic reticulum, ribosomes, mitochondria, Golgi apparatus. The storage materials are volutin, glycogen and fat droplets (Figure 3.2.8).

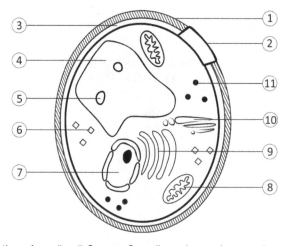

Fig. 3.2.8. Yeast cell structure.

Explanations: 1 – cell wall, 2 – scar, 3 – cell membrane, 4 – vacuole, 5, 6 – spare materials, 7 – nucleus, 8 – mitochondrion, 9 – endoplasmic reticulum, 10 – Golgi apparatus, 11 – ribosome.

The size of the yeast cell ranges from 5 to 10 μm long and 2 to 6 μm wide. Both cell size and shape are important features in the identification of yeast. However, it should be remembered that these parameters may vary depending on the physiological state of the cells, culture conditions and environmental conditions. Yeast cells can have different shapes, often characteristic of different genera: spherical (1) (*Rhodotorula, Torulaspora*); ellipsoidal (2) (*Saccharomyces*); lemon-like (3) (*Saccharomycodes, Kloeckera, Nadsonia, Hanseniaspora*); bottle-like (4) (*Pityrosporum*); cylindrical (5) (*Schizosaccharomyces*) (Figure 3.2.9).

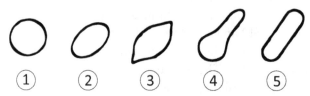

Fig. 3.2.9. Shapes of yeast cells (explanations in the text).

Some yeast spend their entire life cycle as a single cell. Others can produce pseudomycelium, or can form regular mycelium with septa, which is also an important feature in identification.

Yeast are not a homogeneous taxonomic group, they belong to different phyla (most of them to Ascomycota). Their sexual reproduction takes place in a manner characteristic of the phylum to which they belong, with the difference that in the sexual process, vegetative cells of the respective sex types are fused.

Yeast classified as Ascomycota produce a diploid zygote, which is transformed into a sac. After meiosis, haploid ascospores are formed (usually 4). There are no ascogenous hyphae and no fruiting bodies. There is also no distinct dikariophase. The shape of the ascospores is also considered in the identification of yeast.

Yeast belonging to Basidiomycota reproduce sexually, producing basidiospores on the basidia. They also form dikaryotic mycelium, but do not form fruiting bodies.

In addition to sexual reproduction, yeast can also reproduce asexually, i.e., vegetatively:

- by budding, e.g. *Saccharomyces cerevisiae*;
- by transverse division (splitting), e.g. *Schizosaccharomyces pombe*;
- by asexual spores, e.g. blastospores (*Candida albicans*), ballistospores, arthrospores, chlamydospores.

Some yeast reproduce only asexually (e.g. *Candida, Rhodotorula*).

S. cerevisiae baker's yeast are of great practical importance. They are used, among others, in the fermentation process for beer and wine (see Section 4.2.1) and biomass production (see Section 4.1.1), as well as in modern technologies, e.g. for vaccine production. *S. cerevisiae* and *S. pombe* yeast are convenient research models for eukaryotic cells, e.g. *Schizosaccharomyces pombe* have been used in microcalorimetric studies of the toxic effects of androgens on eukaryotic cells.

PRACTICAL PART

Materials and media

1. Microorganisms: *Cyberlindnera jadinii (Candida utilis), Kluyveromyces marxianus, Rhodotorula rosea, Saccharomyces cerevisiae, Schizosaccharomyces pombe, Wickerhamomyces anomalus (Pichia anomala).*

2. Media: wort plates, small sugar range, liquid media with urea and tryptophan, medium with citrate as the only source of carbon.
3. Reagents: Ehrlich reagent, methylene blue, Lugol's iodine, Sudan black B dye.

Aim of the exercise
Comparison of macroscopic and microscopic morphology and selected biochemical properties of different yeast species.

Procedure

1. Prepare a live specimen from a liquid culture. Observe and prepare drawings (magnification 400 times). Pay attention to the shape and size of cells, the presence of spores, mycelium and pseudomycelium.
2. Evaluate the physiological condition of the yeast culture (preparations only for *S. cerevisiae*):
   a) evaluation of survival, dyeing with methylene blue diluted 1:10,000. Give the viability of the yeast [%];
   b) staining of yeast spare materials:
   • glycogen:
     Spread the examined yeast in a drop of Lugol's iodine placed in the middle of microscope slide, cover the preparation with a cover slide and observe under the microscope after 2–3 min; the result of the staining: glycogen grains are dark brown;
   • lipids:
     Prepare a smear of the yeast on the slide, dry it in the air, cover the preparation with Sudan Black B solution and stain for 30 min.
     Then pour off the dye, dry the preparation and rinse with a 50% aqueous ethanol solution. Rinse the preparation with water, dry and examine the slide under the microscope. Staining result: blue-grey lipid drops, light pink cell cytoplasm.

3. Examination of selected biochemical properties of yeast. Assess selected biochemical properties of yeast and complete Table 3.2.4.

Table. 3.2.4. Selected biochemical properties of yeast

Biochemical feature	S. cerevisiae	C. utilis	K. marxianus	P. anomala	R. rosea	S. pombe
Urea decomposition (urease activity)						
Citrate as a carbon source						
Tryptophan decomposition						
Glucose fermentation						
Sucrose fermentation						
Lactose fermentation						
Maltose fermentation						
Raffinose fermentation *						

* – on the basis of the raffinose fermentation capacity, brewer's yeast was divided into top-fermenting and bottom-fermenting yeast (more in Section 4.2.1).

3.3. Microbial biotyping by LC-MS/MS and MALDI-TOF/TOF methods

INTRODUCTION

Biotyping – a procedure for species identification of microorganisms and higher organisms on the basis of analysis, e.g. of tissues or body fluids. Currently, this method can also include the identification of tissues, food contamination or metabolomics "in-situ". Modern techniques of biotyping are a natural consequence of the development of knowledge (mainly achievements of genetic, omic and systemic biology research) and technology (mainly mass spectrometry and NMR).

Currently, various methods of determining the species affiliation of microorganisms are routinely used – both those established many years ago and modern ones. Currently used biotyping methods include:
- microscopic methods – classical, staining, fluorescence, etc. This is usually the initial step in identifying the

microorganism. It is characterized by a short period of time from the moment of obtaining the material to the initial result. It allows the presence of bacteria in the examined material to be confirmed and provides information on bacterial cell morphology;

- selective and differential media – e.g. blood agar, chromogenic agar etc. The basis is the use of media that prefer the growth of particular microorganisms or to induce a colour reaction resulting from the growth of selected microorganisms;
- biochemical tests – enzymatic, biochemical series (e.g. API, EntroPluri test). The basis is to induce a colour reaction as a result of microbial growth on indicator microsubstrates (series of media) containing specific enzyme substrates, growth or other microbial substrates specific for selected microorganisms;
- serological tests – non-cultivated, direct (antigen detection) and indirect (antibody detection) (e.g. ELISA). The tests are based on specific immunological reactions, such as precipitation, agglutination, immunofluorescence or immunoenzymaticity, giving easily measurable results (mainly change and intensity of colour);
- molecular tests – PCR and real-time PCR, e.g. Mtb-RIF, which detect characteristic species-specific nucleic acid sequences;
- phage typing – mainly as biosensors, but also as diagnostic methods to identify bacteria on solid media (growth assessment);
- NMR – is only in its initial phase of use as a biotyping tool;
- mass spectrometry – now commonly used, mainly MALDI MS and MS/MS techniques and recently, also LC-MS/MS (Sections 2.3 and 2.4).

3.3.1. LC-MS/MS biotyping on the example of *Mycobacterium* strains

In the case of LC-MS/MS biotyping, different types of molecules are analysed – from very simple to complex, small or large particles, monomers or polymers. As a rule, the compounds present in many species are avoided – typical biomarkers (proteins, sugars, fatty substances or specific biological polymers characteristic for the microorganism, parasite or disease) are sought and used. In the analysis, both methods with chromatographic separation and without chromatographic column (so-called flow injection analysis – FIA) are used. The techniques of biotyping with the LC-MS/MS method are currently developing – there are not many developed and implemented applications, but the volume of scientific research aimed at using LC-MS/MS technique as a tool for identification of microorganisms or tissues is increasing every year.

Mycolic acids (MAs) are one of the basic building blocks of the cell wall of order Actinomycetales and the suborder Corynebacterineae (Figure 3.3.1). In taxonomic terms, they include many representatives of opportunistic human pathogens and numerous saprophytic organisms, including species from genera: *Dietzia, Gordonia, Nocardia or Rhodococcus*.

MAs are very good biomarkers because they:
- may together constitute 20–40% of the dry matter of the cell wall of these microorganisms;
- are only found in the microorganisms of the suborder Corynebacterineae;
- are characterised by great structural diversity within each family and species.

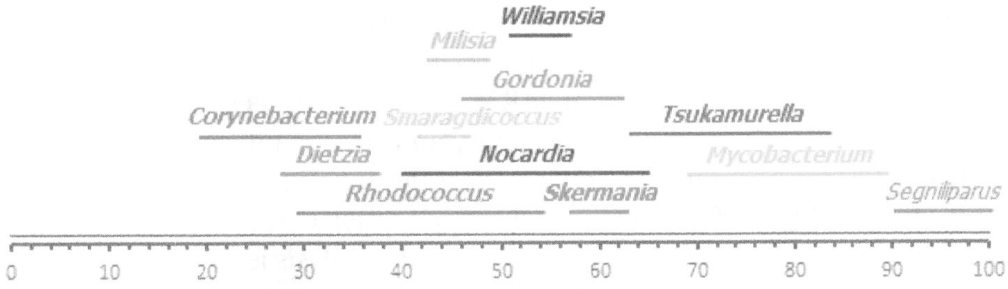

Due to pathogenicity and diagnostic problems, special attention has been paid to MAs with 70–90 carbon atoms in the molecule as potential markers for species identification of the genus *Mycobacterium* – both saprophytic, used in the production of steroid drugs, and pathogenic, which constitute

Fig. 3.3.1. MAs value differentiation within Corynebacterineae suborder.

a significant problem in microbiological diagnostics and treatment of tuberculosis and mycobacteriosis. The structure and location of MAs in *Mycobacterium* is shown in Figure 3.3.2. Tandem MS/MS mass spectrometers with an electrospray ion source (ESI) appear to be most suitable for comprehensive MA analysis. The multitude of scanning modes allows the collection of maximum detailed data about the tested compounds in terms of quality. Tandem MS/MS type mass spectrometers allow assessment of the size of whole molecules, length and structure of individual hydrocarbon chains, degree of saturation and the presence of possible functional groups in the MA structure and to select the most specific markers for biotyping in the final targeted methods.

The basis for analysis in targeted methods based on MRM scanning is the characteristic MAs fragmentation, in which the fragmentation ions from α-alkyl chains – 339.4 m/z, 367.4 m/z and 395.4 m/z, corresponding to chains of length C20, C22 and C24 respectively, are highly visible (Figure 3.3.3).

Based on the experimental data from the MRM method, species-specific signal profiles can be created, which are the basis for the determination of the affiliation of *Mycobacterium* species. This type of data can then be collected and used as a database application or as a reference point for comparative analysis or principal component analysis (PCA).

An example of PCA analysis on data from targeted MRM for *Mycobacterium tuberculosis* (Mtb) and other *Mycobacterium* strains (MOTT, *Mycobateria* other than tuberculosis – outside of the tuberculosis group) is shown in Figure 3.3.4.

The developed LC-MS/MS FIA-MRM methodology is characterised by high sensitivity, selectivity and significant reduction of diagnostic time. It also seems that the only factor limiting the sensitivity of the method is the specification of the LC-MS/MS apparatus itself. The cost of LC-MS/MS analysis is also very competitive in relation to other methods. The analysis time compared to other routinely used Mycobacterium biotyping methods is as follows:

- classical method: 6-week detection + 6-week drug-sensitivity assessment;
- genetic methods and other modern methods: 2–24 hrs detection + 5–7 days to 6 weeks drug-sensitivity assessment;

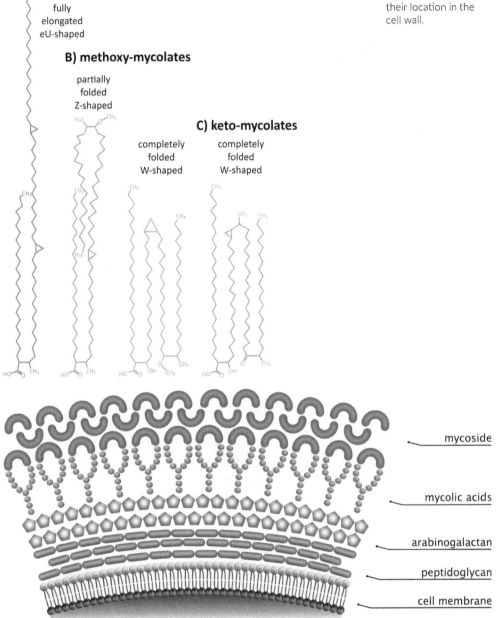

A) α-mycolates

fully
elongated
eU-shaped

B) methoxy-mycolates

partially
folded
Z-shaped

C) keto-mycolates

completely
folded
W-shaped

completely
folded
W-shaped

mycoside

mycolic acids

arabinogalactan

peptidoglycan

cell membrane

Fig. 3.3.2. Main MAs and their forms in *Mycobacterium* and their location in the cell wall.

179

- MS/MS methods:
  - detection variant I: 5–7 days of culture + material preparation and LC-MS/MS analysis (2 h);
  - detection variant II: preparation of direct material + LC-MS/MS analysis (2 h);
  - detection and assessment of drug-sensitivity: 1–7 days of culture + material development and LC-MS/MS analysis (2 h).

Fig. 3.3.3. Sample MS2 MAs spectrum with a mass of 1136 m/z.

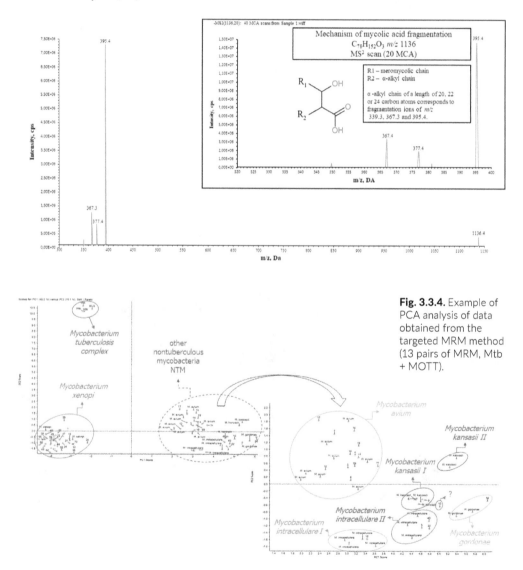

Fig. 3.3.4. Example of PCA analysis of data obtained from the targeted MRM method (13 pairs of MRM, Mtb + MOTT).

The LC-MS/MS-based diagnostic routine test is further developed under the NCBiR project: POIR.01.01.01–00–0798/16 – years 2017–2020.

PRACTICAL PART

Materials and media

1. Reference strains of pathogenic and saprophytic microorganisms: *M. tuberculosis* h37rv, *M. chelonae*, *M. kansasii*, *M. terrae*, *M. xenopi*, *M. smegamatis*, *M. avium*.
2. Media: 7H9 Middlebrook (Difco Laboratories Ltd., West Molesey, UK) with addition of 0.05% Tween 80 (Sigma).
3. Reagents: 25% solution of KOH, 19% solution of HCl, methanol (analytical grade), chloroform (analytical grade), reagents for LC-MS/MS analysis (ultra pure): methanol, ammonium formate, chloroform.
4. Apparatus: liquid chromatograph Agilent 1200 and mass spectrometer QTRAP 3200 (Sciex).

Aim of the exercise
Species identification of Mycobacterium tuberculosis and selected mycobacteria from the MOTT group based on the profile of mycolic acids analysed by a LC-MS/MS method.

Procedure

1. Culture *Mycobacterium* strains on 7H9 Middlebrook medium with 0.05% Tween 80 on a rotary shaker (125 rpm) at 37°C until the end of the logarithmic growth phase.
2. Centrifuge *Mycobacterium* strain culture (3,500 rpm, 10 min, 4°C).
3. Add 2 ml of methanol and shake/vortex for 30 to 60 seconds and centrifuge again (as above).
4. Decant methanol.
5. Add 2 ml 25% KOH in methanol.

6. Incubate 60–90 min at 90°C (screw tubes tightly!).
7. Cool to room temperature and acidify by adding 1.5 ml HCl (19%).
8. Add 2 ml of chloroform and extract for 5–10 min (RM-2 shaker or vortex).
9. Take 1–1.5 ml of the lower phase (chloroform) into a new tube.
10. Dilute the samples with methanol in a minimum ratio 1:1.
11. Prepare autosampler containers with inserts.
12. Take 300–400 μl of the sample and transfer to the inserts in the autosampler containers.
13. Perform the LC-MS/MS analysis using the liquid chromatograph Agilent 1200 and mass spectrometer QTRAP 3200 (Sciex). Carry out the determination by injecting the sample without using a column, using a mixture of methanol : chlorophyll : acetonitrile (20:40:40) with 5 mmol/L ammonium formate directly into an ion source operating in negative polarity mode (FIA-MRM).

Analysis of results

Interpret the results obtained (comparative analysis or PCA) – indicate signals (mycolic acids) characteristic of the species.

3.3.2. MALDI-TOF and MALDI-TOF/TOF biotyping

MALDI-TOF biotyping

Species identification of microorganisms using the MALDI-TOF technique is based on the analysis of pseudo-molecular ions (MS scan) of compounds forming a characteristic "fingerprint" (pattern, profile) as a basis for biotyping. For the analysis it is not necessary to know which molecules we are analysing, but their m/z values and the relative intensity of the measured signals. The whole process is carried out according to the following pattern:

**microbiological material → analysis → profile →
matching → identification**

Species identification is carried out on the basis of MS profiles obtained from molecules found in the microbial outer shells (cell membrane/wall, mainly peptides and low molecular weight proteins) and/or cell extracts (different types of molecules).

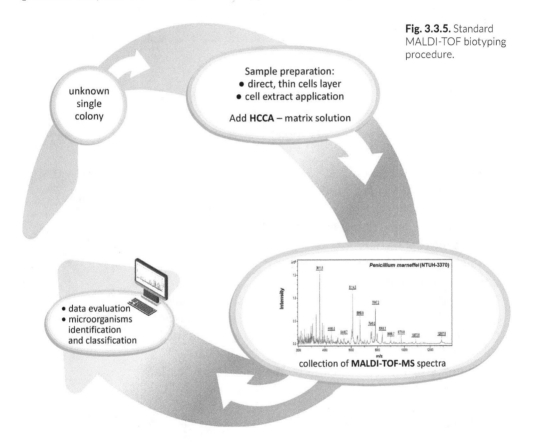

Fig. 3.3.5. Standard MALDI-TOF biotyping procedure.

The software included in the devices (e.g. Biotyper set by Bruker) enables data acquisition, results analysis, importing commercial databases, creating own databases and reporting the results. Generated reports contain information on: organism identified to species, phylum or no match, degree of spectrum match to database and graphical presentation of results.

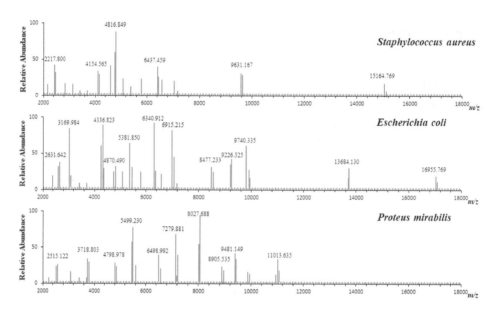

Benefits of the method:

- reliable identification results in a short period of time – 6 to 72 h after obtaining microbial culture or more by other methods compared to a few min by MALDI-TOF MS;
- simplicity of testing;
- no additional tests are required;
- little material required for analysis;
- possibility to identify microorganisms directly in the material from the patient – sputum, swab, biopsy specimen, urine, etc.;
- simple software and operation of the device (no knowledge of mass spectrometry is required);
- automatic transmission and reporting of identification results;
- low operating costs;
- facilitates the organisation of work in the laboratory;
- in the future, the possibility of detecting drug resistance mechanisms (MRSA).

Fig. 3.3.6. MALDI-TOF MS profiles obtained directly from microbial cells (colony material transferred to MALDI plates and subjected to direct analysis).

MALDI-TOF/TOF biotyping

At the moment, the basis is a proteomic analysis carried out according to the scheme: protein isolation → digestion → peptide sequencing. The standard proteomic approach is based on the sequencing of positively ionised (+) peptides from the C-end of the peptide by scanning MS2 in a tandem spectrometer such as MALDI-TOF/ TOF. The method used for microbial identification uses positive (+) and negative (-) ionisation of peptides derivatised with CAF-/CAF+ (or shorter CAF/CAF) (Figure 3.3.7). The CAF/CAF reagent allows chemical activation of fragmentation in the mode of negative and positive ionisation:

- MS/MS (–) fragments derivatised peptide ions → mostly b-series;
- MS/MS (+) fragments derivatised peptide ions → mostly y-series.

Fig. 3.3.7. CAF/CAF method scheme.

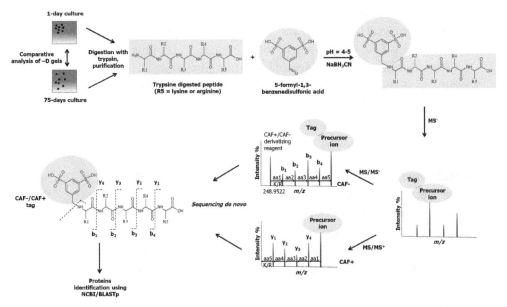

Chemical activation of fragmentation is also used in other applications and occurs in positive ionisation. It is a well-known de novo sequencing tool, but it generates a lot of side effects and mass spectra are sometimes difficult to interpret. The CAF/CAF reagent does not generate side reactions or interferences, allowing for unambiguous and simple mass spectra, and the unique sequences of each organism are the basis for its identification. Reading in two directions of the same sequence – from C-end to (+) and from N-end to (–) – significantly reduces identification errors

(Figure 3.3.8). Simplified mass spectra (only the b(–) and y(+) ions) are analysed in the prototype PROTEIN READER software.

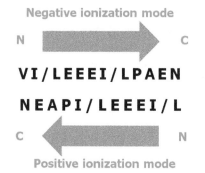

Fig. 3.3.8. Peptide sequencing from C-end (+) and N-end (–).

Advantages of the procedure:
- peptide analysis in negative and positive ionisation;
- sequencing takes into account only b(–) and y(+) ions without additional secondary and side fragmentation;
- clinical applications require effective data quality control systems (MS/MS –/+);
- preparation of the sample for analysis takes about 3 h and can be fully automated;
- basis of biotyping is the genetic sequence;
- software fully meets the analytical requirements (speed, accuracy, reporting, etc.).

Disadvantages of the solution:
- at the moment the solution is at the prototype stage;
- the application targets single cell microorganisms (although the procedure may gradually be extended to other organisms).

PRACTICAL PART

Materials and media

1. Media: Agar (bacteriological plates) depending on microbial growth requirements: universal, blood, wort, other.
2. Reagents: MALDI matrix – α-cyano-4-hydroxycinamic acid (CHCA) (10 mg/ml) in 50% ACN with 0.1% TFA.
3. Apparatus: MALDI-TOF/TOF 5800 system (Sciex).

Aim of the exercise
Species identification of microorganisms on the basis of extracellular molecule profiles using the MALDI-TOF method.

Procedure

1. Inoculate several species of microorganisms on bacteriological plates (reduction culture).
2. Incubate plates at the optimal temperature for microbial growth until visible colonies are obtained.
3. Collect microbiological material from the bacteriological plate (one colony).
4. Transfer material to MALDI plate:
   a) version 1 – apply a thin layer of cells and add a matrix (CHCA);
   b) version 2 – suspend the cells in 10–20 µl of the matrix (CHCA) and apply about 1 µl of the mixture to the MALDI plate.
5. Allow samples to dry completely.
6. Analyse samples in TOF MS positive polarisation mode in the mass range 1,000–15,000 m/z.

Analysis of results

1. Analyse the profiles of extracellular molecules obtained in the MALDI-TOF by comparison and/or PCA.
2. Indicate the main signals that are species-specific.
3. Summarise the results in the form of a table.

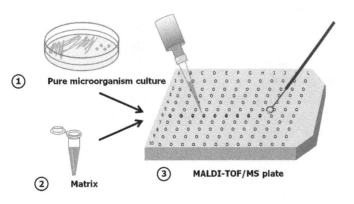

Fig. 3.3.9. Scheme of procedure for MALDI-TOF microbial biotyping.

①   Pure microorganism culture

②   Matrix

③   MALDI-TOF/MS plate

Literature

Books

Cramer, R. (ed.), 2016. *Advances in MALDI and laser-induced soft ionization mass spectrometry*. Springer, Cham, s. 227–246.

Długoński, J. (red.), 1997. *Biotechnologia mikrobiologiczna – ćwiczenia i pracownie specjalistyczne*. Wydawnictwo Uniwersytetu Łódzkiego, Łódź.

Długoński, J. (red.), 2016. *Microbial Biodegradation From Omics to Function and Application*. Caister Academic Press, Norfolk.

Dyson, P., 2019. *Streptomyces*, w: Schmidt, T.M. (red.), *Encyclopedia of Microbiology*, 4th ed. Academic Press, London, s. 334–345.

Kunicka, A., Rajkowska, K., 2012. *Drożdże*, w: Libudzisz, Z., Kowal, K., Żakowska, Z. (red.), *Mikrobiologia techniczna. Tom 1: Mikroorganizmy i środowiska ich występowania*. Wydawnictwo Naukowe PWN, Warszawa, s. 43–59.

Moore, D., Robson, G.D., Trinci, A.P.J., 2011. *Twenty First Century Guidebook to Fungi*. Cambridge University Press, New York.

Nikiforova, M.N., LaFramboise, W.A., Nikiforov, Y.E., 2015. *Amplification-based methods*, w: Kulkarni, S., Pfeifer, J. (red.), *Clinical Genomics*. Academic Press, London, s. 57–67.

Piotrowska, M., Żakowska, Z., 2012. *Grzyby strzępkowe*, w: Libudzisz, Z., Kowal, K., Żakowska, Z. (red.), *Mikrobiologia techniczna. Tom 1: Mikroorganizmy i środowiska ich występowania*. Wydawnictwo Naukowe PWN, Warszawa, s. 60–83.

White, T.J., Bruns, T., Lee, S.H., Taylor, J.W., 1990. *PCR Protocols: A Guide to methods and Application*. San Diego, Academic Press.

Review articles

Adzitey, F., Huda, N., Ali, G.R.R., 2013. *Molecular techniques for detecting and typing of bacteria, advantages and application to foodborne pathogens isolated from ducks.* Biotech. 3, 97–107.

Barka, E.A., Vatsa, P., Sanchez, L., Gaveau-Vaillant, N., Jacquard, C., Klenk, H.P., Clement, C., Ouhdouch, Y., van Wezel, G.P., 2016. *Taxonomy, physiology, and natural products of Actinobacteria.* Microbiol. Mol. Biol. Rev. 80, 1–43.

Das, S., Dash, H.R., Mangwani, N., Chakraborty, J., Kumari, S., 2014. *Understanding molecular identification and polyphasic taxonomic approaches for genetic relatedness and phylogenetic relationships of microorganisms.* J. Microbiol. Methods. 103, 80–100.

Raja, H.A., Miller, A.N., Pearce, C.J., Oberlies, N.H., 2017. *Fungal Identification Using Molecular Tools: a Primer for the Natural Products Research Community.* J. Nat. Prod. 80, 756–770.

Spatafora, J.W., Chang, Y., Benny, G.L., Lazarus, K., Smith, M.E., Berbee, M.L., Bonito, G., Corradi, N., Grigoriev, I., Gryganskyi, A., James, T.J., O'Donnell, K., Roberson, R.W., Taylor, T.N., Uehling, J., Vilgalys, R., White, M.M., Stajich, J.E., 2016. *A phylum-level phylogenetic classification of zygomycete fungi based on genome-scale data.* Mycologia 108, 1028–1046.

Original scientific papers

Butorac, A., Dodig, I., Bačun-Družina, V., Tishbee, A., Mrvčić, J., Hock, K., Dimnič, J., Cindrić, M., 2013. *The effect of starvation stress on Lactobacillus brevis L62 protein profile determined by de novo sequencing in positive and negative mass spectrometry ion mode.* Rapid Commun. Mass Spectrom. 27, 1045–1054.

Kowalski, K., Szewczyk, R., Druszczyńska, M., 2012. *Mycolic acids-potential biomarkers of opportunistic infections caused by bacteria of the suborder Corynebacterineae.* Postępy Higieny i Medycyny Doświadczalnej, 66, 461–468.

Mohamed, M.E., Suelam, I.I., Saleh, M.A., 2010. *The presence of toxin genes of Clostridium perfringens isolated from camels and humans in Egypt.* Vet. Arhiv. 80, 383–392.

Różalska, S., Pałecz, B., Długoński, J., 2008. *Calorimetric detection of the toxic effect of androgens on fission yeast.* Thermochim. Acta 474, 91–94.

Szewczyk, R., Kowalski, K., Janiszewska-Drobinska, B., Druszczyńska, M., 2013. *Rapid method for Mycobacterium tuberculosis identification using electrospray ionization tandem mass spectrometry analysis of mycolic acids.* Diagn. Microbiol. Infect. Dis. 76, 298–305.

Varga, J., Kocsube, S., Toth, B., Frisvad, J.C., Perrone, C., Susca, A., Meijer, A., Samson, R.A., 2007. *Aspergillus brasiliensis sp. nov., a biseriate black Aspergillus species with world-wide distribution.* Int. J. System. Evolut. Microb. 57, 1925–1932.

4
Industrial applications of microorganisms

4.1. Biosynthesis processes

Every single microbial cell is an independent organism. The very small size of microbial cells makes the body surface relatively large in relation to its volume. Consequently, the environment exerts a strong influence on microorganisms, allowing the absorption of nutrients into the cell and the release of metabolic products out of the cell. It also has a significant impact on the rate of metabolism taking place in the cell, including biosynthetic processes, as seen by the rate of biomass doubling, which is much greater in microorganisms than in other organisms (Table 4.1).

Organism	Biomass doubling time
Bacteria	20–120 min
Micromycetes	1–6 h
Algae	2–48 h
Crops like grass and alfalfa	1–2 weeks
Chickens	2–4 weeks
Pigs	4–6 weeks
Cattle	1–2 months

Table 4.1. Biomass doubling time of some microorganisms, plants and animals.

The time needed to obtain the biomass necessary for the technological process and the rate of synthesis of compounds produced in primary and secondary metabolism are the main factors (apart from the biological properties of the compounds themselves) determining the wide use of microorganisms in biotechnological processes.

4.1.1. Microbial biomass production and using

INTRODUCTION

Microbial biomass can be a beneficial source of protein, often described as single cell protein (SCP). Among the microorganisms most commonly used for biomass production, attention should be paid to the yeast *S. cerevisiae*. This species has accompanied humanity for thousands of years, making it possible to obtain products such as bread, wine or beer. Thanks to Pasteur's discoveries, the production of pure yeast cultures began on an industrial scale in the 19th century and despite the passage of time these microorganisms are of fundamental importance in the food industry. It is also important that yeast is positively perceived as a beneficial microorganism. Apart from protein, the yeast biomass contains sugars, lipids, mineral salts, B vitamins and vitamin D. The physiology of *S. cerevisiae* yeast is well known. These unicellular fungi, belonging to the phylum Ascomycota, ellipsoidal or spherical, 5–10 µm in size and reproduce (mainly) vegetatively by budding or by sexual reproduction when conditions are unfavourable to budding (see Section 3.2). During budding, a gradual increasing protrusion (bud) is produced on the surface of the stem cell. After the separation of the daughter cell from the stem cell, a permanent scar remains on both. At the site of the scar the cell cannot form a new bud, and thus the number of scars indicates both the age of the cell and its ability to propagate further. From a technological point of view, the process of scar formation on the surface of the cells limits the use of continuous cultures for the production of baker's yeast biomass. Based on several decades of experience, efficient and cost-effective processes have been developed on an industrial scale to obtain yeast biomass. Currently, various raw materials are used to produce yeast biomass, which are usually by-products of the agri-food industry. The raw material used to produce these microorganisms is beet molasses or cane molasses – a by-product of sugar production, containing about 40% sucrose.

Other popular by-products, such as whey or starch, are not absorbed directly by S. cerevisiae. Industrial yeast production involves several basic stages. It starts with the incubation of a selected strain in pure culture in flasks and then sterile transfer of the yeast to larger and larger reactor tanks, carrying out batch culture each time, until the multiplied yeast are finally placed in a tank for batch or fed-batch culturing. The resulting biomass is separated from the culture medium by centrifugation and cleaned from the culture liquid using filter presses or rotary filters, so that the mycelium dry matter content is about 27–33%. Compressed yeast can then be dried to a dry matter content of 92–96%. The lower the water content, the longer the product is suitable for use (see Section 4.2.3).

Yeast can grow under aerobic or anaerobic conditions, with higher biomass production capacity under aerobic conditions. However, exceeding a certain sugar concentration in some strains of S. cerevisiae results in an increased ethanol synthesis and a decrease in biomass production efficiency. The use of fed-batch culture allows control of the amount of sugar in the tank.

PRACTICAL PART

Submerged, batch culture of *S. cerevisiae* yeast in a bioreactor

Materials and media

1. Microorganisms: *Saccharomyces cerevisiae*.
2. Media: synthetic medium according to Lobos.
3. Apparatus: Labfors 5 bioreactor from Infors HT (Switzerland).

Aim of the exercise
The aim of the exercise is to obtain the biomass of S. cerevisiae yeast. The exercise includes the preparation of the bioreactor for submerged culture, inoculation of the medium and control of the examined process.

Procedure

1. Culture in a bioreactor:
   a) connect the bioreactor filled with one litre of medium, after sterilisation in the autoclave, to the control panel;
   b) calibrate the pH electrode and place it in the tank, start mixing, connect the temperature sensor;
   c) after setting the temperature (28°C), saturate the culture medium with oxygen in order to calibrate the oxygen electrode;
   d) Introduce into the bioreactor, 50 ml of inoculum obtained after 24 h of multiplication in the same medium, on a shaker (140 rpm).;
   e) culture using airflow = 1.0 v/v/m and stirrer speed = 250 rpm;
   f) in case of strong foaming, apply silicone anti-foaming agent;
   g) after half an hour from inoculation, take a culture sample for analysis of the biomass content and glucose concentration in the post-culture liquid;
   h) culture time and frequency of analyses should be agreed with the instructor;
   i) after completing the culture, turn off the bioreactor, remove and wash the electrodes, sterilise the remaining culture in the autoclave, wash the tank.

2. Determination of mycelium dry matter:
   a) centrifuge culture samples (3,000 × g, 10 min) in Falcon tubes of known weight, separate the supernatant, dry test tubes with yeast biomass at 105°C to constant weight;

3. Determination of glucose.
   Preparation of the glucose standard curve for the concentration range 100 ng – 10 µg/ml:
   a) Weigh 10 mg of the glucose standard in a Pyrex tube on a 0.1 mg precision laboratory scale and dissolve in deionised H_2O to 1 mg/ml.
   Determinations should be performed using the LC-MS/MS kit and flow analysis injection technique (without the use of an analytical column);

b) using the HPLC Agilent 1200 Series LC with the QTRAP 3200 (Sciex) mass spectrometer, inject 10 µl of the sample directly (without column). Use water: methanol (40:60 v/v) mixture as the mobile phase, flow rate: 1 ml/min, negative polarisation, analyse for selected fragmentation pairs of MRM 179–59 and 179–89;

c) in the case of test samples, separate the culture medium from the microorganisms (centrifugation 3,000 × g, 10 min) and dilute the supernatant obtained accordingly.

Analysis of results

The report should include the results of the performed analyses. On their basis determine the biomass efficiency coefficient, the rate of biomass gain.

Submerged, continuous culture of *S. cerevisiae* yeast in a bioreactor

Materials and media

As described in the previous section (for submerged, batch culture).

Aim of the exercise
To obtain the biomass of yeast by means of continuous culture in a fermenter, consolidation of calculations of basic parameters characterising the cultures in the chemostat.

Procedure

1. Prepare the bioreactor for experiments and process parameters, as done previously for batch culture.
2. For the experiment, use a synthetic medium as in batch culture.

3. Inoculate 50 ml of 24-h *S. cerevisiae* culture on the same medium with shaking. Connect the container with the additional medium with a sterilised tube through a pre-calibrated peristaltic pump with a bioreactor. Collect the culture fluid from the reactor in the sterilised flask through an installed and pre-sterilised drain system.

4. Carry out the culture at an air flow of 1.0 v/v/m at 28°C and stirrer speed of 250 rpm.

5. When growth is achieved in the growth inhibition phase (use batch culture data), start the medium dosage after calculating (D) from (μmax) from batch culture. Apply a 70% μmax in practice.

6. Complete the process after 24 h.

Analysis of results

Determine the biomass concentration in the culture, the amount of substrate used and productivity.

4.1.2. Characteristics, classification and directions for practical use of surfactants. Screening of *Bacillus* strains capable of lipopeptide surfactant production

INTRODUCTION

Characteristics, classification and directions for practical use of surfactants

Surface active agents of biological origin are commonly known as biosurfactants. Microorganisms, mainly bacteria and yeast, constitute the largest group of organisms capable of synthesising this kind of compounds. In addition to biosurfactants, there are also synthetic surfactants (obtained by processing crude oil compounds) and bio-based surfactants (produced by combining both synthetic and plant-based compounds). Regardless of this division, all surfactants are characterised by an amphiphatic (amphiphilic) structure. This means that the molecules of this group of compounds contain both a hydrophobic and a hydrophilic part, which may be ionic (cationic, anionic, amphoteric) or non-ionic

(Figure 4.1.2.1). The amphiphilic structure renders biosurfactants capable of accumulation at the interface between phases of different polarity. This results in decrease of the surface and interfacial tension. Depending on the chemical structure and the type of adjacent phases (air/liquid; liquid/liquid; solid surface/air or solid surface/liquid), surfactants may be involved in stabilising or destabilising emulsions and foams, wetting, coagulation or dispersion.

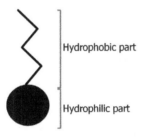

Hydrophobic part

Hydrophilic part

Fig. 4.1.2.1.
Amphiphilic structure
of surfactant molecule
shown as a monomer.

The most important parameters allowing characterisation and comparison of the physicochemical properties of surfactants include: 1) Critical Micelle Concentration (CMC); 2) Surface tension and interfacial tension (values are determined for surfactants at CMC); 3) Hydrophile-Lipophile Balance (HLB) and 4) ability to stabilise (destabilise) emulsions and foams. When the CMC value is exceeded, the surfactant molecules combine and form different assemblies, such as micelles, tubes and bubbles. Further increase of surfactant concentration after achieving CMC does not result in further decrease of surface tension (Figure 4.1.2.2).

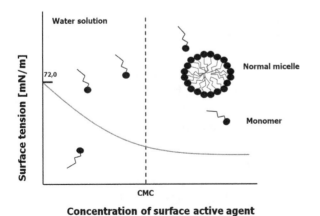

Water solution

72,0

Normal micelle

Monomer

Surface tension [mN/m]

CMC

Concentration of surface active agent

Fig. 4.1.2.2.
Relationship
between surfactant
concentration in
aqueous solution,
surface tension value
and surfactant's
ability to form micelles

Explanation: CMC – Critical Micelle Concentration.

In non-polar fluids, however, after exceeding the CMC value, surfactants form so-called reverse micelles, in which hydrophilic parts of the surfactant are directed inside the micelles, while hydrophobic parts are directed outside of them. Compared to synthetic surfactants, CMC values of microbiological surfactants are usually much lower. In an aqueous environment, surfactants form so-called normal micelles, in which the hydrophilic groups are directed outside the micelle, staying in contact with the aqueous medium (Figure 4.1.2.2). Normal micelles have an ability to solubilize hydrophobic compounds by physically encapsulating them inside the hydrophobic core of the micelles. This process is called micellar solubilization.

Surfactant activity is determined by measuring surface tension at the interface of the air and water surfactant solution (the concentration of which has to correspond with CMC). The interfacial tension is usually determined by using n-hexadecane. Hydrophobicity of the compounds can be compared by using their HLB value. The HLB scale ranges from 1 to 20, and was created by using oleic acid (HLB = 1) and sodium oleate (HLB = 20) as hydrophobicity standards. HLB values are mainly used to compare different synthetic surfactants. The HLB value of biosurfactants does not exceed 6. They form so-called "W/O (water in oil) emulsion" in which the aqueous phase is dispersed in the form of tiny droplets inside a continuous oil phase. Compounds with HLB values in the range from 10 to 18 form an emulsion called O/W (oil in water). The emulsifying activity (E24) can be determined by mixing the aqueous surfactant solution with a hydrophobic liquid. Hydrophobic liquids often used for this purpose include kerosene (mixture of liquid hydrocarbons), hexadecane, vegetable or synthetic oils and various organic solvents (e.g. ethyl acetate). The emulsion content in the sample is observed 24 h after mixing the two liquids together.

Low-molecular biosurfactants are compounds whose molecular weight does not exceed 2000 Da. Most of them significantly reduce surface tension. Surfactin, produced by *Bacillus* strains is an example of such compounds. Surfactin lowers the surface tension from 72 (water/air system characteristic value) to 27 mN/m at concentrations as low as 13–30 mg/l, though it doesn't show high emulsifying activity. On the other hand, high molecular weight biosurfactants are mostly heteropolymers (with molecular weights up to 1,000 kDa) and are often very effective emulsifiers. Such compounds include for example, emulsan, produced by *Acinetobacter calcoaceticus* strains. On the basis of the chemical structure of the hydrophilic part of microbiologically produced surfactants, they are divided into five major classes: 1) glycolipids; 2) lipopeptides and lipoproteins; 3) fatty acids, phospholipids and neutral lipids; 4) complex biosurfactants (e.g., glycolipoproteins) and 5) special biosurfactants including, among others, components of fimbria and envelopes (Table 4.1.2.1). Formulae of selected surfactants are presented in Figure 4.1.2.3.

Table 4.1.2.1. Classification of microbiological biosurfactants taking into account the chemical structure of the hydrophilic part (modified from: Paraszkiewicz, Kuśmierska 2017; Sharma 2018).

Class	Groups/families/examples of compounds	Example of a microorganism
Glycolipids	Rhamnolipids	*Pseudomonas aeruginosa*
	Trehalose lipids	*Nocardia erythropolis*
	Sophorolipids	*Candida bombicola*
Lipopeptides and lipoproteins	Surfactin, iturin, fengycin, putisolvin	*Bacillus* and *Pseudomonas* strains
Fatty acids, neutral lipids, phospholipids	Phosphatidyl ethanolamine	*Rhodococcus erythropolis*
	Corinomycolic acid	*Corynebacterium lepus*
	Spiculisporic acid	*Penicillium spiculisporum*
Polymeric biosurfactants	Emulsan	*Acinetobacter venetianus*
	Liposan	*Candida tropicalis*
	Alasan	*Acinetobacter radioresistens*
	Mannoproteina	*Saccharomyces cerevisiae*

A)

B)

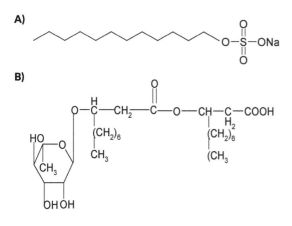

Fig 4.1.2.3. Selected surfactants structures.

C)

Fig 4.1.2.3. cont.

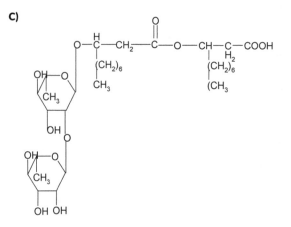

D1)

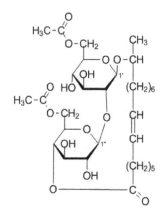

D2)

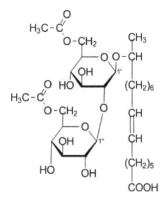

E)

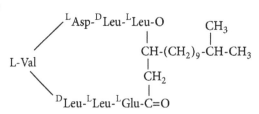

F)

$CH_2-CO-{}^LAsn$ ── ${}^DTyr$ ── ${}^DAsn$

$CH_3-CH_2-CH_2-(CH_2)_9-CH$ .... ${}^LGln$

$NH-{}^LSer$ ── ${}^DAsn$ ── ${}^LPro$

G)

$$CH_3-(CH_2)_n-CH-CH-CO-^LGlu-^DOrn-^DTyr-^DThr-^LGlu-^DAla-^LPro-^LGln-^LTyr-^LIle$$

$$\underset{OH}{|} \qquad \underset{O}{|}$$

Explanations: A) synthetic surfactant – sodium dodecyl sulphate (SDS); B) *P. aeruginosa* monoramnolipid ($C_{26}H_{48}O_9$), molar weight 504 g; C) diramnolipid *P. aeruginosa* ($C_{32}H_{58}O_{13}$) with a molar mass of 650g; D) sophorolipid (D1 – in lactone form and D2 – in acid form) *C. bombicola*; E) cyclic lipopeptide: surfactin *B. subtilis*; F) cyclic lipopeptide: iturin *B. subtilis*; G) cyclic lipopeptide: fengycin *B. subtilis*.

Surfactants are used widely as detergents, emulsifiers or wetting and foaming agents in various industries, including: the petrochemical industry, crude oil recovery, agriculture, food industry, cosmetics industry, pharmaceutical industry, ceramics industry, paper industry, medicine and remediation processes for contaminated environments. Due to their low price, synthetic surfactants are used far more frequently on an industrial scale than the biological ones. Nevertheless, intensive work is being carried out to reduce production costs and commercialise microbial surfactants. This is due to the fact that biosurfactants have a number of advantages over their synthetic counterparts such as higher surface activity, lower toxicity, they are obtained from renewable sources and are more readily biodegradable. Challenges related to the production, isolation and quantitative and qualitative analysis of *Bacillus* biosurfactants (lipopeptides) are discussed in detail in Section 4.1.6.

Cyclic *Bacillus* lipopeptides and methods of analysis of these compounds

Bacillus is a genus of Gram-positive, spore-forming, aerobic or facultatively anaerobic rod-shaped bacteria. These bacteria are often referred to as cosmopolitan microorganisms. They can be isolated from many different types of soils (especially from the rhizosphere) as well as aquatic environments or even from the digestive tract of eukaryotes, and from plants (in which they are most often found as facultative endophytes). Many strains of the genus *Bacillus* show the ability to produce compounds of potential commercial use (including, among others, enzymes, biosurfactants and antibiotics). The biosurfactants produced by *Bacillus* strains are low-molecular secondary metabolites, classified as cyclic lipopeptides. The peptide part of these compounds is produced without ribosome contribution, by Non-Ribosomal Peptide Synthetases (NRPS). This method of biosynthesis results in great structural diversity. Therefore, individual biosurfactants (e.g., surfactin) can occur in the form of different isoforms (characterised by a different sequence of amino acids in the peptide part) and homologues (differing in length of fatty acid residue). Cyclic *Bacillus* lipopeptides (CLPs) are divided into three families: surfactin, iturin or fengycin family. CLPs significantly reduce surface tension, have strong foaming properties and high affinity to erythrocyte membranes (causing their lysis). The surfactin family includes such compounds as surfactin, lichenysin, esperin and pumilacidin. Surfactin is a compound that exhibits antiviral, antibacterial, antithrombotic and antitumor activity. The peptide part of the surfactin is a heptapeptide (it contains 7 amino acids) combined by lactone bond with a β-hydroxy fatty acid of variable length (13 to 15 carbon atoms long). Surfactin is one of the strongest biological surfactants. After reaching its CMC (13–20 mg/l) it reduces surface tension from 72 to 27 mM/m. The iturin family includes such compounds as iturin (A, AL and C isomers), bacillomycin (D, F, L and LC isomers) and mycosubtilin. These compounds contain heptapeptide (which differs in structure from the one in surfactin family) combined with β-amino fatty acid (14 to 17 carbon atoms long). Iturin exhibits antifungal and antibacterial activity. Fengycins are a group of *Bacillus* lipopeptides, notable among which, are fengycin (A, B and S isomers) and plipastatin (A and B isomers). These lipodecapeptides contain eight amino acids connected by an internal lactone ring in the cyclic part of the peptide and a chain of saturated or unsaturated β-hydroxy fatty acid (containing 14 to 18 carbon atoms) in the hydrophobic part. Fengycins have the ability to inhibit the growth of filamentous fungi, however they exhibit weaker hemolytic properties compared to surfactins and iturins. Biosurfactant production can be detected by various methods. Most of them scan the supernatant of the centrifuged liquid culture for the presence of the biosurfactant. Detection of biosurfactants in the supernatant (obtained after centrifugation of liquid culture) is based on measuring of surface tension (using a tensiometer) and evaluation of the emulsifying and foaming activity. In the absence of a tensiometer, rapid screening tests are commonly used in preliminary studies. In the drop collapsing test, a 10 μl drop of the examined supernatant is placed on a hydrophobic surface and left until it dries. Along the examined sample, control samples of 5% water solution of SDS (positive control) and water (negative control) are also placed on the surface. The drops containing biosurfactants collapse, whereas drops not containing surfactants remain stable. After all drops dry, the diameter of the drop is measured (Fig. 4.1.2.4.).

Fig. 4.1.2.4. Drop collapsing test. Quick screening method of detecting the presence of surfactant in the liquid.

- **LOW WETTING ABILITY**
- **HIGH SURFACE TENSION**
- **HIGH WETTING ANGLE**

- **HIGH WETTING ABILITY**
- **LOW SURFACE TENSION**
- **LOW WETTING ANGLE**

Explanations: A) a drop of water; B) a drop of liquid containing surfactant. The drops were placed on a hydrophobic surface.

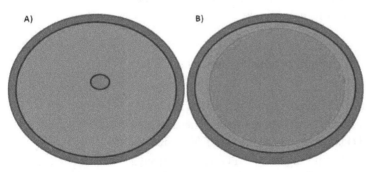

A)   B)

Fig. 4.1.2.5. Oil spread method. Quick screening method of detecting the presence of surfactant in the liquid; bird's-eye view of the oil surface.

Explanations: A) a drop of water; B) a drop of liquid containing surfactant.

The microplate analysis can be used for both qualitative and quantitative evaluation of a sample. 100 µl of a liquid culture supernatant is placed in the 96-well plate wells. The plate is then placed on a sheet of graph paper. While looking through the plate at the graph paper, an observer can notice if the grid of the paper turned more dense while looking through supernatant wells than through those containing water. If it did, the supernatant contains surfactants. This phenomenon is caused by a change in the shape of the drop meniscus.

Another method for surfactant detection is called the oil spread method. Pour 50 ml of distilled water in a large (25 cm in diameter) petri dish, then add 20 µl of oil. The oil coats the water surface. Next, 10 µl of the test sample is applied to the oil surface. If the sample drop is surrounded by an emulsified halo, the sample is considered to contain biosurfactants. The diameter of the halo is directly proportional to the surfactant concentration in the sample (Figure 4.1.2.5). This method is considered to be a very accurate way to detect microorganisms capable of producing surfactants. The kerosene test is most commonly used for detecting the presence of emulsifiers (compounds capable of stabilising emulsions). Sometimes kerosene is replaced by another hydrophobic liquid, e.g. chloroform, diesel or vegetable oil. The test consists of an intensive, several-minute shaking of the sample (with a total volume of 10 ml) containing 4 ml of bacterial culture supernatant and 6 ml of kerosene. This test should be performed in a tightly closed, graduated 20 ml tube. After 24 h, the volume of stable emulsion is read off the scale and the emulsifying activity E24 (%) is determined using the formula: (emulsion volume/sample volume) × 100%.

Use of *Bacillus* strains capable of producing biosurfactants as biopesticides

Plant Growth Promoting Microorganisms (PGPMs) can be distinguished among rhizospheric microorganisms. These include both bacteria and fungi (Plant Growth Promoting *Rhizobacteria* – PGPR and Plant Growth Promoting Fungi – PGPF, respectively). *Rhizobacteria*, included in the PGPM group, have the ability to stimulate the development of plants through a number of different mechanisms. Their influence manifests both in the rapid growth of plant biomass, as well as improvement of the vitality and health status of plants. The mechanisms of promoting plant growth by the bacteria from the above group include the production of enzymes, volatile compounds and antibiotics (which also include biosurfactants, among others from the families of surfactin, iturin and fengycin).

Biosurfactants produced by *Bacillus* strains in many ways contribute to plant health promotion. Surfactin, for example, participates in the formation of biofilm and sliding movement, which helps *Bacillus* cells colonise the surface of root hair cells. As a result of colonisation, an ecological niche is occupied, which prevents the attack and development of phytopathogenic fungi. Moreover, biosurfactants from the surfactin family have an ability to inhibit growth and action of phytopathogenic bacteria and viruses. Iturins are characterised by antifungal properties against many microfungi and yeast (their presence disturbs the osmotic balance of membranes by creating ion channels). A similar effect against microfungi is shown for fengycin.

In recent years, more and more attention has been paid to the use of *Bacillus* strains belonging to *PGPR* promoting plant health as biological agents against plant diseases. These types of agents are characterised by a number of advantages in comparison with their synthetic counterparts. They are characterised by lower toxicity, diverse biological activity and biodegradability. The effectiveness of biological agents depends on such features as:

- ability to persist in the ambient environment;
- ability to form stable consortia;
- ability to transition to an endophytic state;
- being safe for the environment and for proper metabolism of the rhizosphere of a given plant type.

The ability of *Bacillus* strains to produce spores is an additional technological advantage, influencing the possibility of developing a durable and easy to apply biological agent. These types of agents last longer in unfavorable environmental conditions such as drought or nutrients shortage or extreme temperature or pH values.

Bacilli cyclic lipopeptides often act in a synergistic way. This can be observed especially in the following pairs: surfactin and iturin; 2) surfactin and fengycin and 3) iturin and fengycin. Due to this phenomenon, strains with the ability to synthesise surfactin, iturin and fengycin simultaneously have a high application potential (they can be used, for example, as biopesticides).

PRACTICAL PART

Materials and media

1. Soil from the rhizosphere of selected plants; soil contaminated with petroleum-derived compounds.
2. Blood medium: plates containing agar (2%) and fresh sheep blood (5%); universal medium with CTAB-MB complex.
3. LB medium.
4. Reagents for QuEChERS extraction.
5. Surfactin and iturin standards.
6. Reagents for the analysis of lipopeptide biosurfactants by LC-MS/MS technique.
7. Apparatus: liquid chromatograph Agilent 1200 and mass spectrometer QTRAP 3200 (Sciex).

Aim of the exercise
Isolation of microorganisms capable of synthesising surfactants, including strains of the genus Bacillus.

Procedure

1. Inoculation of soil samples on plates with blood medium – containing agar (2%) and fresh sheep blood (5%) and with CTAB-MB medium – universal medium containing 0.0005% acetyltrimethylammonium bromide (CTAB) and 0.0002% methylene blue (MB).

2. Isolation of selected colonies (surrounded by a brightening or emulsion zone, hemolysis zone, blue-coloured substrate zone) in order to obtain pure cultures.
3. Preparation of 48–72 h submerged cultures in Luria-Bertani medium (LB).
4. Detection of the presence of biosurfactants in culture supernatant (surface and emulsifying activity) using the so-called fast screening methods: drop collapsing test, oil spreading method, microplate analysis, kerosene emulsification test.
5. Characteristics of microscopic morphology of bacteria (stained by the Gram method) characterised by the ability to produce biosurfactants.
6. Testing the *Bacillus* strains for the ability to produce lipopeptide class biosurfactants – surfactin and iturin;
   • obtaining 72 h liquid cultures in LB medium;
   • isolation of lipopeptides from liquid cultures by modified QuEChERS method;
7. Separation of the lipopeptides by liquid chromatography, using the Agilent 1200 with a Kinetex C18 column (50 mm × 2.1 mm, 5 µm; Phenomenex, USA) at 40°C. Solvents: water (channel A) and methanol (B) with the addition of 5 mM ammonium formate. Perform the determination in positive polarisation mode using the QTRAP 3200 (Sciex) spectrometer equipped with ESI-type ion source.

Analysis of results

Evaluation of soil taken from the rhizosphere of crop plants and soil contaminated with petroleum-derived compounds as a potential source of surfactant producing microorganisms, including *Bacillus* bacteria producing lipopeptide biosurfactants.

4.1.3. Microbiological production of enzymes from the hydrolase class

INTRODUCTION

The production of enzyme preparations belongs to the most developed branches of biotechnology. On an industrial scale, the production of enzymes (amylolytic preparations) was initiated in the USA in 1894. According to an analysis of the market made by the Reports and Data portal in 2019, the enzyme market in 2017 was 7,082 million USD, with a projected increase to 10,519 million USD in 2024. In 2018 as much as 24% of enzymes produced worldwide were hydrolases.

Most of the enzymes produced on an industrial scale are obtained using microorganisms. Nevertheless, plant and animal tissues can also be a source of enzymes (4 and 8% of total production of enzyme preparations, respectively). For example, germinated cereal grains (so-called malt) are a traditional source of amylolytic enzymes, and tropical plants such as pineapple and papaya are a source of proteases (described as bromelain and papain, respectively). Trypsin and catalase are obtained from bovine liver. The so-called natural rennet – a mixture of proteases capable of hydrolysing milk proteins, can be obtained from the stomach walls of calves. However, it should be emphasised that most of the currently produced rennet cheeses are obtained using genetically modified rennet – secreted from the culture of recombinant strains of *Escherichia coli* K-12, *Kluyveromyces marxianus* var. *lactis* or *Aspergillus niger*.

The usefulness of microbiological methods in enzyme production is determined by a number of factors, the most important of which are:

- microorganisms produce enzymes very quickly and in principle, could represent an unlimited source;
- microbiological methods of enzyme production are relatively cheap, both from the point of view of biosynthesis and further stages of the technological process (separation of the enzyme from biomass, thickening, purification, etc.);
- the use of microorganisms for enzyme production provides independence from seasonal and climatic fluctuations;

- relatively simple techniques are developed (and used) to increase productivity through:
  - optimisation of microbial culture conditions;
  - selection and mutagenisation of industrial strains;
  - use of genetic engineering methods at the cellular (protoplast fusion) and subcellular level.

In addition to increasing the efficiency of enzyme biosynthesis by microbial cells, the use of various methods of selection and improvement of strains allows the development of:

- mutants, so-called constitutive (capable of producing the enzyme in the absence of an inductor);
- strains resistant to repressor activity;
- microorganisms capable of growing under industrial conditions;
- non-toxic or antibiotic strains.

Microbiological production of enzymes most often involves the use of substrates containing natural raw materials and waste from the agri-food industry, such as molasses, potato starch, corn steep liquor, soybean flour, malt marc, distillery grain or spent grain. Most often it is also necessary to add an appropriate inductor. For example, amylase synthesis is induced by the presence of starch, which is a substrate for transformations catalysed by this enzyme. The analogue of substrate or its derivatives can also be an inductor. On the industrial scale the biosynthesis of enzymes in submerged cultures is carried out in bioreactors with a capacity of 10–100 m^3 for about 50 to 150 h (depending on the type of enzyme). Enzymatic preparations can also be obtained during the cultivation of microorganisms in/on solid media (Solid State Fermentation (SSF) method), using rice or wheat bran as substrates. This type of culturing is preferred especially for enzymes produced by micromycetes. For proper growth of the microorganism and the biosynthesis of the enzyme, it is necessary to ensure adequate humidity of the raw material (usually 40–70%). After the completion of the biosynthesis process it is necessary to separate the enzyme from the culture. The basic technique of enzyme separation from the culture conducted in solid medium is water extraction. Submerged cultures should be first subjected to centrifugation or filtration (in order to separate the biomass from the post-culture liquid). Extracellular enzymes are secreted directly from the culture fluid, whereas in the production of intracellular enzymes it is necessary to disintegrate the cell walls beforehand.

Enzymes produced on a large scale can be used in various industries. Hydrolases catalysing the breakdown of bonds by connecting water have found the widest application. Among hydrolases, the most important are amylolytic, cellulolytic, pectinolytic, proteolytic and lipolytic enzymes.

Amylases are enzymes decomposing starch, a polysaccharide built from the residues of D-glucopyranose. This polysaccharide contains two polymers: amylose (unbranched chain, bonds α-1,4) and amylopectin (branched chain, bonds α-1,4 / α-1,6). Amylases of plant origin have long been used in the brewing and spirits industry. Currently, amylases obtained with the use of microorganisms are commonly applied in the food industry. The greatest use is made of amylolytic enzymes produced by both fungal (*Aspergillus oryzae, A. niger, A. avamorii*) and bacterial strains (*Bacillus subtilis, B. cereus, B. polymyxa*). High activity of the above enzymes is also determined in cultures of the following microorganisms: *Lactobacillus, Micrococcus, Pseudomonas, Arthrobacter, Proteus, Penicillium, Mucor, Candida, Rhizopus*.

The amylase group includes α- and β-amylase, glucoamylase, pullulanase, α-D-glucosidase, isoamylase, cyclodextrin glycosyl transferase. α-Amylases (1,4-α-glucan-glucanohydrolases) cleave starch mainly into low-molecular weight dextrins and a small amount of maltose. They cut the α-1,4 bonds and attack the starch molecule from inside. β-Amylases (α-1,4-glucan-maltohydrolases) are enzymes that cleave starch into maltose and small amounts of macromolecular dextrins. *B. licheniformis* and *B. amyloliquefaciens* are most commonly used to produce these enzymes on an industrial scale. The enzymes are stabilised with Ca ions and the commercial product is an aqueous solution containing about 2% proteins or a solid preparation. The production rate of amylases in the logarithmic phase of microbial culture is low and increases when the growth rate of the culture decreases. Synthesis of amylase, like most extracellular enzymes, is subject to catabolite repression. The presence of glucose accelerates the growth of microorganism culture while inhibiting the production of amylases. These amylases (α- and β-) are constitutional and their production is regulated by many genes.

The high demand for amylases is associated with the widespread use of starch in many sectors of the food industry, which primarily include:
- distillation – saccharification of mash using amylolytic preparations saves barley and increases the efficiency of spirit production;
- brewing – the use of amylases allows the use of unmalted raw materials (e.g. barley, rice, corn and wheat meal) for beer production;
- potato and starch industry – production of thickeners, sauce additives, pudding, baby food ingredients;
- baking – accelerated hydrolysis of starch increases the content of sugars used by yeast, which in turn allows better quality of bread to be obtained;
- vegetable and fruit processing – during the production of jams and marmalades.

Additionally, amylases are used in the cosmetics industry – as an addition to washing agents.

After amylases, proteases hydrolysing peptide bonds in proteins are the next important enzymes obtained on an industrial scale. Proteolytic enzymes are characterised by different specificity of action, which determines the nature of the resulting products, and thus their potential areas of use. These enzymes are used primarily in the food industry, e.g.:
- in the dairy industry – to coagulate milk proteins in the cheese making process, to produce casein hydrolysates, to stabilise and improve wettability of milk powder. A separate group here are enzymes of microorganisms developing on the surface of cheeses or in their mass during ripening (blue cheese, smear-ripening cheese);
- in the fish industry to speed up the ripening of salted and pickled herring and the hydrolysis of fish waste (mainly proteases produced by *Aspergillus* sp. fungi);
- during maturation of meat (mainly beef), which shortens the maturation time and improves consistency (delicacy, tenderness), biological value and organoleptic characteristics;

- for the production of protein hydrolysates, e.g. from waste from the fish, meat and leather industries, hydrolysis of keratin waste and to improve the assimilability of feed ingredients;
- in baking to improve the bulkiness and consistency of dough.

Additionally, microbial proteases are used by the household chemistry industry as ingredients in washing powders. The concentration of active protein in washing powders ranges from 0.015 to 0.025%. For this purpose, proteases produced by *B. licheniformis* and *B. amyloliquefaciens* are mainly used. They must meet high requirements and be stable at high temperature, alkaline pH and in the presence of chelating agents. Such enzymes are called alkaline proteases.

Acidic proteases, on the other hand, are used mainly in the medicine and food industry, and their optimum pH is 2–4. They are primarily produced by micromycetes of the genera *Mucor* and *Aspergillus* and strains of *Streptococcus lactis, S. cremoris* and *Lactobacillus helveticus*.

Neutral proteases are relatively unstable, sensitive to pH and temperature changes. They are used in the leather and food industry and obtained on an industrial scale using culture of *B. subtilis, B. cereus, B. megaterium, Pseudomonas aeruginosa, Streptomyces griseus* and *A. oryzae* strains.

Microbial proteases are also used in the medicinal and pharmaceutical industry in hydrolysis processes to obtain preparations with anti-inflammatory, antioxidant, hypoallergenic and antimicrobial effects.

Lipases belong to the esterases that break down triglycerides of higher fatty acids. Since lipids are not soluble in water, lipase reactions take place on the borderline between the aqueous phase (where enzymes are dissolved) and the solid phase (containing fats). The production of lipases mainly involves the use of the fungi from the genera *Rhizopus, Aspergillus* and *Mucor*. These enzymes are especially applied in cheese-making to speed up ripening and improve the aroma of cheeses, as well as to regulate the composition of fatty acids in triglycerides (which is important for example in shaping the nutritional value of margarines). In addition, fungal lipases are part of numerous washing preparations used in household chemistry.

Lytic complexes containing several or more different hydrolytic enzymes are also of great practical importance. The production of lytic complexes uses strains of moulds, mainly belonging to *Trichoderma viride, Penicillium funiculosum*, as well as Actinomycetes and bacteria from genera *Streptomyces, Micromonospora, Cytophaga* and *Bacillus*. The quantitative and qualitative composition of enzymatic complexes depends on the type (structure) of biopolymers used in the culture medium as a source of carbon and energy, which also act as lytic enzyme inductors. Partially or fully purified enzymatic complexes are applied in various industries to decompose natural homo- and heteropolymers. In scientific research they are useful to obtain protoplasts of higher plants and eukaryotic microorganisms as a result of cell wall digestion.

PRACTICAL PART

Comparison of activity of proteolytic enzymes produced by different *Bacillus* strains

Materials and media

1. Microorganisms: 24 h cultures on universal or wort agar slants.
2. Media: liquid universal.
3. Reagents: 2% casein solution; Folin's reagent, 5% TCA, 6% Na_2CO_3. Immediately prior to use, dilute Folin's reagent with distilled water 1:4.
4. Apparatus: centrifuge, spectrophotometer.

Aim of the exercise
Biosynthesis and determination of proteolytic enzyme activity in cultures of Bacillus strains isolated from soil and the B. subtilis strain (from the University of Łódź strain collection).

Procedure

1. Macro- and microscopic characteristics of the strains. Gram staining of strains used in the study.
2. Preparation of 48-h liquid cultures. Rinse the culture obtained on slant with 5 ml of universal medium and transfer to 30 ml of the same medium. Carry out the culture at 30°C on a rotary shaker (130 rpm). After 48 h of incubation, centrifuge the culture (3,500 × g, 15 min, 4°C). Prepare samples to determine the activity of proteolytic enzymes from the supernatant obtained.
3. Determination of proteolytic enzymes activity:
   a) prepare dilutions in 1 ml from the culture liquid obtained after centrifugation according to the scheme:

Reagent	Undiluted sample	2 × dilution 1 ml	5 × dilution 1 ml	10× dilution 1ml	10×dilution 1ml (control)
2% casein 2 ml, heated 28°C	+	+	+	+	+
5% TCA 5 ml	−	−	−	−	+
Incubation 10 min., temp. 28°C – enzymatic decomposition of casein					
5% TCA 5 ml	+	+	+	+	−
Filter through tissue paper funnels into the second series of tubes, collect 1 ml each and transfer to the third series of tubes.					
6% Na$_2$CO$_3$ 4 ml	+	+	+	+	+
Folin's reagent (previously diluted with distilled water) 1 ml	+	+	+	+	+
Incubation 30 min at room temperature					
Extinction with respect to the control sample at wavelength λ = 750 nm					

   b) add 2 ml of 2% casein solution heated to 30°C to each sample. Also add 5 ml 5% TCA to the control tube;
   c) incubate test samples and controls for 10 min at 30°C (stage of casein enzymatic decomposition), then filter through tissue paper funnels into a second series of tubes;
   d) take 1 ml of each sample and transfer to the next (third) series of tubes.;

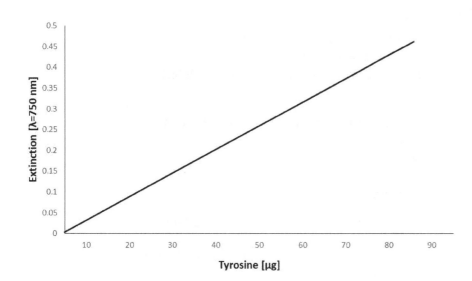

Fig. 4.1.3. Example of a standard curve showing the relationship between absorbance and the amount of tyrosine (Folin's method).

e) add 4 ml of 6% Na_2CO_3 and 1 ml of Folin's reagent diluted with distilled water to all samples in the third series of tubes, then incubate for 30 min at room temperature;

f) read the extinction with respect to the control sample on the spectrophotometer at 750 nm;

g) from the results obtained determine the content of tyrosine released, from the standard curve:

C tyrosine = (A + 0.002)/0.0052
where: A – sample extinction value;

h) calculate the activity of proteolytic enzymes in the post-culture liquid:

enzyme activity (μmol tyr./min) =
= (μg tyrosine ×dilution ×8)/(181 ×10)

where: 8 – conversion factor resulting from the dilution (1 ml sample + 5 ml 5% TCA + 2 ml 2% casein), 181 – molecular weight of tyrosine;

i) determine for all four samples and calculate the average.

217

Analysis of results

On the basis of the data obtained, compare the proteolytic activity of the strains isolated from the soil and the reference *Bacillus subtilis* strain (from the strains collection of IMBiI).

Production and examination of T. viride lytic enzymes activity

Materials and media

1. Microorganisms: *Trichoderma viride, Cunninghamella echinulata, Curvularia lunata.*
2. Media: ZT slants, substrate for *Trichoderma viride* for lytic enzymes production.

Aim of the exercise
Production of lytic enzymes using T. viride culture and evaluation of the activity of the obtained enzyme preparation (obtaining protoplasts of C. echinulata or C. lunata strains).

Procedure

1. Wash off the spores from 10-day cultures of *T. viride* on ZT slants with NaCl physiological solution and transfer to a 1 l conical flask containing 250 ml of medium for *T. viride* (containing autoclaved mycelium of *C. echinulata* or *C. lunata* as an inductor of lytic enzymes).
2. Carry out *T. viride* cultures for 7 days on a shaker at 28–30°C.
3. Filter the culture through the Büchner funnel (connected to a water pump).
4. Add ammonium sulphate to the filtrate, up to 80% saturation (561 g per 1 l of filtrate) and leave at 4°C to precipitate proteins.

5. Centrifuge the precipitate at 4°C, dissolve in distilled water and conduct dialysis in a cold store for 36 h, changing the water several times.

6. Centrifuge the contents of the dialysis bag (3,000 × g, 10 min, 4°C), lyophilise the supernatant and keep it at −20°C in a sealed flask.

7. Control of preparation activity – releasing *C. echinulata* or *C. lunata* protoplasts:

   a) filter *C. elegans* or *C. lunatus* culture on PL_2 medium from the end of the logarithmic growth phase (18–24 h of growth) through a Büchner funnel and rinse 0.8 M $MgSO_4$ in citrate phosphate buffer at pH 4.2;

   b) transfer a sample of wet mycelium (0.5 g) into a 50 ml flask containing 5 ml 0.8 M $MgSO_4$ with a pH of 4.2 and add a 2.5 mg/ml lyophilised *T. viride* lytic complex;

   c) incubate on a rotary shaker, with gentle agitation, at 28 to 30°C.;

   d) after 3; 5 and 17 h of incubation, take samples and count the released protoplasts in the Thoma chamber;

   e) after complete digestion of the mycelium, filter the sample through sterile glass wool and centrifuge the filtrate (1,200 × g, 10 min, 4°C);

   f) collect the film containing protoplasts on the surface of the liquid with a Pasteur pipette, rinse with a suitable osmotic stabiliser and use for further testing.

4.1.4. Polysaccharide biosynthesis

INTRODUCTION

Numerous groups of microorganisms are capable of synthesising various extracellular polysaccharides (homo- and heteropolymers). Of these, more than 20 polysaccharides are produced on an industrial scale and used in the medical, food, chemical and cosmetic industries. Some of the microbiological polysaccharides available on the market are presented in Table 4.1.4.

Table 4.1.4. Selected polysaccharides produced on an industrial scale using microorganisms

Polysaccharide	Chemical structure	Microorganisms
Dextran	$-$Glc 1 $\xrightarrow{\alpha}$ Glc 1$-$ z nielicznymi odgałęzieniami 1 $\xrightarrow{\alpha}$ 4, 1 $\xrightarrow{\alpha}$ 3, 1 $\xrightarrow{\alpha}$ 2	*Leuconostoc mesenteroides, L. dextranicum, Acetobacter* sp., *Streptococcus mutans*
Bacterial cellulose	$-$Glc 1 $\xrightarrow{\alpha}$ Glc 1 1 $\xrightarrow{\beta}$ 4,	*Komagateibacter xylinus*
Xanthan	$-$Glc 1 $\xrightarrow{\beta}$ 4 Glc $-$ $\uparrow$3 α $\vert$ 1 Man 6-OAc $\uparrow$2 β $\vert$ 1 GlcUA $\uparrow$4 β $\vert$ 1 Man 4,6 $-$ O pirogronian	*Xantomonas campestris*
Alginate	4 D-ManA 1 $\xrightarrow{\beta}$ 4 D-ManA 1$-$ 4 L-GulA 1 $\xrightarrow{\alpha}$ 4 L-GulA 1$-$	*Azotobacter vinelandii, Pseudomonas aeruginosa*
Curdlan	$-$Glc 1 $\xrightarrow{\beta}$ 3 Glc $-$	*Alcaligenes faecalis*
Pullulan	$\longrightarrow$ 6(Glc 1 $\xrightarrow{\alpha}$ 4 Glc 1 $\xrightarrow{\alpha}$ 4 Glc) 1 $\longrightarrow$	*Aureobasidium pullulans*
Scleroglucan	$-$Glc 1 $\xrightarrow{\beta}$ 3 Glc $-$ $\uparrow$6 β $\vert$ 1 Glc	*Sclerotium glucanicum*

Explanations: Glc-glucose, GlcUA-glucuronic acid, GulA-guluronic acid, Man – mannose, ManA-mannuronic acid, Ac-acetate.

Dextran

One of the most important polysaccharides produced by microorganisms (from a medical and economic point of view) is dextran. Strains of *Leuconostoc mesenteroides* are most commonly used for its production. These are Gram-positive cocci forming diplococci and short chains belonging to the Lactobacillaceae family, which do not have the ability to move and do not produce spores. These bacteria are relative anaerobes capable of lactic acid heterofermentation, have no ability to move and do not produce spores. The production strains form linear dextran containing up to 95% of 1,6-α bonds and few side branches in 1,3-α and 1,4-α, and occasionally 1,2-α. The molecular weight of the dextrans produced by the strains is highly dependent on the culture conditions and ranges from several thousand to several million.

The enzyme dextranosaccharase (2-glucosyltransferase 1,6-α-glucan:D-fructose, EC 2.4.1.5) is responsible for the synthesis process, which during transglycosylation, transfers with UDP, the glucose residue from sucrose to the acceptor according to the scheme:

$$\text{sucrose} + \text{UDP} \quad \text{UDP} + \text{glucose} + \text{fructose}$$
$$\text{UDP} - \text{glucose} + \text{fructose} \longrightarrow \text{acceptor -glucose} + \text{UDP}$$

The acceptor can be low molecular weight dextran, maltose or sucrose.

The production medium contains sucrose and mineral salts. The optimum pH for enzyme synthesis is 6.5–7.0, while the highest activity of dextranosaccharase is observed at pH 5.0–5.2. The most favourable temperature for Leuconostoc mesenteroides growth is 30°C, but the maximum enzyme production takes place at 23°C and therefore the process of dextran biosynthesis is conducted at room temperature. The production process is carried out without aeration using a batch culture method and the biosynthesis time usually does not exceed 48–72 h. Continuous cultures and technologies using solution and immobilised dextranosaccharase are also developed and used. The yield of production strains in relation to sucrose as a substrate is 40–45%. The increase in yield is limited by the production of by-products with in which the composition is predominated by fructose.

The use of dextran

Raw dextran, the so-called primary, is a mixture of molecules of different sizes, containing 1,000–1 000 000 glucose units. The biological effect of dextran depends on the size of its particles. For therapeutic purposes, the fractions of dextran of smaller weight and the small mass distribution of the molecules are suitable. Two types of dextran are currently used:

- medium-molecular weight 70,000 (50,000–80,000);
- low-molecular weight 40,000 (30,000–45,000).

Dextran with a molecular weight of 40,000–110,000 Da in the form of a 6% solution in 0.85% NaCl or glucose is used as a blood plasma substitute because it creates the same osmotic pressure, is durable and, most importantly, has no toxic or allergic effects and only slightly reduces blood protein levels. Dextrans of different molecular weights are applied in transplantology as components of organ storage solutions as well as components of dental implants. 10% dextran solution is useful as a dressing for wounds (burns, bed sores) that are difficult to heal because it easily absorbs exudate fluids. Dextran forming a complex with iron can be used to treat anaemia. Low-molecular weight dextran becomes an anticoagulant with properties similar to heparin (a substance that prevents venous embolism) after receiving an electrical charge (converted to dextran sulphate). Dextran (modified) can also be an enzyme and cell immobiliser.

Various applications of dextran have made it one of the most important polymers synthesised by microbiological means.

PRACTICAL PART

Materials and media

1. Microorganism: *Leuconostoc mesenteroides* – a strain stored in the form of glycerin "stocks" (Section 1.7).
2. Media: liquid for the biosynthesis of dextran, solid for the cultivation of *Leuconostoc mesenteroides*.
3. Reagents: 25% H_2SO_4, 0.1 M $Na_2S_2O_3$, 1% starch solution, KJ, Luff-Schoorl reagent, composition: (25 g $CuSO_4$

× 5 H_2O, 50 g citric acid, 388 g Na_2CO_3 × 10 H_2O, distilled water to 1 l), 10% trichloroacetic acid, 50% and 96% ethyl alcohol.

Aim of the exercise
Characteristics of bacteria capable of extracellular polysaccharides biosynthesis (using the example of Leuconostoc mesenteroides) and production of dextran on a laboratory scale.

Procedure

1. Observation of pure cultures of *Leuconostoc mesenteroides* on plates.
2. Microscopic observations of Gram stained preparations. Make the preparation from culture on a plate.
3. Evaluation of *L. mesenteroides* culture on liquid medium after 24 h, 2 and 4 days of incubation:
   a) macroscopic observations of the culture. Pay attention to the increase in opacity with the age of the culture;
   b) determination of the pH of the culture by means of indicator papers;
   c) approximate determination of culture viscosity (dextran solutions). Measure the flow time of 5 ml of the culture from a 10 ml pipette;
   d) determination of dry mass of dextran in the culture. Precipitate dextran from the culture with 96% ethanol. Squeeze the obtained ball of dextran, weigh it and then dry it at 50°C to constant weight;
   e) determination of fructose content in culture by the Luff-Schoorl method.
   Weigh 20 g of the culture, transfer quantitatively to a 1 litre volumetric flask and fill up to the mark with distilled water. Then transfer 25 ml of the solution to a 300 ml ground Erlenmeyer flask and add 25 ml of Luff-Schoorl reagent. Bring the liquid in the flask to a boil under a reflux condenser and continue boiling for 10 min on a small flame. After cooling, add 25 ml of

25% H_2SO_4 (add the acid slowly, as the CO_2 release may cause the liquid to flow out of the vessel), add 3 g KJ and titrate with respect to starch with 0.1 M $Na_2S_2O_3$ solution. The blank is 25 ml of water.

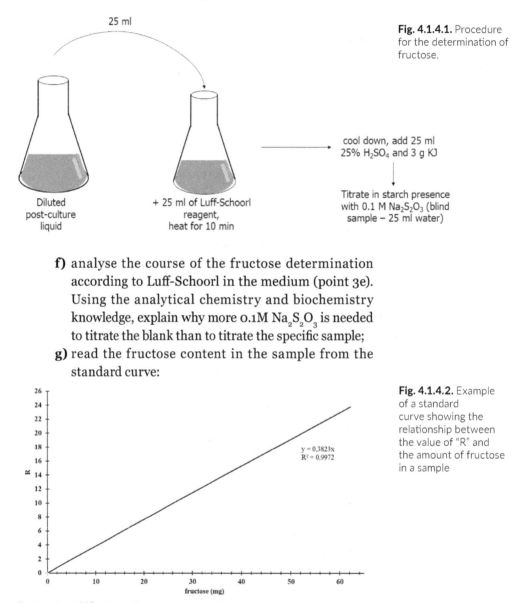

25 ml

Fig. 4.1.4.1. Procedure for the determination of fructose.

Diluted post-culture liquid

+ 25 ml of Luff-Schoorl reagent, heat for 10 min

cool down, add 25 ml 25% H_2SO_4 and 3 g KJ

Titrate in starch presence with 0.1 M $Na_2S_2O_3$ (blind sample – 25 ml water)

f) analyse the course of the fructose determination according to Luff-Schoorl in the medium (point 3e). Using the analytical chemistry and biochemistry knowledge, explain why more 0.1M $Na_2S_2O_3$ is needed to titrate the blank than to titrate the specific sample;

g) read the fructose content in the sample from the standard curve:

Fig. 4.1.4.2. Example of a standard curve showing the relationship between the value of "R" and the amount of fructose in a sample

$y = 0.3823x$
$R^2 = 0.9972$

fructose (mg)

Explanations: "R" – the difference, expressed in millilitres, between the amount of 0.1M $Na_2S_2O_3$ solution needed to titrate the blank and the amount used to titrate the test sample.

224

h) calculate the theoretical and practical efficiency of dextran biosynthesis on the basis of the determined fructose content and the known initial amount of sucrose in the medium, using the equation:

$$n\ C_{12}H_{22}O_{11}\quad =\quad [C_6H_{12}O_6]n\quad +\quad n\ C_6H_{12}O_6$$

saccharose	dextran	fructose
100 g	47,36 g	52,64 g

4. Determination of dextran by turbidimetric method:
a) standard curve preparation:

Fig. 4.1.4.3. Procedure diagram for the standard curve preparation.

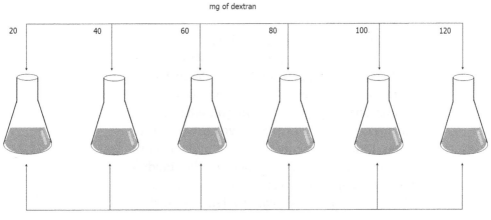

mg of dextran

0.5 ml 10% trichloroacetic acid

Measure 0.5 ml of 10% trichloroacetic acid into 50 ml volumetric flasks, then add dextran in the appropriate amount. Make up to the final volume of 12.5 ml with water. Add 50% ethyl alcohol to the samples, stirring gently until turbidity occurs (about 12.5 ml). After 15 min measure the absorbance on the spectrophotometer at 720 nm. The blank is water. Plot a standard curve.

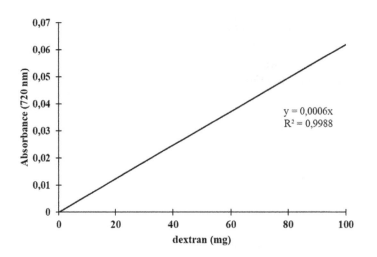

b) determination of the amount of dextran in the test sample: Dilute the contents of the flask with distilled water to about 500 ml (depending on the turbidity). Take 12 ml of the diluted sample, add 0.5 ml 10% trichloroacetic acid and add 12.5 ml 50% ethanol. After 15 min measure the absorbance at 720 nm with respect to water. Calculate the total amount of dextran (in mg) contained in the test sample using the standard curve and taking into account the dilutions. Calculate the theoretical and practical efficiency of dextran biosynthesis process.

Bacterial cellulose

Another important polysaccharide produced by microorganisms is cellulose synthesised by bacteria. The most common strains used for its production include *Komagataeibacter xylinus,* formerly known as *Gluconacetobacter xylinus* or *Acetobacter xylinum*. These are Gram-negative, motionless and non-spore-forming bacteria belonging to the Acetobacteraceae family (see Section 4.3.1). These bacteria are aerobes that synthesise a flexible, yellow or white cellulose membrane, strongly bonded to the cells, which allows them to remain on the surface of the liquid medium to gain access to oxygen.

These bacteria synthesise cellulose microfibrils, which are then secreted outside the cells in the form of a ribbon permanently bound to the cell. Cellulose polymerisation is a multi-stage process:

- sub-elementary cellulose nanofibrils with a diameter of 1.5 nm, composed of 10−15 parallel glucan chains, are secreted by channels located in the bacterial cell wall (50−80 per cell);
- sub-elementary nanofibrils from adjacent channels are bonded by hydrogen bonds and form elementary nanofibrils (1.5 × 3 nm);
- elementary nanofibrils join together by hydrogen bonds to form a beam of 3.0 × 3.8 nm first, then 3.0 × 6.8 nm and 6.0 × 9 nm;
- in the last stage, a cell-bound cellulose ribbon of 40−60 nm width, 10 nm thickness and length up to 10 μm is formed.

Cellulose ribbons from many bacterial cells interweave to form a compact membrane on the surface of the culture medium.

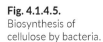

Fig. 4.1.4.5.
Biosynthesis of cellulose by bacteria.

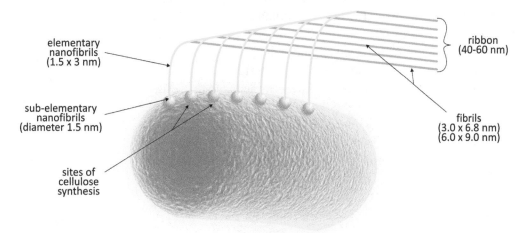

elementary
nanofibrils
(1.5 x 3 nm)

sub-elementary
nanofibrils
(diameter 1.5 nm)

sites of
cellulose
synthesis

ribbon
(40-60 nm)

fibrils
(3.0 x 6.8 nm)
(6.0 x 9.0 nm)

Synthesis of cellulose in bacteria takes place according to the following scheme:

glucose

↓ *glucokinase*

glucose-6-phosphate

↓ *phosphoglucomutase*

glucose-1-phosphate

↓ *UDP-glucose pyrophosphorylase*

uridine diphosphate glucose

↓ *cellulose synthase*

cellulose

The production medium contains glucose and a source of nitrogen, e.g. yeast or corn extract, peptone or ammonium salts. The optimal pH for the synthesis process is 4.5–6.8, while the most favourable temperature for the growth of *K. xylinus* is 26–31°C. Most often the production process is carried out without aeration, in stationary conditions, using the method of batch culture, and the time of biosynthesis does not exceed 7 days. The yield of production strains in relation to glucose as a substrate is from 1 to 5 g of cellulose per litre of medium.

Bacterial cellulose has a number of advantages compared to plant cellulose. The most important are: possibility of obtaining pure cellulose (without hemicellulose, lignin and other impurities), low toxicity and high biocompatibility. In addition, bacterial cellulose fibre is longer and stronger, which increases its mechanical strength and flexibility. Bacterial cellulose can be modified at the synthesis stage and after extraction to the desired shape and thickness, which is an extremely important feature in the production of dressings, as it makes it easier to fit them to specific (injured) body parts. In addition, the bacterial polymer is more porous and has an increased absorbency compared to plant cellulose, which allows it to be soaked with various substances, such as antibiotics. Like plant cellulose, it is easily biodegradable.

Practical application of bacterial cellulose:
- in medicine, it is used as a dressing material ("artificial skin" accelerates the healing of wounds) and as implants, e.g. in the trachea, nasal septum, ears. Since bacterial cellulose dressing can be produced in any shape and size, it is often used in places where it is difficult to place other dressings, e.g. in the groin or interdigital spaces of the hands;

- in environmental protection it is used as a component of filtration materials enabling ultrafiltration of water, sewage treatment or absorption of toxic compounds;
- in the paper industry it is used to improve paper or create paper with special properties;
- in acoustics it is used to produce loudspeaker membranes;
- in cosmetology it is used as a stabiliser of creams and emulsions and as a carrier of active substances in cosmetic masks.

PRACTICAL PART

Materials and media

1. Microorganism: *Komagataeibacter xylinus* – strain stored in the form of glycerol stocks (Section 1.7).
2. Media: liquid for cellulose production and solid for *K. xylinus* culture.
3. Reagents: 1% NaOH, Cross-Bewan reagent (1 g $ZnCl_2$ dissolve in 2.5 ml 37% (concentrated) solution of HCl.

Aim of the exercise
Characteristics of bacteria capable of extracellular polysaccharides biosynthesis (using the example of Komagataeibacter xylinus) and laboratory scale cellulose production.

Procedure

1. Observation of pure *Komagataeibacter xylinus* cultures on plates and slants.
2. Microscopic observations of Gram stained preparations. Make the preparation from culture on a plate.
3. Evaluation of *K. xylinus* culture in liquid medium after 7 days of incubation.
4. Extraction of cellulose from *K. xylinus* liquid culture. Collect the cellulose film from the liquid culture, drain it and place it in a 1 l beaker. Pour 200 ml of 1% NaOH. Boil

for 1 to 1.5 h. Then transfer the cellulose into a 1 l beaker with 200 ml of deionised water and rinse thoroughly in distilled water. Then place the cellulose in Petri dishes (5 cm diameter) and dry at 100°C for about 5–10 min. Weigh the resulting product and calculate the practical efficiency of cellulose biosynthesis.

$$C_6H_{12}O_6 = [C_6H_{10}O_5]n + H_2O$$

glucose   cellulose

180g   162g   18g

5. Properties of bacterial cellulose.
Dissolve the cellulose in Cross-Bewan reagent, and then precipitate with water. At the same time, do a comparative test with filter paper (vegetable cellulose). Write the conclusions in a notebook.

4.1.5. Antibiotic biosynthesis using tetracyclines as a study model

INTRODUCTION

Antibiotics are the products of the specific (secondary) metabolism of actinomycetes (especially from the genus *Streptomyces*), fungi (*Aspergillus, Penicillium, Cephalosporium*) and specific bacteria (mainly from the genus *Bacillus*). These substances are widely used in chemotherapy because they act selectively, at low concentrations, bactericidally or bacteriostatically on pathogenic microorganisms; some also exhibit anticancer effects, are food preservatives or growth promoters for farm animals, and are widely used in biochemical and molecular biology research.

Antibiotics can be classified according to their antimicrobial spectrum (disruption of transport through cytoplasmic membranes, inhibition of murein and protein synthesis, DNA function, respiration and oxidative phosphorylation), taxonomic affiliation of the producer, biosynthesis pathway and chemical structure (β-lactam, aminoglycosidic, macrolide, peptide, tetracyclines and others).

Tetracyclines are a group of chemical compounds made up of 4 carbocyclic rings, which differ in their substituents at 5, 6 and 7 carbon (Figure 4.1.5.1) and mainly affect the pharmacokinetics of the drug (solubility and removal of the drug from the body). The dimethylamine group, found at C-4, is primarily responsible for the antibacterial properties of tetracyclines.

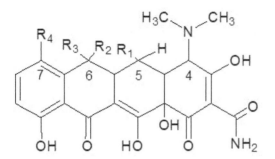

Fig. 4.1.5.1. Scheme of chemical structure of tetracyclines.

Over 1,000 substances belong to this class of antibiotics. Tetracyclines most commonly used in human and animal therapy are shown in Table 4.1.5.1.

Table 4.1.5.1. Relevant Chemical substituents of the most important tetracyclines

Name of tetracycline	Substituents				Producer strain
	R_1	R_2	R_3	R_4	
Tetracycline	H	OH	CH_3	H	S. aureofaciens
Aureomycin (7- Chlortetracycline)	H	OH	CH_3	Cl	S. aureofaciens
Terramycin (5- Oxytetracycline)	OH	OH	CH_3	H	S. rimosus
Declomycin (6- dimethyl-7-chlorotetra-cycline)	H	OH	H	Cl	S. aureofaciens
Vibramycin (6- deoxy-5-hydroxy-tetracycline)	OH	H	CH_3	H	semi-synthetic
Minocycline (7- dimethylamino-6-demethyltetracycline)	H	H	H	$N(CH_3)_2$	semi-synthetic
Metacycline (6- deoxy-6-dimethyl-6-demethyl-6-methylene--5-hydroxytetracycline)	OH	$=CH_2$	$=CH_2$	H	semi-synthetic

Tetracyclines are broad-spectrum antibiotics acting on Gram-positive, Gram-negative, *Rickettsia, Coxiella, Chlamydia, Mycoplasma* bacteria and protozoa. The mechanism of their action is to block protein biosynthesis by breaking the codon-anticodon connexion between tRNA and mRNA. As tetracyclines bind to the S7 protein of the 30S subunit of the ribosome, the effect is to inhibit the aminoacyl-tRNA binding to the ribosomal acceptor.

The following strains *of Streptomyces* are used on an industrial scale:

- *S. aureofaciens* for the production of tetracycline, chlorotetracycline and their 6-dimethyl derivatives;
- *S. rimosus* for synthesis of oxytetracycline;
- *S. viridifaciens* for formation of tetracycline and chlorotetracycline (on a smaller scale);
- *S. psamoticus* for manufacture of 6-dimethyltetracycline and 6-dimethylchlorotetracycline (on a limited scale).

In all cases, these are repeatedly improved mutants and recombinant strains, which usually come from the soil. Industrial strain spores are stored in a lyophilised form. The multiplication of the strains is carried out on slants at a temperature of 26–27°C for 7 days, where, while maintaining the activity of the strain, there is limited multiplication of vegetative cells and abundant sporulation. The inoculation material is subcultured in 4–5 stages lasting 24–36 h in shaking flasks and then the production bioreactor is inoculated with 5–10% of the culture material. The culture media have a composition developed for each type of tetracycline, strain used and culture stage. During the whole process it is necessary to use intensive aeration, anti-foaming agents and very precise temperature control. Biosynthesis takes 5–7 days. Tetracyclines are accumulated extracellularly and precipitated in the form of Ca and Mg salts. After cooling to 5–10°C, the post-culture suspension is acidified with sulfuric or oxalic acid in order to dissolve these salts, coagulate the protein and facilitate the separation of mycelium. Isolation of tetracycline antibiotics from the post-culture (after sediment separation) can be carried out by extraction, precipitation, sorption and ion exchange methods.

PRACTICAL PART

Oxytetracycline synthesis

Materials and media

1. Microorganisms: *Streptomyces rimosus* strain producing oxytetracycline, strains of actinomycetes (as potential antibiotic producers), isolated during the exercise described in Section 1.2.4. Reference strain *Bacillus cereus* NCIB 8145.
2. Media: potato slants – for the multiplication of *S. rimosus* strain producing oxytetracycline, plating medium IM-I, multiplication medium IM-II, production medium IM-III, plates with compensating, primary and inoculation agar for the determination of oxytetracycline content in the culture medium by biological method, broth medium for the multiplication of reference *B. cereus* NCIB 8145 strain.
3. Apparatus: fermenter Labfors 5 (Infors HT Switzerland), centrifuge, spectrophotometer, water bath, liquid chromatograph Agilent 1200 and mass spectrometer QTRAP 3200 (Sciex).

Aim of the exercise
Analysis of the course of oxytetracycline biosynthesis in fed-batch actinomycetes culture.

Procedure

1. Wash off actinomycetes conidia with physiological saline from the potato slant culture.
2. Transfer the conidia suspension into a 100 ml flask containing 25 ml of medium IM-I. Culture on a shaker for 48 h at 28°C.
3. Take 5 ml of culture and transfer to a 100 ml flask containing 25 ml of medium IM-II. Incubate on a shaker for 18–24 h at 28°C.

4. Transfer 25 ml of actinmycetes culture from medium IM-II to a bioreactor containing 700 ml of medium IM-III.
5. Set the mixing at 200 rpm, air flow 1 v/v/m (air/medium/min) and 28°C.
6. During the culture, measure in real time the level of oxygenation, pH and the amount of CO_2.
7. After 36 h of incubation, start the substrate pump, introduce 0.3 g/l/h of glucose (initial concentration 300 g/l).
8. Carry out biosynthesis for 5 days at 28°C, with aeration 0.5–1.0 v/v/m (0.5–1.0 l air/l medium/min).
9. During the culture, take samples to estimate the antibiotic content.
10. Determine the standard curve of the antibiotic:
    a) weigh 50 mg of oxytetracycline and dissolve in 44.4 ml of 0.01 M HCl, resulting in a standard solution containing 1,000 IU/ml;
    b) prepare a series of dilutions of oxytetracycline solution containing 1,000 IU/ml according to Table 4.1.5.2.

Table 4.1.5.2. Preparation of series of dilutions of oxytetracycline solution

Sample no	Oxytetracycline content in the sample [µg/10 ml]	Volume of reagents (in ml) used for preparation				
		Control samples		Appropriate samples		
		0,01 M HCl	Oxytetracycline solution at concentration 1,000 IU/ml	0,01 M HCl	Oxytetracycline solution at concentration 1,000 IU/ml	$FeCl_3$ 0,05%
1	250	9.75	0.25	4.75	0.25	
2	500	9.50	0.50	4.50	0.50	
3	750	9.25	0.75	4.25	0.75	
4	1000	9.00	1.00	4.00	1.00	5 ml each
5	1250	8.75	1.25	3.75	1.25	
6	1500	8.50	1.50	3.50	1.50	
7	1750	8.25	1.75	3.25	1.75	

c) leave the samples containing diluted antibiotic at room temperature for 10 min. Add 5 ml of 0.05% $FeCl_3$ to the appropriate samples. To induce the colour reaction of the sample, leave all samples at room temperature again for 10 min and then measure the absorbance (E) on the spectrophotometer at 490 nm;

d) plot a standard curve on the basis of the obtained E values.

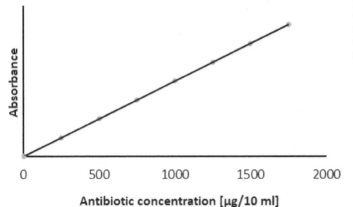

Fig. 4.1.5.2. Standard curve showing the relationship between oxytetracycline amount and absorbance.

Absorbance

0   500   1000   1500   2000

Antibiotic concentration [µg/10 ml]

11. Determine the oxytetracycline content in the culture medium by chemical method:
   a) Centrifuge 50 ml of culture (2,000 × g, 15 min, room temperature);
   b) Take 20 ml of supernatant and acidify with 5 M HCl to pH 1.8–2.0.;
   c) prepare samples of the following composition:

Test sample	control sample
1 ml filtrate	1 ml filtrate
ml 0.01M HCl	9 ml 0.01M HCl
ml 0.05% FeCl$_3$	

   d) after preparation, leave the samples for 10 min at room temperature and read the absorbance on the spectrophotometer at 490 nm. Read the oxytetracycline content in the culture medium from the standard curve.

12. Determine the oxytetracycline content in the culture medium by biological method:
   a) Prepare 8 plates with the medium for antibiotic determination: pour into a Petri dish (9 cm in diameter) successively: 15 ml of the agar compensating medium, 12 ml of the primary medium and 4 ml of the inoculation medium containing *B. cereus* NCIB 8145, prepared as in point 12b;

b) preparation of the inoculation medium:
take 0.1 ml from 24 h *B. cereus* NCIB 8145 culture on
the broth and transfer to 4 ml of inoculation medium,
previously liquefied in a boiling water bath and cooled
to +50°C. Mix thoroughly and pour onto a plate
containing the primary medium;

c) after 1 h, place 1 cylinder on the surface of the medium,
in the middle of the plate and press it to the depth of
the upper layer of the agar;

d) introduce 0.5 ml of the test sample undiluted, 5 times
diluted, 10 times diluted and reference solutions
of known antibiotic concentration (1, 5, 10, 20, 50,
100 µg/ml) into the cylinders (one plate, one cylinder);

e) after 24 h of incubation at 37°C, measure the bacterial
inhibition zone around the cylinders. Prepare an
antibiotic concentration diagram for the inhibition
zone and then read the antibiotic content in the
biological sample from the diagram;

f) determine the oxytetracycline content of the culture
medium using QuEChERS extraction and analysis
with LC-MS/MS. To 10 ml of the culture medium,
devoid of bacterial biomass, add 10 ml of acetonitrile
and a salt mixture containing: 2 g $MgSO_4$; 0.5 g
NaCl; 0.5 g $C_6H_5NaO_7 \times 2\ H_2O$; 0.25 g $C_6H_6Na_2O_7$
$\times$ 1.5 H_2O. Vortex and centrifuge the samples.
Take the upper organic layer into Eppendorf tubes
and perform chromatographic and spectrometric
analysis. Separate the antibiotics using the liquid
chromatograph Agilent 1200 and Kinetex C18 column
(50 mm $\times$ 2.1 mm, 5 µm; Phenomenex, USA), at 40°C,
solvents: water (channel A) and acetonitrile (B) with
0.1% formic acid, and then, in positive polarisation
mode, perform the determination using QTRAP 3200
(Sciex) spectrometer equipped with an ESI-type ion
source.

Analysis of results

Evaluate changes in antibiotic content during culture.

4.1.6. Production of bacterial lipopeptides

INTRODUCTION

Bacterial lipopeptides show antibacterial and antiviral properties. The antimicrobial effects of surfactin and iturin have been described many times. However, despite several decades since their discovery, biosurfactants are not widely used, and the main reason limiting the use of lipopeptides is the high cost of their production in bioreactors. Nowadays, reactor vessels with mechanical mixing are most often used for *B. subtilis* batch cultures synthesising surfactin. It is important to monitor glucose and surfactin levels in the culture. A decrease in the surfactant concentration was observed in cultures after the microbial consumption of glucose, which means that the surfactant can be used by bacteria as a carbon source. On average, between 100 and 350 mg of biosurfactant is obtained from one litre of synthetic or semi-synthetic medium, depending on the bacterial strain used. It has also been shown that the addition of Fe ions and maintaining a pH of 6.3–6.7 can increase the amount of surfactant even to 3,000 mg/l. Moreover, it has been observed that the surfactin synthesis efficiency is closely related to foam production. Chemical elimination of the foam negatively affected the synthesis of biosurfactant. On the other hand, successive removal of the foam with the surfactin contained in it allows up to 1,700 mg/l of biosurfactant to be obtained.

PRACTICAL PART

Materials and media

1. Microorganisms: *Bacillus subtilis* KP7 strain.
2. Media: medium LB.
3. Apparatus: bioreactor Labfors 5 (Infors), liquid chromatograph Agilent 1200 and mass spectrometer QTRAP 3200 (Sciex).

Aim of the exercise
Learning about the technique of submerged culture of microorganisms on the example of Bacillus subtilis culture capable of efficient synthesis of surfactin.

Procedure

1. *B. subtilis* culture in the "Labfors 5" fermenter from Infors with a total capacity of 3.6 dm³ carried out as a batch culture:

   a) Prepare *B. subtilis* KP7 bacterial inoculum on medium LB – 20 ml. The cultures should be kept for 24 h on a shaker at 28°C, 160 rpm;

   b) prepare 1 dm³ of medium;

   c) fill the fermenter with medium, place it in an autoclave and sterilise;

   d) after removing the fermenter from the autoclave, place it on the test stand and connect all necessary media, calibrate the electrodes;

   e) Introduce 20 ml of inoculum into the cooled medium. Carry out the culture at an air flow of 1.0 v/v/m and a stirrer speed of 200 rpm;

   f) in case of strong foaming, use mechanical foam removal from the system;

   g) determine the optical density of the culture at 620 nm at 0, 6, 12, 24, 36 and 48 h;

   h) take 30 ml of culture medium from the fermenter at selected time points.

2. Determination of lipopeptides:

   a) the first stage of lipopeptides isolation from biological samples is their separation from the culture medium. In order to separate the bacterial biomass from the culture medium, centrifuge with MPM centrifuge (5,000 × g, for 10 min);

   b) then transfer the supernatant into 50 ml Falcon tubes;

   c) add 10 ml of acetonitrile to the tubes prepared above;

   d) Weigh the salt mixture containing:

   - 2 g $MgSO_4$;
   - 0.5 g NaCl;
   - 0.5 g $C_6H_5NaO_7 \times 2\ H_2O$;
   - 0.25 g $C_6H_6Na_2O_7 \times 1.5\ H_2O$.

   e) add the salt mixture to the homogenate obtained above;

   f) then shake the contents of the tubes intensively for one min;

   g) centrifuge the prepared tubes at 4,000 rpm, at 4°C, 7 min, in order to obtain two separated phases;

h) take the upper organic layer into Eppendorf type tubes for chromatographic analysis coupled with tandem mass spectrometry. Separate the lipopeptides using the liquid chromatograph Agilent 1200 and Kinetex C18 column (50 mm × 2.1 mm, 5 μm; Phenomenex, USA) at 40°C, solvents: water (channel A) and methanol (B) with 5 mM ammonium formate, and then perform the determination in positive polarisation mode using QTRAP 3200 (Sciex) spectrometer equipped with ESI-type ion source;

i) after the end of cultivation, turn off the bioreactor and wash it thoroughly.

Analysis of results

Evaluate the effect of foaming on lipopeptide production.

4.1.7. Citric acid biosynthesis

INTRODUCTION

Citric acid (2-hydroxypropane-1,2,3-tricarboxylic acid) is one of the most important products in the group of organic acids obtained by biotechnological means. The human body is capable of metabolising L isomers of organic acids, while racemic mixtures containing both L and D isomers of this group of compounds are obtained by chemical synthesis. Therefore, organic acids, including citric acid, used in food, pharmaceutical and cosmetic industries must come from biological sources. Citrus fruits, including lemons, oranges and pineapples, are a natural source of citric acid. These fruits were initially the only raw material from which citric acid was obtained.

$$H_2C\text{-}COOH$$
$$|$$
$$HO\text{-}C\text{-}COOH$$
$$|$$
$$H_2C\text{-}COOH$$

citric acid

Currently, about 99% of the world's citric acid production is based on the use of microbial technology.

In the food industry, this compound is used in a variety of meat, fruit and vegetable products, confectionery, carbonated drinks and dairy products (including processed cheese). Citric acid performs a variety of functions in the above products, the most important of which are lowering the pH value and stabilising the product, limiting enzymatic oxidation processes and providing the desired sensory qualities. It prevents sucrose crystallisation, has a positive effect on the stability of emulsions and limits the activity of many oxidation-reduction enzymes. The synergistic action of citric acid and other antioxidant compounds is also used. The addition of citric acid to jellies and jams allows suitable conditions for pectins to act as gelling agents. The presence of this compound positively influences the taste and smell of various food products (especially beverages). Moreover, it is added to some white wines and ciders in order to maintain a light colour and prolong the shelf life. Citric acid inactivates heavy metal ions and thus has a protective effect for vitamin C.

The pharmaceutical industry uses citric acid as an additive to various preparations, because it accelerates the dissolution of active ingredients of drugs. It is also used as an anticoagulant and oxidation inhibiting (stabilising) factor in vitamin preparations. In cosmetic products, the presence of citric acid primarily ensures the required pH of the product and reduces oxidation processes. Citric acid and its salts are used as cleaning agents, removing metal compounds or buffering agents in household chemistry products, textile, paper and ceramic industries.

Highly efficient, genetically modified strains of filamentous fungi *Aspergillus niger* and A. wentii are used for the biological production of citric acid. Such strains are obtained, among others, through the use of mutagenisation and selection of wild strains isolated from plant material and soil. Citric acid production is carried out by either by surface or submerged methods in various types of bioreactors, e.g. in stirrer or air-lift types. Spectrophotometric and chromatographic methods are used to determine the citric acid content in post-culture liquids.

The main raw material used to produce citric acid is molasses (a by-product of the sugar industry), containing about

40% sucrose, or pure sucrose. Fractions from oil distillation, containing aliphatic hydrocarbons with chain length of C10-C20 can also be used as a raw material for citric acid biosynthesis (such method was used e.g. in Japan). In this case, the yeast of *Candida* genus was acid producer. The product of biosynthesis using alkanes as a raw material, apart from citric acid, is a significant amount (up to 40%) of isocitric acid. In 2015, global production of citric acid exceeded 2 million tonnes, with the dominant share (59%) from Chinese factories.

The EMP pathway and the Krebs cycle fully function during the mycelium growth phase in the carbohydrate medium. In the second stage of cultivation, the so-called idiophase (phase of citric acid accumulation), changes in the activity of some of the enzymes of the Krebs cycle in the mycelium are noted: the activity of citrate synthase multiplies with almost complete inhibition of cis-aconitate hydratase and isocitrate dehydrogenase. The Krebs cycle is therefore almost completely blocked at the citric acid transformation stage. Formation of further amounts of citric acid needs reactions other than the Krebs cycle, providing precursors to its synthesis. Such a typical supporting (anaplerotic) reaction is the carboxylation of pyruvic acid to oxaloacetic acid. Two pyruvic acid molecules formed by EMP undergo transformations with the participation of coenzyme A: one in the process of oxidative decarboxylation to acetyl-CoA, the other as a result of carboxylation to oxaloacetic acid:

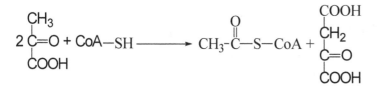

The two intermediate compounds formed, using citrate synthase, are condensed to citric acid with the release of CoA. The balance of this process given by Meyrath is as follows:

$$5\ C_6H_{12}O_6 + 6\ O_2 \rightarrow 4\ HOOC\text{-}CH_2\text{-}C(OH)(COOH)\text{-}CH_2\text{-}COOH$$

900 g   768 g

$$+\ 6\ (CH_2O) + 8\ H_2O$$

180 g

The expression (CH_2O) on the right-hand side of the equation shows the mass of mycelium produced during the growth phase – it is so substantial that it must be included in the process balance. According to this equation, the theoretical yield of the citric acid biosynthesis process is 85.3% with respect to hexose.

The amount of mycelium produced in this process is, in practice, slightly less than the quantity that would result from the Meyrath equation. This is because Meyrath assumed in the equation that at the stage of acid accumulation, full coupling of electron transport with oxidative phosphorylation and ATP formation takes place, which is then used in cellular anabolism. In fact, during the acid accumulation phase, there is an alternative route of electron transport without the classical respiratory chain and without the phosphorylation process. This "energy idle" electron transport route is not accompanied by anabolism reactions and further microbial mass gain. The inhibition of ATP synthesis in the acid accumulation phase is of great regulatory importance as it does not inhibit the activity of phosphofructokinase, the key enzyme of the EMP pathway. Thus, the EMP pathway is still active in the idiophase and provides precursors for citric acid synthesis.

PRACTICAL PART

Materials and media

1. Microorganisms: *Aspergillus niger* strains.
2. Media: sucrose, molasses.
3. Apparatus: liquid chromatograph Agilent 1200 and mass spectrometer QTRAP 3200 (Sciex).

Aim of the exercise
Comparison of selected industrial strains of Aspergillus niger in terms of biomass production and citric acid biosynthesis efficiency.

Procedure

1. Morphological characteristics of *Aspergillus niger* strains used in industrial practice:

   a) observation of plate cultures (conducted on wort medium with 2% agar) after 2, 4 and 7 days of incubation at 28°C. Describe the macroscopic and microscopic morphology of the mycelium and the microscopic morphology for the different stages of growth: hyphae, conidial group, conidia (magnification 600 times):

2. Laboratory attempts to biosynthesise citric acid using different media and strains:

   a) cultures of 100 ml volume, carried out on a molasse medium using the surface method in volumetric conical 300 ml flasks;

   b) cultures on sucrose medium carried out with surface and deep method under shaken conditions (120 rpm) at 30°C.

3. Fermentation sample analyses.

   After 7 days of biosynthesis conducted with the surface method and after 5 days of the process conducted with the deep method, determine:

   a) presence of oxalic acid as a by-product of fermentation:

   - treat a drop of medium with a saturated solution of $CaCl_2$ on a watch slide. In the presence of oxalic acid, Ca oxalate crystals appear;

   b) mycelium dry matter content:

   - place the mycelium from the surface culture on a Büchner funnel and rinse it several times with distilled water, then dry it with tissue paper and place it on a Petri dish of known weight. Dry the mycelium to a constant mass, initially at about 50°C and then at 105°C;

   - place the mycelium from the deep (shaken) culture on filter paper in the Büchner funnel, drain and rinse several times with water, continue as with the mycelium from surface culture;

243

c) citric acid content, using:
- titration of 5 ml of post-culture liquid with an aqueous solution of 0.25 M NaOH with respect to phenolphthalein (1 ml of 0.25 M NaOH corresponds to 0.016 g citric acid),
- mass spectrometry method using the HPLC Agilent 1200 Series LC with the QTRAP 3200 (Sciex) mass spectrometer. Inject directly (without column) 10 µl of the sample. Use a flow rate of 0.7 ml/min, water: methanol mixture (70:30 v/v) as a moving phase, spectrometer operating in negative polarisation mode, analyse for selected fragmentation pairs of MRM 191–111 and 191–87 (for more information see Section 2.3). Prepare a citric acid standard curve;

d) on the basis of the results obtained (citric acid content and mycelium dry matter), calculate the yield of the biosynthesis of the tested acid and compare it with the theoretical value obtained from the Meyrath equation.

Analysis of results

Calculate the theoretical and practical efficiency of citric acid production.

4.2. Fermentation processes

Fermentation processes are among the changes that have long been used by man, long before Pasteur laid the foundations of industrial microbiology. During the fermentation of sugars, microorganisms produce compounds with a low molecular weight and a relatively simple chemical structure, such as methanol, ethanol, acetone, butanol, which can also be produced by chemical means. The choice of technology (biological or chemical) is primarily determined by the profitability of production. Nevertheless, it must be remembered that fermentation, especially lactic acid and ethanol fermentation, produces numerous food products and stimulants with valuable nutritional and/or organoleptic properties that can only be obtained with the participation of microorganisms, e.g. fermented milk products, cheese, alcoholic beverages.

In industry, the term "fermentation" is often also used to describe other processes (technologies) in which microorganisms are used, even though these are biosynthesis or biotransformation processes, such as "citric fermentation" or "acetic fermentation". The book adopts terminology consistent with the principles used in biochemistry and therefore citric acid biosynthesis and acetic acid production by microorganisms are discussed in Sections 4.1.7 and 4.3.1, respectively.

4.2.1. Winemaking and brewing

INTRODUCTION

Wine and beer – products known from time immemorial – are the result of alcoholic (ethanol) fermentation carried out by yeast in fruit must (in the case of wine) or malt wort (in beer production). The yeast used in these industries belongs to the *Saccharomyces* genus (see Section 3.2.5). They are not typical fungi because they do not form mycelia – they are unicellular microorganisms, which is why they are called yeast-like fungi for the distinction. Different yeast types with specific technological characteristics are used in the industrial process. The term "type" is not a taxonomic term, but a typically technological one and characterises a given yeast culture in terms of its suitability for production, e.g. wine, with specific taste and aromatic qualities. Winemaking uses different types of yeast belonging to the species *S. cerevisiae* v. *ellipsoideus* (elliptical variety), while brewing usually uses *S. carlsbergensis* (synonym *S. uvarum*).

Yeast can grow under both aerobic and anaerobic conditions. Oxygen dissolved in the medium plays a role in regulating the direction of cellular metabolism. Under aerobic conditions, an intensive growth of yeast cell mass takes place, while under anaerobic conditions, fermentation with ethanol production dominates. This phenomenon, known as the Pasteur effect, consists of the fact that under aerobic conditions, the Krebs cycle and respiratory chain are activated in yeast cells, transferring electrons to oxygen with simultaneous coupling with ATP synthesis. Triphosphosphoradenosine acid is the main effector inhibiting the activity of phosphofructokinase in the EMP pathway. Increased concentration of ATP inhibits glycolysis, while intensive use of ATP in anabolic reactions accelerates the glycolysis process. Therefore, the use of glucose in the synthesis process under aerobic conditions is a very economical solution. However, under anaerobic conditions, phosphofructokinase, aldolase and pyruvic kinase of the EMP pathway are inducted, thus accelerating glycolysis. The pyruvate formed in the absence of the Krebs cycle and the respiratory chain, undergoes enzymatic decarboxylation to acetaldehyde, which under these conditions (anaerobic process) is an electron acceptor, reducing to ethanol. Under anaerobic conditions, the efficiency of cellular material synthesis in relation to glucose is low, while intensive ethanol formation takes place. The fragment of cellular metabolism discussed here can be presented by the following equation:

$$\underset{\substack{\text{COOH} \\ | \\ \text{C}=\text{O} \\ | \\ \text{CH}_3}}{} \xrightarrow{\text{decarboxylase}} CO_2 + CH_3-\underset{\substack{\text{H} \\ \diagdown \\ \text{O}}}{C} \xrightarrow{\substack{\text{NADH-H}^- \quad \text{NAD}^- \\ \diagdown \quad \diagup}} CH_3-CH_2-OH$$

The complete alcoholic fermentation balance is as follows:

$$\underset{180\,g}{C_6H_{12}O_6} \quad \rightarrow \quad \underset{92\,g}{2\,CH_3-CH_2-OH} + \underset{88\,g}{2\,CO_2}$$

The equation does not take into account a small part of the substrate used to build yeast cells. Taking this into account, as well as the possibility of creating small amounts of by-products, the practical efficiency of the alcoholic fermentation process ranges from 94–96% of the theoretical efficiency.

In addition to the wine and brewing industries, yeast is used in the distillery industry, where the final product is pure spirit, used for both food and industrial purposes. In addition, *S. cerevisiae* yeast is used in the production of baker's yeast, where the final product is a compressed mass of live yeast cells (for more information see Sections 1.3 and 4.2.3).

The wine production scheme involves several basic stages. The basic raw material for wine production is grape, but other fruits such as apples, pears, cherries, redcurrants, plums, gooseberries, raspberries, blackberries, strawberries or currants (so-called fruit wines) can also be used. The fruits are crushed, then the resulting juice (called must) is separated from the fruit residue (white wines) or left with the fruit residue (red wines). The grapes themselves contain yeast, but since sulphur compounds are added to the must, most of them are eliminated. The obtained juice is supplemented with selected types of yeast with resistance to low pH, high alcohol content (>10%) and sulphur compounds. The fermentation process begins with the supply of oxygen to help the yeast multiply and is then conducted under anaerobic conditions. The main stage of fermentation lasts 4–8 days. The next stage of wine production, ripening, begins after the fermented must has been separated from the sludge collecting at the bottom of the container.

A popular alcoholic beverage is also cider, obtained from fermented apple juice (not flavoured with sugar), with an alcohol content of 1.2–8.5%.

In contrast to wine production, the technology of brewing beer is more complicated. Beer is a beverage, the production of which mainly uses barley, hops, water and yeast. In addition to barley, other cereals can also be used. Beer can be divided into two groups: bottom-fermentation (dominant group) and top-fermentation beer. Bottom-fermentation beers (generally called lager, from the German lagern, i.e., store) are obtained by using yeast that forms floccules and falls to the bottom of the tank (*Saccharomyces uvarum*, previously called *Saccharomyces carlsbergensis*). Most top-fermented beers are obtained using S. cerevisiae, accumulated on the surface of the wort, e.g., Ales, Porters etc.

The basic raw material for beer production is barley. It is characterised by high starch and amylase content. Proper preparation of barley largely determines the overall beer making process. After cleaning, the barley grain is soaked in water, the germination process begins (malt production), during which mainly amylolytic, proteolytic and cellulytic enzymes are activated. The malt is then dried and the germs are separated (de-germination). The crushed malt is mixed with water. The mashing process begins at a high temperature (>50°C). When setting the temperature ranges, optimum conditions for the activity of α-amylases (70°C) and β-amylases (60–65°C) are taken into account, as well as the pH, which is also important for the activity of proteolytic enzymes (pH 5.2–5.5). In the next stage, the hot mash (low temperature increases viscosity) is filtered. The resulting solid phase (spent grain) can be used as feed, while the wort is cooked with the added hop cones (brewing process). The resulting wort, devoid of hoppers, is quickly cooled. In the next stage the wort is fermented with the use of yeast. Then, if bottom-fermentation yeast is used, the beer undergoes lagering at a temperature of about 2°C, in the meantime another fermentation of the remaining sugars takes place and the beer gains taste. After a few months the beer is filtered and packed. If top-fermentation yeast is used, however, the ageing process is much shorter and the beer can be enriched with various additives, then purified, pasteurised and packed. The different stages of beer production vary according to the type of beer.

PRACTICAL PART

Winemaking

Materials and media

1. Microorganisms: *Saccharomyces cerevisiae, Saccharomyces uvarum* (types used in winemaking).
2. Media: liquid wort, apple must. Raw materials with known and predetermined sugar composition.

Aim of the exercise
Characteristics of wine yeast, acquaintance with wine production technology.

Procedure

1. Fermentation test using different types of wine yeast:

   a) setting of fermentation samples:
   - determine the extract content of the must;
   - sweeten the must with sucrose to a total of 25% extract;
   - complete the must with ammonium phosphate in the amount of 0.2 g/l;
   - pasteurise the prepared must and after cooling, inoculate with pure yeast culture (4% mother yeast);
   - close the samples fermentation pins, seal with paraffin, weigh every 24 h for 7 days of main fermentation;
   - on the basis of CO_2 loss, make a diagram expressing the dynamics of fermentation during 7 days;
   - determine the alcohol content of the wine by distillation and the actual extract;
   - calculate the theoretical and practical yields from the balance of sugar and alcohol produced, compare the results obtained for individual types of wine yeast.

   b) microbiological examination of prepared wines:
   - determination of the number of microorganisms in the wine by means of membrane filters: filter 50 ml of wine through a disinfected filtering kit. Remove the membrane from the filter and place it on solidified medium in a Petri dish (wort with agar). Thermostat at 28°C. After 48 h, analyse the growth of the microflora, microscoping the individual colony types (drawings) and give the total number of microorganisms in 1 l of wine. Compare the results with the indications of the standard;

   c) thermostat test of wine:
   - Fill two tubes with wine, then close one with a lignin stopper and the other with a rubber stopper. Insert the test tubes into the heaters (at 25°C) and observe any changes (turbidity, sediment or a skin on the surface) during two weeks. In case of changes, prepare microscopic preparations from the samples and indicate the type of microflora (drawings).

Brewing

Materials and media

1. Microorganisms: *Saccharomyces cerevisiae,
Saccharomyces carlsbergensis* (types used in brewing).
2. Media: liquid wort.

Aim of the exercise
Microbiological evaluation of beer production stages

Procedure

1. Comparison of bottom and top-fermentation yeast:
   a) macro- and microscopic observations of cultures on the wort. Pay attention to the characteristic cell arrangement at low microscope magnification (150 times);
   b) test for the rate of yeast fall in a water suspension (flocculation): take some bottom yeast suspension into one tube, some top yeast suspension into another tube, shake thoroughly with water and observe the difference in the nature of the suspension and the rate of yeast fall. The yeast suspension should not be very dense, so that is possible to observe turbidity in the form of a dusty, floating top yeast suspension and a flocculent, dropping bottom yeast suspension in the semi-transparent liquid during mixing. Then acidify the sample of bottom yeast with 5% H_2SO_4, shake it and observe a change in the rate of falling and disintegration of the floccules:

Water suspension	Nature of turbidity	Falling rate
Top fermenting yeast		
Bottom fermenting yeast		
Bottom fermenting yeast after acidification		

   c) Herzfeld test (raffinose fermentation):
   Place a suspension consisting of 10 ml of 1% raffinose solution in yeast water + 1 ng of yeast in a single-armed tube. After 24 h of incubation at 30°C, determine the

amount of CO_2 produced. The bottom yeast ferments the whole raffinose, while the top yeast only 1/3.

2. Microbiological control of beer production:
   a) microscopic observations of seed yeast (the so-called mother yeast). Cheque the viability and nutrition of the yeast;
   b) inoculate seed yeast from an appropriate dilution on solidified wort medium with the addition of 5 mg % actidone to inhibit the growth of the yeast.
   Count bacterial colonies after 2 days of incubation at 25°C. Provide the result as the number of bacteria in 1 g of seed yeast;
   c) microbiological analysis of beer from different production phases:

Sample	Density °Blg	Approximate number of yeast cells	% of budding cells	Presence of infections
Yeast cake				
Wort after joining with yeast				
Beer 1–2 days of the main fermentation				
Beer 6–8 days of the main fermentation				
Beer from the lagering room				
Ready-made beer				
Infected beer				

Calculate in the direct preparation, assuming the preparation height of 0.01 mm.

4.2.2. Practical use of lactic acid bacteria

INTRODUCTION

Lactic acid bacteria belong to the *Lactobacillaceae* family and are very diverse in morphology. Besides cocci arranged in chains (*Lactococcus, Leuconostoc*) or tetrads (*Pediococcus*), there are also short or long bacilli in chain form (*Lactobacillus*).

These bacteria are Gram-positive, non-sporulating, immobile, relatively anaerobic, mostly without catalase and respiratory chains. They are very demanding in terms of growth medium components. In addition to simple carbon sources (simple sugars), they require many amino acids as well as B vitamins for growth. They do not absorb nitrogen from nitrate salts or ammonium nitrogen. Therefore, the natural habitat of these bacteria are: milk, plants, plant remains and mucous membranes of mammals, but they are not parasites. They do not occur in water or soil. This group of bacteria contains both meso- and thermophiles. The common feature of the bacteria of the *Lactobacillaceae* family is the production of significant amounts of lactic acid (2-hydroxypropionic acid) from simple carbohydrate compounds. Left-handed D(–) or right-handed L(+) lactic acid or a racemic mixture of these acids is produced. In the human body only the form of L(+) lactic acid is metabolised, therefore it is important what enantiomer is produced by bacteria.

Differential carbohydrate metabolism in these bacteria is the basis for the division into two subgroups: homofermentative and heterofermentative. Homofermentative bacteria, which include *Lactococcus lactis, Pediococcus acidilacci, Lactobacillus plantarum, Lactobacillus delbrueckii* subsp. *bulgaricus,* lead a highly homogeneous fermentation with an efficiency of not less than 1.8 mole of lactic acid from 1 mole of hexose, which is more than 90% of the theoretical efficiency.

In homofermentative lactic acid fermentation bacteria, in the absence of a respiratory chain, under anaerobic process conditions, the electron acceptor is pyruvate formed by EMP. The enzyme responsible for this reaction is lactic acid dehydrogenase, whose coenzyme is NAD. This process can be presented by the following equation

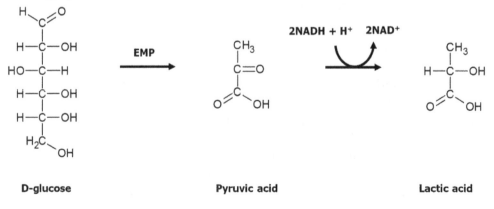

D-glucose   Pyruvic acid   Lactic acid

Oxidation (dehydrogenation) of NADH + H+ in this reaction enables a further process of hexose conversion to lactic acid, because the oxidised form of NAD can participate in the oxidation reaction of 3-phosphoglyceric aldehyde to 3-phosphoglyceric acid in the EMP pathway and thus a further pool of pyruvic acid is supplied.

The metabolism of simple carbohydrates in heterofermentative lactic acid bacteria is different. These bacteria do not have L-enzyme aldolase splitting 1,6-diphosphofructose into two trioses, so there is no EMP pathway. In turn, they have 6-phosphoglucose dehydrogenase and 6-phosphogluconate dehydrogenase, as well as further enzymes of the hexose monophosphate pathway (HMP), which enables them to metabolise sugars in this pathway. As a result of such metabolism, the products of hexose fermentation in heterofermentative lactic acid bacteria are: lactic acid, ethanol or acetic acid, carbon dioxide.

The scheme of hexose metabolism in heterofermentative lactic acid bacteria with HMP pathway is presented below.

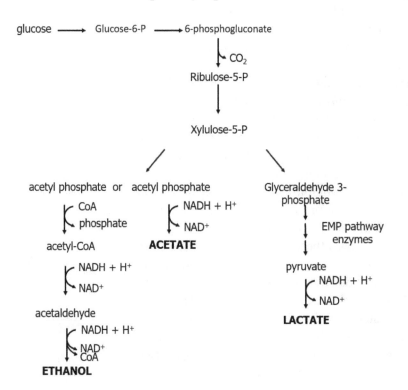

Most often, heterofermentative lactic acid bacteria produce acetic acid, lactic acid and CO_2. For example, *Leuconostoc mesenteroides* produces ethanol instead of acetic acid. Another type of lactic heterofermentation is known, which is found in *Lactobacillus bifidus*, an obligate anaerobe present in the intestinal microbiota. The metabolism of hexose in this microorganism is carried out without the release of CO_2 according to the following reaction:

$$2\,C_6H_{12}O \rightarrow 2\,CH_3CH(OH)COOH + 3\,CH_3COOH$$

The use of lactic acid bacteria is very wide and is mainly related to the food industry. In the improvement and preservation of plant material, lactic acid fermentation is used in the preparation of vegetable silages (cabbage, cucumbers, beets) as well as feed silages for cattle. This process, often carried out with the use of indigenous microflora, involves lactic acid bacteria of the genera *Lactococcus, Pediococcus, Leuconostoc*, as well as ones belonging to *Lactobacillus brevis, L. plantarum, L. fermenti* (for further information see Section 4.2.3).

The second large branch of the food industry, where lactic acid bacteria are used, is the dairy and cheese industry. Pure cultures of lactic acid bacteria, i.e., *Lactococcus lactis, L. diacetilactis, L. cremoris, L. thermophilus, Leuconostoc citrovorum, L. mesenteroides, Lactobacillus delbrueckii* subsp. *bulgaricus, L. brevis, L. helveticus, L. plantarum* and other, are used in the production of butter, cottage cheese and hard cheese and dairy drinks such as yoghurt, kefir, kumis, etc.

A mixed bacterial microbiota, which includes lactic acid bacteria is also used in the production of fermented meat products (e.g. salami) (see Section 4.2.3). In addition, lactic acid bacteria are used in the biosynthesis of dextran (the synthesis of dextran with *Leuconostoc mesenteroides* is discussed in Section 4.1.4), nisin – a polypeptide antibiotic used in the food industry (*L. lactis*), as well as in the production of medicinal and probiotic preparations administered orally to improve the composition of intestinal microbiota, especially after an antibiotic treatment.

Probiotic strains of species *Lactobacillus acidophilus, L casei, L. johnsonii, L. fermentum, L. rhamnosus, L. plantarum,*

L. reuteri, L. salivarius, L. paracasei, L. delbrueckii subsp.
bulgaricus, Streptococcus thermophilus, Bifidobacterium lactis,
B. longum and *B. breve* are particularly useful. Probiotics have
a double effect, preventing / reducing intestinal colonisation
by pathogenic microorganisms and preventing inflammatory
reactions in host tissues. Some strains, used in acute rotaviral
diarrhoea, ulcerative colitis or *Helicobacter pylori* infection, had
a positive effect on the treatment process.

A separate field of biotechnology is the production of
lactic acid for food and industrial purposes using lactic acid
bacteria. Although the chemical synthesis of lactic acid is
nowadays economically comparable with the microbiological
process, biological lactic acid is preferred for food purposes.
Thermophilic, homofermentative bacteria of the genus
Lactobacillus are used in lactic acid biotechnology:

- for plant raw materials: *L. delbrueckii and L. leichmanii*;
- for raw materials of animal origin (e.g. whey): *Lactobacillus delbrueckii* subsp. *bulgaricus.*

In industrial practice, most often molasses, often sucrose, or
starch hydrolysates are used.

The lactic acid produced is neutralised with $CaCO_3$ and after
the fermentation is recovered from Ca lactate with sulfuric
acid. The lactic acid solution, after separation of gypsum (the
common name for Ca sulphate) is purified, thickened to 50% and
is marketed in this form. This process can be presented using the
following formulae:

$$C_{12}H_{22}O_{11} + H_2O \rightarrow 2\ C_6H_{12}O_6 \rightarrow 4\ CH_3CH(OH)COOH + 2\ CaCO_3 \downarrow$$

sucrose $\qquad$ hexoses $\qquad\qquad\qquad\qquad\qquad \downarrow$

$$2\ (CH_3CHOHCOO)_2Ca + 2\ H_2O + 2\ CO_2 \uparrow$$

$$2\ (CH_3CHOHCOO)_2Ca + 2\ H_2SO_4 \rightarrow 4\ CH_3CH(OH)COOH + 2\ CaSO_4$$

The practical performance of this process is in the range of
80–90% of the theoretical performance.

255

PRACTICAL PART

Materials and media

1. Microorganisms: *Lactobacillus delbrueckii* subsp. *bulgaricus, Lactobacillus delbrueckii, Lactococcus lactis*.
2. Media: skimmed milk, medium for lactic acid production with sucrose and molasses.
3. Reagents: 0.1 M NaOH, 0.5 M Na_2CO_3, 0.5 M, H_2SO_4.

Aim of the exercise
Characteristics of lactic acid bacteria and microbiological lactic acid production on a laboratory scale.

Procedure

1. Observation and description of morphological features of lactic acid bacteria in commonly available dairy products, including strains with probiotic properties.
   Prepare the direct preparation (magnification 600 times) and the preparation fixed with gentian violet, and observe under immersion.
2. Use of solid media with indicator to distinguish colonies of lactic acid bacteria. Describe the bacterial growth on solid medium (yeast water + glucose) with $CaCO_3$.
3. Production of lactic acid in milk by industrial strains of Lactococcus lactis and *Lactobacillus delbrueckii* subsp. *bulgaricus*:
   a) inoculate 5 drops of pure culture of these bacteria into the tubes containing 10 ml of sterile milk;
   b) after 5 days of incubation at 28°C (*L. lactis*), 48°C (*L. delbrueckii* subsp. *bulgaricus*), determine the acidity of fermentation trials by titration with 0.5 M NaOH in respect to phenolphthalein. Provide the result in grams of lactic acid per 100 ml of culture.
4. Fermentation trials on industrial substrates: inoculate clean *L. delbrueckii* culture in a sterile manner in the amount of 1.5% on the medium:

    a) with sucrose;

    b) with molasses (molasses diluted to 16°Blg).

        Place the fermentation trials in a thermostat at 48°C. After 7 days of fermentation, determine % of lactic acid.

5. Determination of lactic acid content and process efficiency.

Reactions observed:

$$(CH_3CHOHCOO)_2Ca + 2Na\ 2CO_3 = 2CH_3CHOHCOONa + Na_2CO_3 + CaCO_3$$

$$2CH_3CHOHCOONa + Na_2CO_3 + 2H_2SO_4 = 2CH_3CHOHCOOH + 2Na_2SO_4 + CO_2 + H_2O$$

The yield of lactic acid should be expressed in % relative to the theoretical yield of fermented sugar.

Analysis of results

Calculate the theoretical and practical efficiency of lactic acid production.

4.2.3. Use of microorganisms in the baking industry and for the production of fermented meat and vegetable products

INTRODUCTION

The Neolithic Revolution (the beginning of which is estimated to be about 10,000 BC) consisted in changing the way people got their food. The nomadic lifestyle combined with hunting and gathering was replaced by agriculture, based on a sedentary lifestyle, and methods facilitating plant cultivation and animal breeding were gradually developed. This way of obtaining food proved to be very effective and led to an excess of food. Some of this food was spoiled or was subject to natural conservation processes – resulting from the development of microorganisms releasing significant amounts of lactic acid or ethanol into the environment. The oldest documented traces of food preservation based on microbial growth concern the production of acidified (fermented) fungi, soy sauce (China, 4,000-3,000), wine and beer (Asia, Europe, Africa, 3,000–2,000 BC). By 1,000 BC, the basic principles for making cheese, bread (using yeast and sourdough), pickling vegetables, fish and obtaining fermented dairy and meat products were developed. These principles were constantly improved up to the present day.

Fermented food production in the 21st century

The production of fermented food is a very important branch of the food industry (this type of food represents about 1/3 of global consumption). In some regions of the world, the share of fermented products is up to 40% (by weight) of an individual's diet. The highest production levels are reached by cheese (15 million tonnes), yoghurt (3 million tonnes), fermented mushrooms (1.5 million tonnes), fish and soy sauce, beer (1,000 million hl) and wine (350 million hl).

Classification of fermented products

There are many criteria on the basis of which fermented food is classified. Due to the type of raw material used for fermentation, products are divided into:
- vegetable (including cabbage, cucumbers, beets, garlic, peppers, carrots, celery, onions, patissons, tomatoes, mushrooms, olives);
- of animal origin (milk, meat, fish, crustaceans – so-called seafood);
- mixed – containing fermented plant and animal raw materials (e.g. some kimchi variations).

Another way of classifying fermented food includes division into products:
- consumed as a meal (e.g. yoghurt, salami, bread);
- for consumption, but most often used as a flavouring (e.g. sour cream);
- used only as sauces (soy sauce).

The following products are distinguished in the next division:
- containing live microorganisms (e.g. yoghurt, cheese);
- free from live microorganisms (e.g. bread, wine, soya sauce);
- products where the microorganisms were only used in the initial stages of raw material preparation (e.g. coffee, cocoa).

There is also a distinction between products obtained as a result of development:
- lactic acid bacteria;
- yeast;
- mixed populations of fermenting microorganisms (e.g. kefir);
- mixed populations containing both fermenting and non-fermenting microorganisms (including strains of microscopic fungi and bacteria).

The changes in fermented products include mainly:
- prolongation of durability (protection against the development of unfavourable microorganisms is mainly due to the lowering of pH under the influence of lactic acid or ethanol);
- the product gains new organoleptic characteristics, consisting of a change in taste, smell and/or texture (e.g. wine, cocoa, Balkan yoghurt);
- increased content of vitamins, bacteriocins, probiotic microorganisms and nutrient bioavailability;
- removal of toxins (e.g. cassava).

In Poland, cabbage and cucumbers are among the most popular vegetables eaten in a pickled form. Pickling of these vegetables is a complex process. It requires the addition of a NaCl, which limits the development of undesirable microorganisms, favours the development of lactic acid bacteria and promotes the passage of components contained in plant tissues to the external environment (which is the growth medium for developing bacteria). Sugars from cabbage juice (and in the case of cucumbers pickling added in the form of sucrose solution) are broken down by fermentation into lactic acid, which preserves vegetables. Additionally, others, secreted by bacterial metabolites give the finished products different taste, smell and health characteristics.

Sauerkraut

In Poland, a variety of white cabbage (*Brassica oleracea* L. var. *capitata* L.) is usually used for pickling. The technological process includes the following stages:

- bleaching – storing cabbage for several days in a dark room in order to reduce the level of chlorophyll and sulphur compounds which adversely affect the organoleptic characteristics of the final product;
- chopping of cabbage (as well as carrots, which are a frequently used flavouring supplement and improve the course of milk fermentation);
- placing the chopped cabbage in a container, adding NaCl (2.25–6% of cabbage volume);
- crushing the vegetables (to release the cellular juice) and strong whisking of the successive layers added, in order to reduce oxygen content;
- covering the surface with a weighted cutting board (this treatment prevents the development of aerobic microorganisms), leaving space through which gaseous products produced during fermentation can be extracted;
- placing the container with the cabbage at 18–20°C.

The significance of NaCl addition:
- accelerates the softening of leaves and the outflow of cellular juice, which, by filling the spaces, helps to remove air (oxygen) and is a source of simple sugars, easily metabolised by lactic acid bacteria;
- limits the development of the unfavourable microbiota responsible for decay processes.

The fermentation process involves three stages:
- preliminary (turbulent) fermentation carried out at 20°C for 2–5 days; initially oxygen consuming microorganisms develop (e.g. coliform bacteria); then the environment is controlled by lactic acid bacteria (mainly *Leuconostoc mesenteroides* and *Lactococcus lactis* strains); as a result of their activity, large amounts of carbon dioxide, lactic acid (up to 1%) and other organic acids (e.g. succinic, formic, propionic), ethanol are produced; the resulting products lower the pH of the environment to about 4;
- fermentation carried out at 20°C for 2–5 days and at 18°C for 14–27 days of fermentation; during this time strains of *Lactobacillus brevis*, *L. plantarum*, *L. pentoaceticus* as well as *Pediococcus* bacteria develop intensively; lactic acid

content increases to 1.8%, pH decreases to 3.5 and ethanol level reaches 0.4%.

- after-fermentation and storage at a couple of degrees above freezing.

One of the causes of spoilage of sauerkraut stored at too high a temperature, may be the development of microscopic filamentous fungi and yeast on the product surface. These microorganisms consume organic acids, which leads to a gradual increase in pH, and thus promotes the development of spoilage microorganisms. Additionally, volatile compounds with an unpleasant smell may be formed or pectinolytic enzymes causing softening in prepared food products, may be produced.

Pickled cucumbers

Washed cucumbers are placed tightly in a container, then so-called brine is added, containing 6–8% NaCl, 1–2% sucrose and spices. Unlike cabbage pickling, the diffusion of cucumber cellular juice into the brine is very slow, no turbulent fermentation is observed, and there is no preliminary dominance of *L. mesenteroides*. Cucumber pickling requires 22–23°C room temperature in the first 24 h. Then the process is carried out for 40 days at 12°C or for 21 days at 18°C. At home, pickled cucumbers can be stored for up to 4 months in room temperature and for up to 10 months in a fridge.

The process consists of three stages:
- so-called mixed fermentation, lasting up to 7 days, during which various microorganisms develop; at the end of this phase, lactic acid bacteria gain dominance, the drop in pH value is slow;
- (lactic) fermentation, lasting 10–14 days, characterised by intensive development of *L. plantarum, L. brevis* and *Pediococcus* strains;
- saturation of the solution with lactic acid (pH 3.3; lactic acid 0.8–1.2%), inhibition of lactic bacteria growth due to total exhaustion of sugars;
- fungi of the genus *Pichia* (formerly *Mycoderma*) or the *Dipodascus* fungi (belonging to the family *Saccharomycetaceae*, formerly described as *Oomyces* or

Geotrichum) begin to develop on the surface. Since they decompose lactic acid (which causes a gradual increase in pH), the production of pickled cucumbers should be finished at this stage.

Fermented meat products

In Europe, raw maturing sausages (e.g. salami and pepperoni) are produced mainly in Italy, Hungary and Germany. The microorganisms used in the production of this type of sausage cause the meat to become more stable (prolonged shelf life) and change the colour, taste, smell and consistency of the product. Metabolic properties and technological importance of microorganisms used in the production of fermented meat products are presented in Table 4.2.3.1.

Table 4.2.3.1. Microorganisms used in the production of fermented meat and their functions

Group of microorganisms	Metabolic properties and technological importance
Lactic acid bacteria (*Lactobacillus sake, L. plantarum, L. curvatus, Pediococcus acidilactici, P. pentosaceus*)	Decomposition of carbohydrates by lactic homo- and heterofermentation pathway. As a result of lactic acid production, the pH drops below 5.3 after 2–3 days, and stabilises between 4–8 days (pH = 5). At low pH, the growth of rotting bacteria is inhibited and protein gelation occurs (a process which has a beneficial effect on the consistency and compactness of meat). Additional products of lactic fermentation (including acetic, formic, propionic, pyruvic acids, ethanol, propanol, acetoin, diacetyl and butandiol) participate in the formation of taste and smell.
Reducing and aromatising bacteria (*Staphylococcus carnosus, S. xylosus, Micrococcus varians, Kocuria, Streptomyces griseus*)	Staphylococcus strains produce nitrate reductase, catalysing the reduction of nitrates (V) to nitrites (III) and then to nitric oxide, which reacts with myoglobin to form red nitrosylomyoglobin. *Micrococcus* strains produce catalase, which by decomposing peroxides prevents fat rancidity and adverse changes in colour. In addition, this group of microorganisms secrete various enzymes involved in the transformation of proteins and fats, which affects the production of aroma and taste compounds.
Yeast *Debaromyces hansenii, Candida famata, C. utilis, Pichia, Rhodotorula, Torulospora, Yarrowia lipolytica*)	If they are added to the stuffing, they develop mainly at the edge; they can also be a component of surface microbiota. Using oxygen, they inhibit the effect of peroxidases that destroy the red colour of meat, and secrete proteases and lipases that affect the taste and aroma of sausages.
Moulds – surface microbiota (*Penicyllium nalgiovense, P. chrysogenum, P. candidum*)	The mycelium forms an air layer on the surface of the sausage link. It takes part in the gradual oxidation of lactic acid (a beneficial process because the product is ultimately less acidic). Various proteases and lipases are produced that affect the taste and smell of the product. The development of mycelium reduces the risk of surface contamination by undesirable microorganisms and supports uniform drying of the sausage link.

Salami

A typical salami recipe includes the use of: 4 kg ham, 1 kg lard, NaCl (2.6–4.5%), sugar (0.3–1%), herbs and spices (including sweet peppers, garlic, pepper), nitrates and starter cultures (containing various groups of microorganisms described in Table 4.2.3.1). Table salt inhibits the growth of spoilage bacteria and promotes the development of lactic acid bacteria. Nitrates have a bacteriostatic effect and are also a substrate for nitrate reductase produced by denitrifying microorganisms. Then, in the presence of nitrites, myoglobin is transformed into nitrosylomyoglobin, giving the product an intense pink colour.

The antimicrobial effect of nitrites increases with a decrease in the environmental pH (e.g. at pH 5.0 the required level in meat is about 80 mg/kg). One of the most important reasons for using nitrites is to inhibit the development of *Clostridium botulinum* strains. Nitrites also have an antioxidant effect, inhibiting the breakdown of unsaturated fatty acids. Thus, they increase the durability and taste profile of the finished product. The addition of carbohydrates is necessary for the development of lactic acid bacteria, inhibiting the decay of meat proteins. In industrial practice, carbohydrate mixtures characterised by a different rate of decomposition by lactic acid bacteria (e.g. glucose, lactose and sucrose) are most commonly used. Spices determine the taste and smell, some of them also have antioxidant properties. The fat contained in meat and added in the form of lard affects such features as: consistency, taste, smell, colour in cross-section and tenderness, assimilability and stability of nutrients. The qualitative and quantitative composition of microbiota in the stuffing of raw sausages is very varied and largely depends on the conditions in which the raw meat was obtained and its contamination. During the production process, the primary microbiota in raw sausages is exchanged in the first period of maturation for the technologically desired one, i.e., the acidifying, denitrifying and aromatising microorganisms. Microbial starters are usually ready-made commercial preparations sold as lyophilizates or frozen products. The addition of a starter at a level of 10^6 -10^7 CFU of meat ensures quick control of the environment and correct fermentation. Starters added to meat stuffing should not contain fungi (moulds) developing on the surface of casings. The

263

characteristic smell of salami type meats is the result of production of: 1) fatty acids, aldehydes and ketones obtained from the decomposition of lipids; 2) short peptides and amino acids (especially glutamic and α-aminobutyric acid); 3) ribonucleotide derivatives; 4) certain products of lactic acid heterofermentation; and 5) if smoking is used, also the compounds contained in the smoke (mainly polyphenols).

PRACTICAL PART

Part I
Materials and media

1. Media: MRS, Sabouraud medium with yeast extract and chloramphenicol, Czapek-Dox medium (plates).
2. Fermented maturing sausage covered by mould, sauerkraut, bell pepper, beet, NaCl, sucrose, garlic, bay leaf, allspice, pepper, rye or wheat-rye sourdough bread, other types of bread.
3. Reagents: Methylene Blue Loeffler, reagents for the Gram staining.

Aim of the exercise
Preparation of selected fermented products (pickled bell peppers, beetroot sourdough and sour rye soup). Isolation of microorganisms from fermented meats with fungal growth.

Procedure

1. Isolate microorganisms from sauerkraut juice and samples of fermented cured meats covered by mould, using plates with different microbiological media. Culture at 28°C for 3–5 days.
2. Prepare the pickled bell peppers – wash the peppers, cut them up, remove the core. Peel the garlic, crush the bay leaves. Boil water with NaCl and allspice. Arrange the

peppers tightly in jars (in order to limit the air content) and add the garlic and bay leaves. Pour cooled fluid (3% water solution of NaCl). In addition, prepare samples in which the brine is replaced by water and without spices. Cover the jars, incubate at 25–28°C for 3 days. Then incubate at 18°C for 7–14 days.

3. Prepare beetroot sourdough – necessary raw materials: 1.5–2 kg of red beet (*Beta vulgaris* L. subsp. *vulgaris*), 2 l of water, rye wholemeal bread crust, 2 teaspoons of NaCl, a few cloves of garlic, allspice, a teaspoon of sugar, microorganisms present in a slice of wholemeal sourdough bread are used as a starter.

   Cut the peeled beets into thick cubes; peel the garlic and crush it with a knife; preparation of brine: pour water into the pot, add NaCl; add spices to the bottom of the scalded jar, place beets in the jar, pour cooled brine (the liquid should cover beets thoroughly); put the cap on the jar, but without screwing it on – this will allow the outlet of gaseous fermentation products; leave the sample prepared in such a way at temp. 20–22°C for 5–7 days. Observe daily (measure pH). For the first 3 days a layer of white foam may form on the surface of the liquid, which should be removed with a scalded spoon, the beets can be very gently mixed. After 3 days, if bread was used as a starter, it should be thrown away; none of the beets should stick out above the water surface, so the brine must be refilled. The incubation should be completed after 7 days.

4. Obtain soured rye flour (rye flour sourdough) – (0 h): mix 225 g of wholemeal rye flour (type 2,000) with 225 g of boiled water, place the sample in a glass flask, incubate at 20–25°C for 24 h. Take 112 g of the sample from the previous day (24 h) and combine with 112 g of wholemeal rye flour (type 2000) and 112 g of boiled water, then incubate at 20–25°C for 24 h. Repeat the process (end after 5 days of the process).

5. Carry out a test to distinguish between baked goods produced by acidic fermentation (lactic acid bacteria and yeast) and those produced with acidifying agents (e.g. lactic acid), as follows:

Add 50 ml of water to 5 g of bread crumb, wait 3 min, then grind in a mortar and set aside for 2–3 h. After the indicated time, take 5 ml of the suspension into a measuring cylinder and add 0.5 ml of Loeffler Methylene Blue.

Mix the sample thoroughly and set aside for 5 minutes. Then take a drop of the suspension on the slide and microscope. In case of doubt, perform Gram staining of the uncoloured bread crumb suspension. Dry the sample, fix it with ethanol (70–96%) for 30 min, stain (results of staining the individual components of the bread: yeast and bacteria – dark blue; the covering parts of the grain – intense blue; proteins – blue-green; swollen starch grains – light blue or blue pink).

Analysis of results

Draw conclusions about the tolerance of lactic acid bacteria to NaCl and the use of microorganisms for dough loosening during bread production.

Part II
Materials and media

1. Media: MRS and Sabouraud media with yeast extract and chloramphenicol (plates).
2. Materials: pickled pepper, beetroot sourdough, soured rye flour (including commercial product).
3. Reagents: Loeffler Methylene Blue, carbolic fuchsin by Ziehl-Nielsen.
4. Materials: pH-metre, homogeniser, Thoma chambers.

Aim of the exercise
Characteristics of organoleptic features and microbiological analysis of selected fermented products.

Procedure

1. Evaluate the organoleptic characteristics (smell, pH and consistency) and carry out microscopic analysis of fluids from samples of pickled peppers, beetroot sourdough and soured rye flour.

2. Perform macro- and microscopic observations of microorganisms isolated from ready-made food products (including sauerkraut juice, pickled cucumbers and beetroot sourdough). Evaluate the presence of fungi and bacteria in food products and the usefulness of the applied media. Filamentous fungi – make a live preparation, use lens magnification from 5 to 40 times, other microorganisms – make a preparation fixed and stained with Gram method, use lens magnification 100 times.

3. Carry out microbiological analysis of the soured rye flour by determining the number of lactic acid bacteria and yeast by direct method (according to PN-A74102/1999). Procedure: suspend the test sample of 1 g in sterile distilled water and homogenise for 2 min. (If homogenisation is performed in a mortar, gradually add 50 ml of water to the sample, and after obtaining a homogeneous suspension and adding the remaining 50 ml of water shake for another 10 min. In case of significant rye flour content add additionally 1 ml of acetone). Then take 10 ml of the sample and add 5 drops of Loeffler Methylene Blue and 2 drops of carbolic fuchsin according to Ziehl-Nielsen. Mix thoroughly and incubate at 75°C for 4 min. Apply the sample to the Thoma chamber and microscopy. Determine the number of: a) bacteria and b) yeast according to the following formulae:

   a) $N = n \times 250 \times 1{,}000 \times 100$,

   b) $N = n \times 10 \times 1{,}000 \times 100$,

   where: N – number of microorganisms in 1 Gram of the sample and n – average number of cells in 1 large square; 100 – results from sample dilution.

4. Carry out a microbiological analysis of the soured rye flour by determining the number of lactic acid bacteria and yeast by an indirect method (inoculation), as follows: Dilute the sample to be examined: 10^1, 10^2, 10^3 and inoculate 3 plates from each solution. To determine the

number of lactic acid bacteria, inoculate the sample on MRS solid medium and incubate at 30°C for 48–72 h. In the case of yeast, inoculate the sample on Sabouraud solid medium with yeast extract and chloramphenicol and incubate at 25°C for 72–96 h. After incubation, determine the number of microorganisms in 1 ml of the sample, taking into account the dilution and volume of inoculated sample.

Analysis of results

Draw conclusions concerning production conditions, organoleptic characteristics and the participation of microorganisms in the production of selected food products.

Part III
Materials and media

1. Microorganisms: *S. cerevisiae* yeast (commercial preparations of compressed and lyophilised baker's yeast), yeast biomass obtained as described in Section 4.1.1;
2. Materials: wheat flour (2 kg), rye flour (1 kg), salt (1 kg), oil.
3. Reagents: 0.1 M NaOH, phenolphthalein, 10% water solution of sucrose, 10% water solution of maltose.
4. Equipment: 100 ml flasks, plugs with fermentation tubes, beakers, weighing containers, bowls, knives, dough forms with crossbars, heaters (20°C and 28°C), refrigerator, dryer (60°C and 110°C).

Aim of the exercise
Characterization of selected metabolic abilities and technological features of baker's yeast.

Procedure

1. Determine and compare selected characteristics of compressed and lyophilised yeast:

a) colour, smell, taste, weight (weigh the package of yeast and then compare with the declared net weight on the package);

b) consistency of baker's yeast (it should present a strict, acceptable external surface with no smell of decomposing protein);

c) appearance of a water suspension of yeast (place 1 g of yeast in a test tube, add water and mix; after 5 min describe the appearance of the suspension – it should be homogeneous, without lumps and flocculations on the bottom of the vessel);

d) dry mass of baker's yeast (weigh 1 g of yeast with an accuracy of 0.001 g and then place in weighed and dried weighing vessel; dry the sample to constant mass for 2 hours at 60°C and then at 105°C for 1 h);

e) acidity of yeast (weigh 10 g of yeast, add 50 ml of water and mix; titrate the resulting suspension with 0.1 M NaOH in the presence of phenolphthalein (give the result per 100 g of yeast).

2. Mark the time when the dough is rising. Add the suspension obtained from 5 g of yeast and 160 ml of table salt solution (2.5%) to 280 g of flour (at 35°C). Write down the time of adding the yeast. Next, mix the dough for 5 min, then place it in a pre-prepared and greased tin heated to 35°C. Hang a crossbar over an evenly distributed dough and place it in the thermostat (35°C). When the dough touches the crossbar (1st shoot), take it out of the machine, note the time, then press for 1 min and put it back into the tin, hang the crossbar and move it into the machine until it touches the crossbar (2nd shoot), note the time. Repeat the steps (3th shoot). The result of the dough rising time are the times of individual shoots and their sum.

3. Determine the saccharolytic and malactic activity of the yeast. Spread 500 mg of yeast calculated on dry matter in 10 ml of water (at 35°C). Add 10 ml of 10% sucrose solution or 10% maltose solution, respectively, to the obtained suspension. Place the samples in flasks tightly closed with a stopper and equipped with a fermentation tube. Weigh the flasks and then place them in a heater (35°C) for 1 h and weigh again.

From the mass differences before and after fermentation, calculate the mass of the CO_2 released. The saccharolytic/malactic activity should be given in ml secreted by 100 mg of yeast dry matter per 1 h (1 mol CO_2 = 44.0 g, which corresponds to 22.4 ml).

Analysis of results

Compare the experimental data established for compressed and lyophilised yeast and draw conclusions about the required technological characteristics of yeast for dough loosening.

4.2.4. Asian food obtained by the use of microorganisms

INTRODUCTION

Traditional fermented food is a product of biotechnological processes that involve the natural bacterial flora that inhabits fresh food. It is one of the most practical and economical methods of preservation and at the same time enhances the organoleptic and nutritional qualities of fresh food (see Section 4.2.3). Especially in developing countries, where the use of refrigeration equipment is not always possible, the fermentation process is widely used and is crucial because fermentation extends the shelf-life of food, improves its nutritional value and reduces the risk of foodborne diseases.

The use of mixed microbial cultures in food preparation is quite common in the Western world, but much more popular in the Orient. The use of fermentation processes for food preservation has been known in these countries for 4,000–3,000 years B.C. Fermented food is very popular in Japan, Indonesia, India, Pakistan, Thailand, Taiwan, China and Korea. Nowadays, fermented oriental food is a delicacy all over the world and its health-promoting effect is widely known.

Only selected oriental food products that use microorganisms for production will be discussed in this section. Table 4.2.4.1 shows examples of oriental fermented food.

Product	Microorganisms involved in the process	Country of origin
kimchi	*Lactobacillus kimchii*	South Korea
koji	*Aspergillus oryzae*	Japan
miso	*Aspergillus oryzae, Pediococcus halophilus Saccharomyces rouxii*	Japan
natto	*Bacillus subtilis*	Japan
nata de coco	*Acetobacter xylinum*	Philippines
sake	*Aspergillus oryzae, Staphylococcus, Bacillus, Lactobacillus*	Japan
tempeh	*Rhizopus oligosporus*	Indonesia

Koji is a product used to prepare traditional fermented food in Japan. Literally translated, koji means "grain covered with yeast culture". In practice, it is a culture created by growing different fungi on boiled grains or legumes in a warm, humid place. Most often, the mixture of steam-cooked rice and mould spores is placed on a large porous plate or in a wooden container through which, air with the right temperature and humidity, flows. During the growth, koji forming fungi produce many enzymes, including amylases, proteases, lipases as well as tannins. The enzymes produced, hydrolyse starch, proteins and fats to their components, such as dextrin, glucose, peptides, amino acids and fatty acids. Simple products obtained in this way are substrates (nutrients) for yeast and bacteria, leading to subsequent stages of fermentation. Depending on the intended use, different species of fungi are used to produce koji. *Aspergillus oryzae* is most commonly used. Koji prepared using this species can then be used to produce sake or miso. In addition, other species of the genus *Aspergillus* are used, including *A. sojae, A. usami, A. awamori, A. kawachii*, as well as *Rhizopus* spp., *Monascus* sp., *Mucor* sp. and *Absidia* sp.

Sake is a non-distilled beverage from Japan containing 13 to 16% alcohol. The original sake was called kuchikami and was produced by chewing rice, acorns, chestnuts and millet in the mouth, which were then spat out into a special vat. The first sake was made around 3rd century BC. In the first stage, the rice used to produce sake is ground or polished to provide *Aspergillus oryzae* with adequate access under the outer layer of grain (Figure 4.2.4.1).

Table 4.2.4.1.
Examples of fermented oriental food

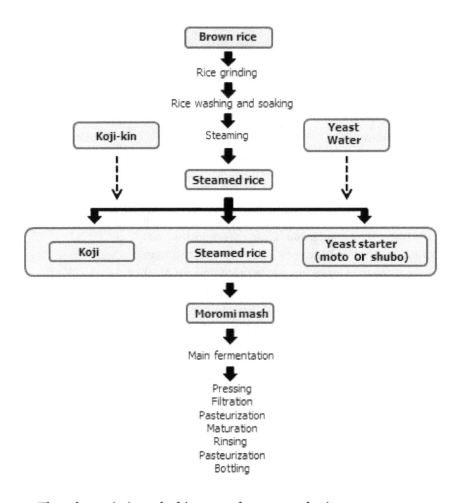

Fig. 4.2.4.1. Scheme of sake production

Then the grain is soaked in water from several minutes to several hours. Then the rice is steamed. Part of it goes directly to the fermentation tank, and part of it is inoculated with fungal spores and moved to a special room with increased humidity and temperature. For about 36 to 45 h, the ripening sake is constantly stirred and controlled and then combined with the remaining rice and water. The yeast, using simple nutrients obtained by the action of enzymes from the mould, carries out alcoholic fermentation and bacteria such as *Bacillus, Staphylococcus* and *Lactobacillus* stabilise the product. After the fermentation, the mash (moromi) is pressed, filtered and pasteurised (unpasteurised sake is called namazake). The last stage of sake production is ageing, which takes about 6 months.

Miso is one of the most characteristic dishes of Japanese cuisine, eaten in the form of a soup (so-called miso-shiru) or used as a seasoning for other dishes. Miso was first mentioned about 700 years BC in the ancient Chinese text of Syurai, but its roots can go back as far as several thousand years, when the fermentation of soy and rice was used on a daily basis by Buddhist monks. Traditionally, the process, depending on the type of miso, takes from several months to several years and is carried out in cedar barrels with lids protected by a pile of stones. The exception is a white rice miso (shiro), the production of which takes about 1 week.

Miso is produced by the slow fermentation of carefully selected and appropriately cleaned soya beans, rice or barley, with the addition of salt and using *A. oryzae* cultures (Figure 4.2.4.2).

Fig. 4.2.4.2. Scheme of miso production

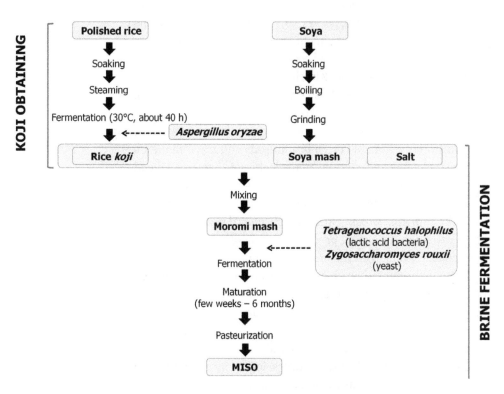

In order to speed up the process, *Tetragenococcus halophilus* lactic acid bacteria and *Zygosaccharomyces rouxii* yeast are added to the boiled, crushed soybean combined with koji. Using this method, about 300–400 kg of miso can be obtained from

100 kg of soya and 100 kg of rice. Miso contains high levels of protein and is rich in vitamins (e.g. vitamin B_2 from soybeans and vitamin B_{12} from lactic acid bacteria), amino acids, organic acids, minerals and dietary fibre. Research suggests that people who eat miso soup on a daily basis are less likely to develop stomach cancer. This may be due to the fact that fatty acids such as oleic and linoleic acids and their esters, which are found in miso, inhibit the proliferation of some cancers cell. Consumption of this food can also lower blood cholesterol levels, act as an anti-atherosclerosis agent and delay cell ageing.

Kimchi is the name given to various traditional fermented vegetables, which are symbols of Korean culture. Kimchi is mainly made from napa cabbage (*Brassica pekinensis*) and radishes with an addition of other vegetables and spices such as garlic, green onion, ginger, red pepper, mustard, parsley, carrots and salt. South Koreans consume more than 18 kg of kimchi per person per year and many believe that their country's rapid economic growth is partly due to the consumption of this dish. Kimchi has a high fibre content with a small number of calories. It is rich in vitamin A, thiamine (B_1), riboflavin (B_2), and one serving of this specialty provides more than 50% of the recommended daily dose of vitamin C and carotene.

Additionally, kimchi is rich in minerals and trace elements such as Ca and Fe and contains many lactic acid bacteria. Due to its nutritional properties, kimchi is on the list of five "healthiest foods in the world". According to some studies, kimchi can reduce the risk of atherosclerosis, cancer and bacterial infections. It can also slow down the ageing process, lower cholesterol levels and have a positive effect on the immune system.

Natto, a fermented product made from soya beans cultured from *Bacillus subtilis* popular in Japan for over 400 years, has been recently gaining more and more popularity in the West. Natto, like other fermented Eastern cuisine specialties, has a high nutritional value and is easily digestible. Natto also has an antibacterial effect. In ancient times, food poisoning was very common in Japan and was treated with natto. Today, about 3/4 of the Japanese population consumes natto at least once a week and half consume natto on average once every three days. To prepare it, the soya beans, after being rinsed and soaked in water, are steamed and then inoculated with *Bacillus subtilis* spores. Since the original source of natto bacteria was rice straw, traditionally prepared soya beans were covered with rice straw without the need for additional introduction of bacteria. Nowadays, specially prepared starter cultures of bacteria are most often used for this purpose. During about 20 h of fermentation, the starch and soya proteins are broken down into a mixture containing amino acids, vitamins and enzymes. Natto also contains saponins and isoflavones, which come from soya bean, as well as the fibrinolytic enzyme, vitamin K_2 and dipicolinic acid produced by the bacteria. After about 15 h of fermentation, natto becomes viscous as a result of the formation of poly-ɤ-glutamic acid. This component, combined with vitamin K_2, has a beneficial effect on the condition of the skeleton, stimulating Ca absorption and preventing osteoporosis. The fibrinolytic enzyme, called nattokinase, has the ability to dissolve blood clots formed in the lumen of blood vessels, which are a common cause of a heart attack or stroke. Dipicolinic acid, found in natto, has an antibacterial effect against selected strains of *E. coli* and *Helicobacter pylori*.

Tempeh is a traditional Indonesian fermented dish, known all over the world for its positive effect on health and special nut and mushroom flavour. Tempeh is usually prepared from soya, but it can also be made of beans, sunflower seeds, peanuts, peas, rice or wheat. Traditional tempeh is made from boiled and peeled soybeans, which are dried and then inoculated with *Rhizopus oligosporus* mould spores. The inoculated seeds are incubated at 30–31°C in perforated containers until they are highly overgrown and bound by mycelium (about 24 h). The most valuable components of tempeh are ɤ-aminobutyric acid, reducing blood pressure and isoflavones, alleviating the symptoms of menopause and acting against atherosclerosis.

Nata de coco is a native Filipino delicacy produced by fermentation of coconut cream or milk for 7 days with bacteria capable of cellulose production. In order to prepare nata de coco, the coconut pulp is ground and pressed to obtain the so-called coconut cream. The cream is inoculated with a starter culture containing the species *Acetobacter*, transferred to a large surface tray and covered with gauze. The tray is incubated for about a week, under static conditions. During this time, *Acetobacter* produces a cellulose layer, which is then collected, cut and ready for consumption.

PRACTICAL PART

Materials and media

1. Microorganisms: reference strains of *Aspergillus oryzae, Acetobacter xylinum, Rhizopus oligosporus, Saccharomyces cerevisiae* and microorganisms isolated from food products.
2. Food products: tempeh, miso, rice.

Aim of the exercise
Micro- and macroscopic analysis of microorganisms used in the production of selected articles classified as oriental food. Preparation of sake.

Procedure

1. Presentation of video showing the production of miso, tempeh, sake, nata de coco and natto and presentation of selected food products.
2. Macro and microscopic observations: *Aspergillus oryzae, Acetobacter xylinum* and *Rhizopus oligosporus*.
3. Production of sake:
   a) grind the rice grains in a mortar and then rinse them;
   b) boil the grains and divide them into two parts;
   c) spread *A. oryzae* mould on a mass of boiled and cooled grains. Place the material prepared in a room with increased temperature and humidity for 48 h, mixing several times;
   d) mix the second portion of boiled rice with water and a portion of *S. cerevisiae* yeast, then leave without mixing for several weeks at 15°C.;
   e) press the post-fermentation mass mechanically to obtain the original fermentation solution. Leave the solution to sediment and filtrer through charcoal to give colour and enrich the taste;
   f) sake prepared in this way undergoes the process of pasteurisation and then about six months of ripening. At this stage, water is added to the solution in order to reduce the alcohol concentration from 20% to about 16%.

4.3. Biotransformation processes

INTRODUCTION

Microorganisms are characterised by the ability to transform and use many exogenous organic compounds. In these processes, in addition to typical catabolic reactions providing energy and precursors to the biosynthesis of new compounds, the transformations occurring at the edge of normal metabolism and having no significant impact on the growth and reproduction of microorganisms can be distinguished.

These are usually one-stage transformations, characterised by high regio- and stereospecificity and are called biotransformation or bioconversion.

Many biotransformations of hydrocarbons, carbohydrates, amino acids, antibiotics, terpenes, alkaloids, prostaglandins, steroids are used in practice. Derivatives obtained as a result of bioconversion have often changed or expanded biological properties, so they can be applied directly, e.g. as pharmacologicals, or serve as a substrate for the production of other valuable chemicals. Biotransformation processes can also be used as intermediate steps in the synthesis of optically active compounds, or bioconversion allows for their separation from the racemic mixture. Among a very large number of biotransformations of importance in biotechnology, the bioconversion of ethanol and sorbitol will be presented, as well as steroids from compounds of more complex chemical structure.

4.3.1. Biotransformation of ethanol and sorbitol

Acetic acid bacteria belonging to the family *Acetobacteraceae* play an important role in the food industry and certain chemical and pharmaceutical industries (e.g. bacterial cellulose production, Section 4.1.4). Although for chemical purposes, acetic acid is obtained by dry wood distillation or chemical synthesis, domestic vinegar is still produced using microbiological technology from raw materials of plant origin. The ethanol contained in the distiller's raw materials or rectificate (spirit vinegar) and in light white or red wines (wine vinegar) is bio-converted. The acetic acid thus obtained is not secreted from the "fermentation" medium but forms an integral part of the finished product called "table vinegar". Acetic acid obtained chemically is not allowed in the production of vinegar for human consumption. Another important economic characteristic of acetic acid bacteria is the ability to oxidise simple sugars to the corresponding monocarboxylic (-one-) acids and sometimes even further to these acids, keto-derivatives.

These abilities of acetic acid bacteria have also been put into practice. Gluconic acid, produced using *Gluconobacter*, has been applied in the food industry, and the δ-lactone of this acid is used in baking. Calcium gluconate has found an application in the treatment of allergic diseases or in case of general deficiency of Ca^{2+} ions in the body; similarly, ferric gluconate is recommended in case of this element deficiency. Sodium gluconate is widely used in the production of washing liquids and washing powders, where it replaces environmentally hazardous polyphosphates. Sodium gluconate is also applied in the construction industry as an additive to concrete mortars in the construction of foundations with increased strength. In addition, Gluconobacter bacteria are useful in the biological-chemical method of vitamin C manufacture. The bioconversion of sorbitol to sorbose, an intermediate compound in the production of vitamin C, is conducted with the participation of these bacteria.

Acetic acid bacteria belong to the obligate aerobes, these are Gram-negative or Gram-changing bacilli, either ciliated or not ciliated, occurring singly, in two or in short chains. These bacteria are characterised by bold or excessively elongated involution forms.

The *Acetobacteraceae* family includes two genera: *Acetobacter* and *Gluconobacter*. The bacteria of both genera have the ability to oxidise ethanol to acetic acid, but only *Acetobacter* can further oxidise acetic acid to CO_2 and H_2O, and they can also oxidise lactic acid; as they have the complete Krebs cycle enzymes. Unlike *Acetobacter, Gluconobacter* does not oxidise acetate and lactate to CO_2, because they do not have Krebs cycle enzymes, but they are characterised by high "ketogenic" activity – they oxidise glucose to gluconic acid and then to keto-derivatives: 2-keto-gluconic, 5-keto-gluconic, and some species of bacteria even up to 2,5-diketo-gluconic acid. Bacteria from the genus *Gluconobacter* also have the ability to oxidise polyols such as glycerol for dihydroxyacetone or sorbitol for sorbose. *Acetobacter* genus has a peritrichal ciliation, while *Gluconobacter* genus have a polar one. *Acetobacter* spp. have the ability to grow on a medium containing nitrogen in the form of ammonium salts, while Gluconobacter require organic forms of nitrogen in the medium.

The above metabolic properties of these bacteria can be presented by the following reactions (Figure 4.3.1).

Fig. 4.3.1. Metabolic transformation of acetic acid bacteria

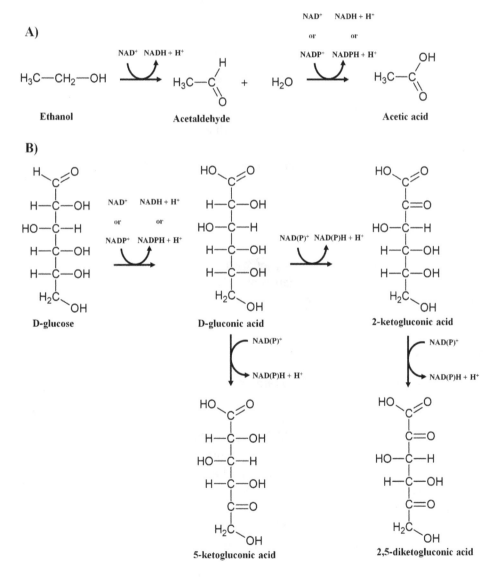

The numerous dehydrogenase coenzymes participating in these oxidising processes are reproduced in oxidised form in the respiratory chain, which both *Acetobacter* and *Gluconobacter* bacteria have in a complete or slightly simplified composition.

PRACTICAL PART

Materials and media

1. Microorganisms: *Acetobacter schüzenbachii, Gluconobacter suboxydans, Gluconobacter* sp.
2. Media: 5°Blg wort slants with 3% ethanol or 3% sorbitol, medium for *Gluconobacter*: for testing ketogenic properties, semi-synthetic medium with glucose, semi-synthetic medium with sorbitol.
3. Reagents: sodium alginate, carrageenan.
4. Marc vinegar.
5. Beech chips from a vinegar generator.

Aim of the exercise
Determination of the capacity of industrial Acetobacter and Gluconobacter strains to oxidise mono- and polyols (ethanol and sorbitol).

Procedure

1. Macro- and microscopic observations of a pure culture of *Acetobacter schüzenbachii* bacteria, used in the generator method of vinegar production.
   a) Culture on a wort slant (5°Blg) with 3% ethanol (30°C). Stain the fixed preparation with gentian violet, observe under immersion. Pay attention to the occurrence of involutional forms.
   b) Observations of the microflora of beech chips taken from the vinegar generator.
   Prepare an imprint preparation of beech chips. After fixation, stain the preparation with gentian violet

and observe under immersion; pay attention to the occurrence of involution forms.

2. Testing of the marc vinegar from a vinegar generator:
   a) centrifuge about 50 ml of marc vinegar at g = 8,000, prepare a microscopic preparation from the sediment, fix, stain with gentian violet and observe under immersion;
   b) Titrate 5 ml of the marc vinegar against phenolphthalein with 0.25 M NaOH solution. Express the result in g of acetic acid in 100 ml of the liquid.

3. Macro- and microscopic observations of pure culture of *G. suboxidans* bacteria – used in the production of gluconic acid:
   a) culture on a wort slant (5°Blg) with 3% sorbitol (28°C). Continue as above in point 1a.
   b) examination of the ketogenic properties of bacteria: Carry out a 3-day culture of *G. suboxydans* at 28°C with shaking on a medium with yeast extract with 5% glycerol. Add a few drops of 20% $CuSO_2$ solution to 2 ml of culture and heat. The appearance of orange Cu_2O precipitate indicates the reduction of bivalent copper by the dihydroxyacetone formed in the medium.

4. Examination of gluconic acid formation in shaken cultures of *G. suboxydans*. Carry out a shaken culture in a semi-synthetic medium containing 8% glucose, at 28°C. Examine the content of the resulting gluconic acid by titrating 5 ml of a 0.25 M NaOH sample against phenolphthalein. Provide the result in g of gluconic acid per 100 ml of sample after 24, 48, and 72 h of the process.

5. Immobilisation of *Gluconobacter* sp. cells in alginate or carrageenan:
   a) carry out bacteria culture on a shaker, in 100 ml of semi-synthetic medium with the addition of 10% sorbitol, for 48 h at 30°C;
   b) after completing the culture, centrifuge the bacteria, wash twice with sterile saline solution, centrifuge and suspend in sterile distilled water in 0.1 of the original volume of the culture medium;
   c) continue as in Section 1.3.5 (Practical part C, points 1b-d).;

    d) transfer the immobiliser beads into a flask containing 100 ml of semi-synthetic medium with sorbitol and place on a shaker at 30°C;

    e) carry out an identical test with free cells of *Gluconobacter* sp. at the same output density;

    f) after 3 days of the process, determine by a reduction (Luff-Schoorl) or colorimetric (with DNS) method the content of the resulting sorbose in both the immobilised and free bacteria samples (control). Proceed as in Section 1.3.5 (Practical part C, point 1h).

4.3.2. Biotransformation of steroids

Steroids are chemical compounds containing the cyclopentano-perhydrophenanthrene (steroid) system, presented on Figure 4.3.2.

Fig. 4.3.2. Scheme of steroid compounds structure

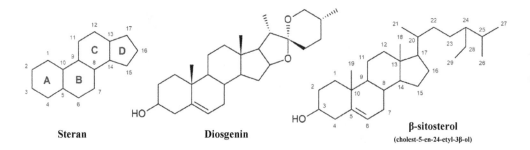

Steran   **Diosgenin**   **β-sitosterol**
(cholest-5-en-24-etyl-3β-ol)

Currently, more than 1,000 different steroid compounds formed during natural metabolic changes in the cells of micro- and macroorganisms have been characterised. Sterols, bile acids, corticosteroids and estrogens are distinguishable due to the type of substituents at carbon atoms in positions 10 and 17 of the steroid molecule, (Table 4.3.2).

Table 4.3.2. Types of substituents at C-10 and C-17 in different steroids

Name of compounds	Substituents at C-10 and C-17	
	C-10	C-17
Estrogens	-H	-H
Androgens	-CH₃	-H
Corticosteroids	-CH₃	-CH₂CH₃
Bile acids	-CH₃	-CH(CH₃)CH₂CH₂COOH
Sterols	-CH₃	-CH(CH₃)CH₂CH₂CH₂CH(CH₃)₂

Steroids also include derivatives whose side chain has cycled. The chemical structures of the most important steroids are shown in Figure 4.3.3.

Fig. 4.3.3. Chemical structures of some steroids.

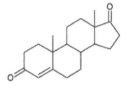

Androstenedione

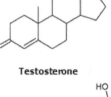

Testosterone

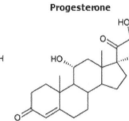

Progesterone

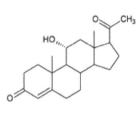

11α-Hydroxyprogesterone

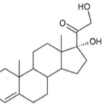

Cortexolone

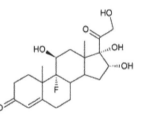

Epihydrocortisone

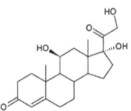

Hydrocortisone

Prednisolone

9a-fluoro-16a-hydroxyhydrocortisone

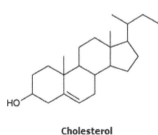

Cholesterol

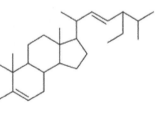

Stigmasterol

Steroid compounds are common in the world of plants and animals. These substances are transformed by many species of both higher organisms and microorganisms belonging to Eukaryotes and Prokaryotes. Of the huge variety of transformations a steroid substrate can undergo, only some of the reactions are important in the biotechnology of steroid drugs and include: the degradation of the sterol side chain (see Subsection 1. 6) (soil bacteria), 11-α-, 11-δ-, 17-α-hydroxylation (filamentous fungi), 16-α-hydroxylation (actinomycetes), dehydrogenation between 1 and 2 carbon atoms in the steroid molecule (*Corynebacterium*), reduction of the ketone group to hydroxyl one at 17 androgen carbon (yeast). In medical practice, about 300 steroid drugs are currently used in a total amount of over 1 million tonnes with a value of around 10 billion US dollars, which puts this group of pharmaceuticals in second place (after antibiotics) in terms of commercial value.

One of the most important properties of steroids is their anti-inflammatory effect. This feature is related to the presence of the hydroxyl group at carbon 11 of the steroid molecule. Filamentous fungi (e.g. *Curvularia lunata, Cunninghamella elegans, Monosporium olivaceum*), capable of 11-hydroxylation, are used for the production of hydrocortisone and related compounds, widely used in medicine, e.g. in the treatment of rheumatoid arthritis, allergic diseases, adrenal insufficiency and many eye and skin diseases.

Steroid monooxygenases (steroid hydroxylases), belonging to the class of oxidoreductases, are responsible for the hydroxylation process. In fungi, these enzymes are associated with the membranes of the endoplasmic reticulum, while in bacteria they occur in the cytoplasm.

During hydroxylation, the oxygen molecule is decomposed into two atoms, one of which is introduced into the C-H bond of the substrate, and the other with $NADPH+H^+$ or $NADH+H^+$ participation is reduced to water according to the scheme:

$$\text{steroid-H} + NAD(P)H+H^+ + O_2 \rightarrow \text{steroid-OH} + NAD(P)^+ + H_2O$$

Hydroxylation efficiency is determined by conditions conducive to high enzyme content in the cell (inductive enzymes) and ensuring optimal concentration of the compounds involved in the reaction – steroid substrate, cofactors, and oxygen.

Induction of the hydroxylation system is usually carried out during biomass multiplication by adding a steroid substrate or other steroids of similar structure, at a concentration of 0.1–0.5 g/l. During the synthesis of steroid monoxygenases, the optimal value of oxygenation, i.e., the so-called DOT (dissolved oxygen tension), is 10–15%. A higher oxygenation level of 24–30% is more advantageous during bioconversion. Due to the competition for reduced cofactors between biosynthetic processes and hydroxylases, full expression of monooxygenases occurs only in the stationary phase of microbial growth.

Steroids are poorly soluble in water, which is significantly conditioned by the type of substitutes in the steroid molecule, especially at C-17 (cholesterol solubility – 1.8 mg/l, testosterone – 23 mg/l). Before being added to the medium, steroids are dissolved in organic solvents (methanol, ethanol, dimethylformamide, dimethylsulphoxide). The size and shape of the crystals precipitated in the medium depends on the type of solvent used, which determines the dissolution of the steroid substrate during culture (steroid biotransformation). The bioconversion of steroids is also influenced by the cell wall, which limits the access of the steroid substrate to the cell interior. Partial or complete deprivation of cell sheathing favours the efficiency of the transformation. Fungal protoplasts show almost four times more ability to hydroxylate steroids than the mycelium from which they are released.

Immobilised systems are also used in the biotransformation of steroids. Immobilisation of cells in gels enables their repeated use in bioconversion, facilitates separation from the medium and reduces toxic effects of organic solvents on microorganisms. With longer use of immobilised organisms, it is necessary to add easily metabolised organic compounds to the environment, enabling regeneration of the hydroxylation system.

PRACTICAL PART

Materials and media

1. Microorganisms: strains of micromycetes *Cunninghamella echinulata* and *Curvularia lunata*.
2. Media: ZT slants, PL_2 medium

3. Reagents: sodium alginate, 0.2 M $CaCl_2$, steroid substrate – cortexolone, product standards – hydrocortisone and epihydrocortisone, internal standard pregn-4-en-11α-ol-3,20-dione.

4. Apparatus: liquid chromatograph Agilent 1200 and mass spectrometer QTRAP 3200 (Sciex).

Aim of the exercise
Verification of the ability of selected strains of micromycetes to hydroxylate corticosteroids by free and immobilised mycelium.

Procedure

1. Multiplication of biomass and induction of hydroxylation system of fungi.
   Wash the spores from a 7–10 day fungal culture on ZT medium with 5 ml of PL_2 medium and transfer the suspension to a 0.2 l flask containing 5 ml of PL_2 medium. After 24 h of incubation on a rotary shaker, at 28°C, add 30 ml of PL_2 medium and culture for the next 24 h on a shaker in a 0.5 l flask with the addition of cortexolone as an inductor (0.1 g/l).

2. Biotransformation using free mycelium.
   Add a steroidal substrate – cortexolone (dissolve 10 mg in 0.5 ml 96% ethanol) to a 24 h induced culture and conduct biotransformation for 24 h under the same conditions as in case of mycelium multiplication. Then add the internal standard (pregn-4-en-11α-ol-3,20-dione), homogenise (MISONIX) until the mycelium break down and extract three times with dichloromethane.

3. Biotransformation using an immobilised mycelium.
   Filter the induced culture, obtained as in point 1 above, wash with sterile distilled water and mix 1 g of wet mass with 5 ml of sterile sodium alginate solution. Drip the suspension with a syringe into 0.2 M $CaCl_2$. Rinse the beads with sterile distilled water and transfer to a 250 ml flask containing 20 ml distilled water and 0.5 g/l steroid

substrate – cortexolone. Incubate on a rotary shaker at 28°C. After 24 h, filter the mycelium immobilised in the gel, rinse with sterile distilled water and place in the same environment with a new portion of steroid substrate. Extract the filtrate as below in point 4. After a few cycles, when the mycelium activity is half that of the initial activity, incubate the gel beads in PL_2 medium, at 28°C, on a shaker for 6 h and use again for bioconversion in water (conditions as above).

4. Extraction and chromatographic analysis of steroids. Extract the filtrate with dichloromethane, three times, using 15 ml each time, with the internal standard in the separator. Dehydrate the extract with anhydrous sodium sulphate, filter and evaporate to dryness. Then analyse the samples by chromatography using the HPLC-MS/MS set (QTRAP 3200) with the Eclipse XDB-C18 column (150 × 4.6 mm, 5 μm), eluting with a methanol: water mixture 50:50, flow rate 1 ml/min.

Analysis of results

1. On the basis of chromatographic analyses, assess the ability of tested strains to cause 11-hydroxylation of cortexolone.
2. Compare the activity of free and immobilised mycelium. Evaluate the change of activity of mycelia immobilised in alginate gel over time and the possibility of hydroxylating system regeneration.

Literature

Books

Asai, T., 1968. *Acetic Acid Bacteria. Classification and Biochemical Activities*. University of Tokyo Press, Baltimore.

Bednarski, W., Reps, A., 2003. *Biotechnologia żywności*. Wydawnictwa Naukowo-Techniczne, Warszawa.

Bednarski, W., Fiedurek, J., 2007. *Podstawy biotechnologii przemysłowej*. WNT, Warszawa.

Bergey, D.H., Krieg, N.R., Holt, J.G., 1984. *Bergey's Manual of Systematic Bacteriology*. Williams and Wilkins, Baltimore.

Długoński, J., 1993. *Protoplasty Cunninghamella elegans (Lendner) jako model w badaniu 11-hydroksylacji steroidów*. Wydawnictwo Uniwersytetu Łódzkiego, Łódź.

Długoński, J. (red.), 1997. *Biotechnologia mikrobiologiczna – ćwiczenia i pracownie specjalistyczne*. Wydawnictwo Uniwersytetu Łódzkiego, Łódź.

Gómez-Pastor, R., Pérez-Torrado, R., Garre, E., Matallana, E., 2011. *Recent advances in yeast biomass production*, w: Matovic, M.D. (red.), *Biomass – Detection, Production and Usage*. InTech Publisher, Rijeka, s. 201–222.

Ledakowicz, S., 2011. *Inżynieria biochemiczna*. WNT, Warszawa.

Paramithiotis, S., 2017. *Lactic Acid Fermentation of Fruits and Vegetables*. CRC Press, Taylor and Francis Group, Boca Raton, FL.

Paraszkiewicz, K., 2016. *Biosurfactant Enhancement Factors in Microbial Degradation Processes*, w: Długoński, J. (red.), *Microbial Biodegradation: From Omics to Function and Application*. Caister Academic Press, Norfolk, s. 167–182.

Paraszkiewicz, K., Długoński, J., Trzmielak, D., 2016. *Application of Recent Omics Achievements in Bioremediation Processes Illustrated by Progress in Microbial Surfactants Commercialization*, w: Długoński, J. (red.), *Microbial Biodegradation: From Omics to Function and Application*. Caister Academic Press, Norfolk, s. 219–232.

Szewczyk, K.W., 2003. *Technologia biochemiczna*. Oficyna Wydawnicza Politechniki Warszawskiej, Warszawa, s. 197–218.

Viljoen, B.J., Heard, G.M., 2014. *Saccharomyces cerevisiae*, w: Batt, C.A., Tortorello, M.L. (red.), *Encyclopedia of Food Microbiology*. Elsevier Science Publishing, San Diego.

Yu, J., Porter, M., Jaremko, M., 2013. *Generation and utilization of microbial biomass hydrolysates in recovery and production of poly(3-hydroxybutyrate)*, w: Matovic, M.D. (red.), *Biomass Now – Cultivation and Utilization*. IntechOpen. https://www.intechopen.com/books/biomass-now-cultivation-and-utilization/generation-and-utilization-of-microbial-biomass-hydrolysates-in-recovery-and-production-of-poly-3-hy (access: 19.06.2020).

Review articles

Bielecki, S., Kalinowska, H., 2008. *Biotechnologiczne nanomateriały*. Post. Mikrobiol. 47 (3), 163–169.

Caulier, S., Nannan, C., Gillis, A., Licciardi, F., Bragard, C., Mahillon, J., 2019. *Overview of the antimicrobial compounds produced by members of the Bacillus subtilis group*. Front Microbiol. 26, 302.

Ciriminna, R., Meneguzzo, F., Delisi, R., Pagliaro, M., 2017. *Citric acid: Emerging applications of key biotechnology industrial product*. Chem. Cent. J. 11, 22.

Contesini, F.J., Melo, R.R., Sato, H.H. 2018. *An overview of Bacillus proteases: From production to application*. Critical Reviews in Biotechnology 38, 321–334.

Cousin, F.J., Le Guellec, R., Schlusselhuber, M., Dalmasso, M., Laplace, J.M., Cretenet, M., 2017. *Microorganisms in fermented apple beverages: current knowledge and future directions*. Microorganisms 5 (3), 39.

Di Cagno, R., Coda, R., De Angelis, M., Gobbetti, M., 2013. *Exploitation of vegetables and fruits through lactic acid fermentation*. Food Microbiology 33, 1–10.

Długoński, J., 1994. *Monooksygenazy steroidowe drobnoustrojów i ich znaczenie w biotechnologii leków*. Post. Mikrob. 33, 147–159.

Donova, M.V., 2018. *Microbiotechnologies for steroid production*. Microbiology Australia, 39, 126–129.

Donova, M.V., Egorova, O.V., 2012. *Microbial steroid transformations: current state and prospects*. Appl. Microbiol. Biotechnol. 94, 1423–1447.

Dufresne, A., 2017. *Cellulose nanomaterial reinforced polymer nanocomposites*. Curr. Opin. Colloid Interface Sci. 29, 1–8.

Fenster, K., Freeburg, B., Hollard, C., Wong, C., Rønhave Laursen, R., Ouwehand, A.C., 2019. *The production and delivery of probiotics: a review of a practical approach*. Microorganisms 7 (3), 83.

Geyer, U., Heinze, T., Stein, A., Klemm, D., Marsch, S., Schumann, D., Schmauder, H.P., 1994. *Formation, derivatization and applications of bacterial cellulose*. Int. J. Biol. Macromol. 16, 343–347.

Hać-Szymańczuk, E., Roman, J., 2009. *Charakterystyka drobnoustrojów wchodzących w skład kultur starterowych i ich wykorzystanie w przetwórstwie mięsa*. Postępy Techniki Przetwórstwa Spożywczego 2, 131–135.

Kanse, N.G., Deepali, M., Kiran, P., Priyanka, B., Dhanke, P., 2017. *A review on citric acid production and its applications*. Int. J. Curr. 6, 5880–5883.

Kim, D.H., Chon, J.W., Kim, H., Seo, K.H., 2019. *Development of a novel selective medium for the isolation and enumeration of acetic acid bacteria from various foods*. Food Control. 106, 106717.

Lavefve, L., Marasini, D., Carbonero, F., 2019. *Microbial ecology of fermented vegetables and non-alcoholic drinks and current knowledge on their impact on human health*. Adv. Food Nutr. Res. 87, 147–185.

Marco, M.L., Heeney, D., Binda, S., Cifelli, C.J., Cotter, P.D., Foligné, B., Gänzle, M., Kort, R., Pasin, G., Pihlanto, A., Smid, E.J., Hutkins, R., 2017. *Health benefits of fermented foods: Microbiota and beyond*. Curr. Opin. Biotechnol. 44, 94–102.

McMurtrie, E.K., Johanningsmeier, S.D., Breidt, F. Jr., Price, R.E., 2019. *Effect of brine acidification on fermentation microbiota, chemistry, and texture quality of cucumbers fermented in calcium or sodium chloride brines*. J. Food Sci. 84, 1129–1137.

Mnif, I., Ghribi, D., 2015. *Review lipopeptides biosurfactants: mean classes and new insights for industrial, biomedical, and environmental applications.* Biopolymers 104, 129–147.

Monteiro de Souza, P., Magalhães, P., 2010. *Application of microbial α-amylase in industry. A review.* Brazilian Journal of Microbiology 41, 850–861.

Murooka, Y., Yamshita M., 2008. *Traditional healthful fermented products of Japan.* J. Ind. Microbiol. Biotechnol. 35, 791–798.

Paraszkiewicz, K., Kuśmierska, A., 2017. *Biosurfaktanty drobnoustrojów (cz. 1–2).* Journal of Health Study and Medicine 1, 57–75, 77–92.

Paraszkiewicz, K., Moryl, M., Płaza, G., Bhagat, D., Satpute, S.K., Bernat, P., 2019. *Surfactants of microbial origin as antibiofilm agents.* Int. J. Environ. Health Res. 2019, 1–20.

Poutanen, K., Flander, L., Katina, K., 2009. *Sourdough and cereal fermentation in a nutritional perspective.* Food Microbiology 26 (7), 693–699.

Randez-Gil, F., Córcoles-Sáez, I., Prieto, J.A., 2013. *Genetic and phenotypic characteristics of baker's yeast: Relevance to baking.* Annu. Rev. Food Sci. 4 (1), 191–214.

Rezac, S., Kok, C.R., Heermann, M., Hutkins, R., 2018. *Fermented foods as a dietary source of live organisms.* Front Microbiol. 9, 1785.

Saleha, T., Syed Faheem, A.R., Ummar, A., 2018. *Tetracycline: classification, structure activity relationship and mechanism of action as a theranostic agent for infectious lesions. A mini review.* Biomed. J. Sci. and Tech. Res. 7, 5787–5796.

Show, P.L., Oladele, K.O., Siew, Q.Y., Zakry, F.A.A., Lan, J.Ch.-W., Ling, T.Ch., 2015. *Overview of citric acid production from Aspergillus niger.* Front. Life Sci. 8, 271–283.

Strnad, S., Satora, P., 2016. *Mikrobiologiczne aspekty produkcji kiszonej kapusty*, cz. 1. Przem. Ferm. i Owoc.-Warzyw. 60 (7–8), 31–33.

Strnad, S., Satora, P., 2016. *Mikrobiologiczne aspekty produkcji kiszonej kapusty*, cz. 2. Przem. Ferm. i Owoc.-Warzyw. 60 (9), 31–32.

Tamang, J.P., Watanabe, K., Holzapfel, W.H., 2016. *Review: diversity of microorganisms in global fermented foods and beverages*. Front Microbiol. 7, 377–405.

Terpou, A., Papadaki, A., Lappa, I.K., Kachrimanidou, V., Bosnea, L.A., Kopsahelis, N., 2019. *Probiotics in food systems: significance and emerging strategies towards improved viability and delivery of enhanced beneficial value*. Nutrients 11 (7), 1591.

Varjani, S.J., Upasani, V.N., 2017. *Critical review on biosurfactant analysis, purification and characterization using rhamnolipid as a model biosurfactant*. Bioresour. Technol. 232, 389–397.

Wojdyła, T., Wichrowska, D., 2014. *Wpływ stosowanych dodatków oraz sposobów przechowywania na jakość kapusty kiszonej*. Inż. Ap. Chem. 53, 424–426.

Original scientific papers

Bernat, P., Długoński, J., 2012. *Comparative study of fatty acids composition during cortexolone hydroxylation and tributylin chloride (TBT) degradation in the filamentous fungus Cunninghamellaa elegans*. Int. Biodeter. Biodegr. 74, 1–6.

Bernat, P., Szewczyk, R., Krupiński, M., Długoński, J., 2013. *Butyltins degradation by Cunninghamella elegans and Cochliobolus lunatus co-culture*. J. Hazard. Mater. 15, 277–282.

Bernat, P., Paraszkiewicz, K., Siewiera, P., Moryl, M., Płaza, G., Chojniak, J., 2016. *Lipid composition in a strain of Bacillus subtilis, a producer of iturin A lipopeptides that are active against uropathogenic bacteria*. World J. Microbiol. Biotechnol. 32, 157.

Elsayed, E.A., Omar, H.G., El-Enshasy, H.A., 2015. *Development of fed-batch cultivation strategy for efficient oxytetracycline production by Streptomyces rimosus at semi-industrial scale*. Braz. Arch. Biol. Technol. 58, 676–685.

Eyini, M., Rajkumar, K., Balaji, P., 2006. *Isolation, regeneration and PEG-induced fusion of protoplasts of Pleurotus pulmonarius and Pleurotus florida*. Mycobiology 34, 73–78.

Jacek, P., Kubiak, K., Ryngajłło, M., Rytczak, P., Paluch, P., Bielecki, S., 2019. *Modification of bacterial nanocellulose properties through mutation of motility related genes in Komagataeibacter hansenii ATCC 53582.* N. Biotechnol. 52, 60–68.

Kędziora, K., Juda, K., Walisch, S., 1982. *Porównanie metod przechowywania konidiów przemysłowych szczepów A. niger.* Przem. Ferm. i Owoc.-Warzyw. 6, 24–27.

Ong, J., Shatkin, J.A., Nelson, K., Ede, J.D., Retsina, T., 2017. *Establishing the safety of novel bio-based cellulose nanomaterials for commercialization.* NanoImpact. 6, 19–29.

Paraszkiewicz, K., Bernat, P., Kuśmierska, A., Chojniak, J., Płaza, G., 2018. *Structural identification of lipopeptide biosurfactants produced by Bacillus subtilis strains grown on the media obtained from renewable natural resources.* J. Environ. Manage. 209, 65–70.

Paraszkiewicz, K., Bernat, P., Siewiera, P., Moryl, M., Sas-Paszt, L., Trzciński, P., Jałowiecki, Ł., Płaza, G., 2017. *Agricultural potential of rhizospheric Bacillus subtilis strains exhibiting varied efficiency of surfactin production.* Scientia Horticulturae 225, 802–809.

Patyra, E., Kwiatek, K., 2016. *Analytical procedure for the determination of tetracyclines in medicated feedingstuffs by liquid chromatography-mass spectrometry.* J. Vet. Res. 60, 35–41.

Płaza, G., Chojniak, J., Rudnicka, K., Paraszkiewicz, K., Bernat, P., 2015. *Detection of biosurfactants in Bacillus species: genes and products identification.* J. Appl. Microbiol. 119, 1023–1034.

Ramamoorthy, V., Govindaraj, L., Dhanasekaran, M., Vetrivel, S., Kumar, K.K., Ebenezar E., 2015. *Combination of driselase and lysing enzyme in one molar potassium chloride is effective for the production of protoplasts from germinated conidia of Fusarium verticillioides.* J. Microbiol. Methods. 111, 127–134.

Ryngajłło, M., Jacek, P., Cielecka, I., Kalinowska, H., Bielecki, S., 2019. *Effect of ethanol supplementation on the transcriptional landscape of bionanocellulose* producer *Komagataeibacter xylinus E25*. Appl. Microbiol. Biotechnol. 103, 6673–6688.

Sanderson, H., Ingerslev, F., Brain, R.A., Halling-Sørensen, B., Bestari, J.K., Wilson, C.J., Johnson, D.J., Solomon, K.R., 2005. *Dissipation of oxytetracycline, chlortetracycline, tetracycline and doxycycline using HPLC-UV and LC/MS/MS under aquatic semi-field microcosm conditions.* Chemosphere 60, 619–629.

Sharma, R., 2018. *Production, characterization and environmental applications of biosurfactants from Bacillus amyloliquefaciens and Bacillus subtilis.* Biocatalysis and Agricultural Biotechnology 16, 132–139.

Teh, M.Y., Ooi, K.H., Teo, S.X.D., Bin Mansoor, M.E., Lim, W.Z.S., Tan, M.H., 2019. *An expanded synthetic biology toolkit for gene expression control in Acetobacteraceae.* ACS Synth. Biol. 8, 708–723.

Vieira, É.D., Andrietta Mda, G., Andrietta, S.R., 2013. *Yeast biomass production: A new approach in glucose-limited feeding strategy.* Braz. J. Microbiol. 30, 551–558.

Zhang, N., Jiang, J.C., Yang, J., Wei, M., Zhao, J., Xu, H., Xie, J.C., Tong, Y.J., Yu, L., 2019. *Citric Acid Production from Acorn Starch by Tannin Tolerance Mutant Aspergillus niger AA120.* Appl. Biochem. Biotechnol. 188, 1–11.

Żęcin, J., Walisch, S., Kaczmarowicz G., 1979. *Modyfikacja metody spektrofotometrycznej oznaczania kwasu cytrynowego w cieczach fermentacyjnych.* Przem. Ferm. i Owoc.-Warzyw. 23 (11), 29.

Websites

https://www.reportsanddata.com/report-detail/enzymes-market (access: 2.12.2019).

https://www.youtube.com/watch?v=tE1-Q0wskfk (access: 2.10.2019).

https://www.youtube.com/watch?v=5Y3giTpXrh8 (access: 2.10.2019).

https://www.youtube.com/watch?v=OxG_SoQ-bAg (access: 2.10.2019).

5

Microorganisms in environmental and human health protection

5.1. Revitalisation of degraded urban green areas

INTRODUCTION

In the context of contemporary problems of urban development (such as density of buildings in the city centre, low air quality, emission of pollutants, high air temperatures, etc.) there is a growing conviction that green areas of cities are important elements determining the health status of the inhabitants. In large cities, these areas are often an important link of natural resources, offering a range of ecosystem services such as support, supply, regulatory, cultural and other services. Green areas include recreation parks, boulevards, as well as ornamental greenery. For a few years now, most of them (damaged and neglected) have been gaining new shine thanks to the Act on revitalisation of 9 October 2015, which treats revitalisation not only as a strictly urban process, but also as a social one. It is assumed that the revitalisation of green areas may improve the health condition of the city's residents and revitalise neglected parts of public space. In the currently revitalised green areas, the number and species composition of plants often increases, historical vegetation arrangement and their specific variety are reconstructed, water purification systems are established (water reservoirs performing not only decorative but also retention functions), natural and secondary elements of equipment (furniture made of native and recycled materials) and technical infrastructure are used, e.g. surfaces permeable to rainwater, historical buildings are renovated depending on new needs, including architectural details about the history of the place/city.

Revitalisation of public space in large cities (such as Warsaw, Kraków, Lodz or Katowice) is often a long-term and complex process. However, it is an important opportunity for a given city to significantly improve the possibilities of tourism development and the comfort of living of its inhabitants, concurrently increasing social security. Therefore, the revitalisation of individual urban facilities may have a positive impact on the importance of areas directly adjacent to the revitalised ones. It may also have a positive long-term effect and lead to the return of many residents from suburban areas to the city centre (e.g. the assumptions of the Municipal Revitalisation Programme of Łódź). Therefore, the case study on revitalisation of green areas may constitute a tool for building a new image of the city, especially in the context of the proecological development.

5.1.1. Interdisciplinary research in urban revitalisation: work stages

Natural revitalisation of the area is a complex process and should be based on interdisciplinary research in the field of biotechnology and environmental microbiology as well as landscape architecture (including ecology and dendrology). Combination of these specialties can give a broader view of the problem of urban pollution elimination. This step-by-step process is presented in Figure 5.1.1 and in the research description below.

The first stage of the discussed process should consist in determining the substantive and spatial scope of revitalisation. First of all, what research will be included in revitalisation, how extensive microbiological, biotechnological, chemical, horticultural, geoengineering and design analyses will be conducted (Figure 5.1.1). Secondly, whether it will concern one or more facilities and how it will affect the development of the natural system in a given city (the so-called city's green infrastructure). As a rule, information on the objects of the planned revitalisation can be found in the municipal planning documents (municipal revitalisation programme, study of land use conditions and directions, eco-physiographic study, etc.).

The next step is to select the research case study, i.e., the place (green area object) where the analysis towards revitalisation is to be conducted. It can be a place such as a greenery, promenade or a city park (city square with vegetation), located in the city centre near compact buildings (e.g. old city tenement houses,

large buildings), bordering with road and pedestrian roads or
service and communication facilities, which may have a potential
impact on the pollution of the area (gas station, transformer
station, landfill of various materials etc.). In terms of size (area),
the facility may be small (about 1–2 ha), but should often be used
by residents or tourists.

Fig. 5.1.1.
Methodology of
revitalisation activities

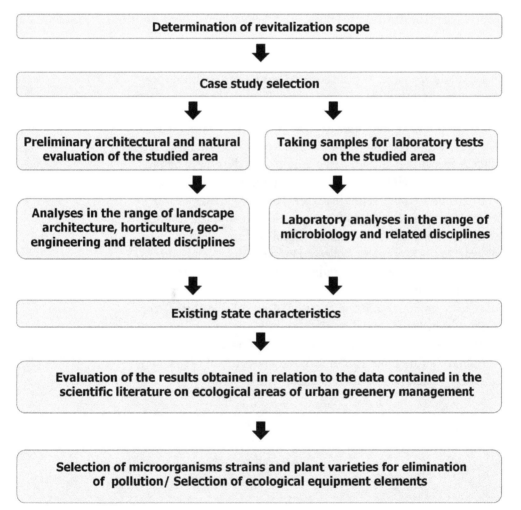

The revitalisation of the selected area will serve not only
the general visual improvement of the object, but much more
broadly – improvement of its natural condition. Such aspects as

soil characteristics, the quantitative and qualitative composition of soil microbiota, the number and spatial distribution of plant species, and elements of land equipment will be analysed. Therefore, the research issues are connected with many tasks indicating interdisciplinary research. Many specialists from the borderline of environmental and industrial microbiology, landscape architecture and ecology should be involved in this process, and the work carried out should include field, laboratory and individual research.

5.1.2. Field research in the scope of landscape architecture and related disciplines

The field research may consist in the execution of the so-called simplified land inventory. All elements of development should then be recognised, including those indicating the type of passive recreation (benches, tables, field café, amphitheatre, etc.) or active (sports ground, running tracks, bicycle paths, gym, etc.), vegetation (estimated species composition), road infrastructure with types of surface and photographic documentation of the object. On the basis of the collected information it is possible to determine the degree of degradation of the object (destruction of development elements), types of leisure activities (observation of user behaviour) and the level of safety of the object. For comparison, a similar elaboration can be made for a green area that has recently been revitalised – it can be an object from a different latitude. An example of the inventory of revitalised greenery objects is presented in Table 5.1.2.1.

Table 5.1.2.1. Sample inventory table of a green area object

No.	The aspect examined	Elements under consideration (examples)	Status as of the day (date)	
			degraded object	revitalised object
I	Leisure facilities	Water reservoir with a clearing / beach	X	X
		Children's playground	X	X
		Sports ground	X	X
		Café	X	X
		Field game garden	X	X
		Field gym	X	X
		Amphitheatre		X
		Leisure Garden Other...		X
II	Plant diversity	Coniferous trees	6 species	6 species
		Deciduous trees	56 species	60 species
		Coniferous shrubs	5 species	5 species
		Deciduous shrubs	42 species	47 species
		Herbal plants	1 species	1 species
		Climbers	1 species	2 species
III	Road infrastructure (surfaces))	Bituminous surface	X	
		Ground surface	X	
		Surface made of granite cubes	X	X
		Paving surface	X	X
		Rinsed paving stone surface		X
		Mineral surface		X
IV	Level of object degradation (damage)	Recreational equipment	No	No
		Road infrastructure elements (surfaces)	Yes (30%)	No (0%)
		Other development elements (e.g. fencing) Other...	No(0%)	Yes (5%)
V	Leisure activities	Education	X	X
		Festivals/concerts	X	X
		Sports games	X	X
		Walking and contemplation	X	X
		Picnic on the grass		X
		Games and plays Other...	X	X
VI	Facility security level	Area under monitoring surveillance	Yes (5%)	Yes (5%)
		Number of offences recorded	1	b.d.
		Illuminated roads	Yes (75%)	Yes (75%)
		Fencing	Yes	Yes
		Opening hours Other...	6-20/22*	6-20/22*

* During the period: 1 April–30 September; X – positive answer.

5.1.3. Laboratory research in biotechnology, microbiology, environmental chemistry and related disciplines

Laboratory tests are designed to determine the quality of soil from both microbiological and chemical perspectives. The number of samples taken from a degraded green area depends on its size, shape and expected distribution of contamination, and in the case of an area of 1–2 ha, it usually ranges from 20 to several dozen soil samples. In the course of microbiological analysis, first of all, attention is paid to the presence of microorganisms capable of efficient elimination of PAHs, heavy metals and other toxic contaminants found on the basis of chemical tests. The presence of microorganisms in the soil that stimulate phytoremediation processes in different plant species is equally important. The microorganisms that inhabit the plant rhizosphere are of particular importance in this process. Microorganisms, developing on the surface of the root system and inside plant tissues, often promote biochemical activity and the ability of plants to eliminate toxic contaminants, and limit the development of phytopathogens (pathogenic strains of bacteria, fungi and viruses). The characteristics and microbiological analysis of soil, including areas contaminated with toxic substances, are further discussed in Sections 1.2.2–1.2.4.

Soil chemical analysis should include measurement of pH and content of toxic xenobiotics, including PAHs and heavy metals, especially As, Pb, Cd and Ni. Soil pH, which is an indicator of the concentration of hydrogen ions in the soil solution, provides information on the potential for plant growth, assimilation of nutrients and elimination of contaminants. Under the influence of various factors, including human activity, the soil pH may change significantly, hence the need to control it simultaneously with the analysis of pollutants and the planning of planting in the analysed area. It should also be borne in mind that urban green areas include not only parks or greenery, but often also significant forest areas (e.g. the Łagiewnicki Forest in Łódź, which is a remnant of the former primaeval forest), where the pH of forest soils is usually lower than that of cultivated soils.

The pH is measured in KCl solution, which allows consideration, of not only the concentration of hydrogen ions in the soil solution (pH measured in water), but also hydrogen ions poorly bound to the soil fraction. The ranges of pH for different soils are given in Table 5.1.3.1.

pH KCl of soils			
Arable		Forest	
Strongly acidic	< 4.0	Very strongly and strongly acidic	< 3.5
Acidic	4.1–4.5	Strongly acidic	3.6–4.5
Medium acidic	4.6–5.0	Acidic	4.6–5.5
Slightly acidic	5.1–6.0	Slightly acidic	5.6–6.5
Neutral	6.1–6.5	Neutral	6.6–7.2
Slightly alkaline	6.6–7.0	Slightly alkaline	7.2–8.0
Medium alkaline	7.1–7.5	Alkaline	> 8.0
Alkaline	>7.5		

Table 5.1.3.1. Ranges of pH KCl depending on the type of soil (own elaboration on the basis: Kabała, Karczewski 2017)

The qualitative and quantitative analysis of PAHs and heavy metals in soil allows assessment of whether, and to what extent, the acceptable standards contained in the Ordinance of the Minister of the Environment of Poland from 1 September 2016 have been exceeded, as well as to what extent it is advisable to plant on a given area, plants with high phytoremediation capacity and to enrich soil microbiota with microbial strains that participate in the elimination of pollutants.

The condition of plants is reflected in the chlorophyll content in their leaves. A decrease in its content indicates the action of abiotic stress factors, e.g. the presence of toxic but also biotic substances in the soil and air, e.g. infections caused by bacterial and/or fungal pathogens. Measurement of chlorophyll content in leaves can be carried out non-invasively, directly in the field, using portable (pocket) cameras, the operation of which is based on the recording of chlorophyll a fluorescence, i.e., remission of light energy absorbed by energy antennas of the plant's photosynthetic apparatus. The results of chlorophyll content are given in CCI units (Chlorophyll Content Index). This technique is also very useful for quick monitoring of changes taking place in plants during phytoremediation and during fungal and bacterial infections.

5.1.4. Total score and summary of research

Before formulating recommendations for the revitalisation of a given area, it is necessary to evaluate the results obtained with reference to the regulations in force as well as the data contained in the scientific literature concerning ecological trends in urban greenery management. In recent years, an intensification of the ecological trend in urban greenery management has been observed, which is reflected, for example, in the creation of ecological theme parks, pocket parks, development of post-industrial areas, former landfills, gravel pits, mine heaps, etc., for recreational purposes.

A review of publications and scientific studies in this field shows that for revitalised areas, in addition to the standard planting of urban ornamental plants (barberries, birches, ornamental acers, oaks, acacia robins, lime trees, etc.), it is more and more often recommended to plant specially selected species (varieties), resistant to increasingly difficult urban conditions, caused, among others, by the presence of dust pollutants – suspended particulate matter (PM 2.5 and PM 10) and contamination of soil and air with toxic substances, including PAHs. Plants with phytoremediation properties which are conducive to the natural elimination of toxic pollutants are also used here, and thus contribute to the improvement of environmental health conditions (Table 5.1.4.1).

Table 5.1.4.1. Characteristics of selected plant species useful for planting within polluted urban green areas

Species	Distance between plants [m]	Properties and applications	Natural conditions	
			Soil type	Insolation
Ground cover plants				
Hedera helix	5	Evergreen, easy to spread, resistant to low temperatures, very good phytoremediation capacity: reduces dust (especially PMx) and other pollutants, e.g. NOx, it is recommended to plant them under the tree canopy, on retaining walls, trellises and trusses (green walls)	Slightly acidic to alkaline soils	Shadow or half-light
Vinca minor	12	Evergreen, used mainly in shady places under trees	Very low soil requirements	Sun

Species	Distance between plants [m]	Properties and applications	Natural conditions	
			Soil type	Insolation
Euonymus fortuni	7	Evergreen, good phytoremediation properties, for use on green roofs and as the basic covering species under tree canopy	All types of soils with high humidity	Sun or half-light
Deciduous shrubs				
Buxus sempervirens	2	Evergreen, good photoreduction properties, the plant absorbs most of the pollutants (especially PMx dust), for hedge forming, only used in the buffer (insulation) zone of the area	All types of soils	Shadow or half-light or sun
Cornus mas	6	Good phytoremediation properties (absorbs the most contaminants (especially NOx), suggested use in the buffer zone	Fertile soils, rich in nutrients, with neutral to slightly acidic pH	Sun or half-light
Physocarpus opulifolius "Diabolo" and "Luteus"	3	The plant tolerates low temperatures and drought, for hedge formation, good phytoremediation properties (especially NOx reduction), suitable for hedges, but not in a buffer zone	Dry and humid soils	Sun or half-light
Ribes alpinum "Schmidt"	2	Plant resistant to the competition of other plants, good phytoremediation properties (especially NOx reduction), perfect for shaping hedges in the pedestrian zone of the object, not recommended for planting in a buffer zone	Slightly acidic or neutral soils	Sun or half-light or shade
Tamarix gallica	3	Large size plant, resistant to drought and salinisation of soil, good phytoremediation properties in the reduction of PMx and NOx, for planting in a buffer zone	Tolerant plant, good growth on light and permeable soil	Sun
Coniferous shrubs				
Taxus baccata	3	Evergreen, for use on hedges, average phytoremediation properties, proposed use in buffer zone	Tolerant plant, grows best on neutral to acidic soils	Sun or half-light or shade
Microbiota decussata	3	Evergreen, reduces dust pollution, proposed use in the buffer zone	Acidic or slightly acidic soils	Shade or half-light
Pinus mugo	2	Evergreen, reduces dust pollution (especially NOx), proposed application in the buffer zone	Low soil requirements, inhabits extreme environments, acidic and alkaline soils	Sun or half-light

Notes: PMx – particulate matter; NOx – nitrogen oxides.

The next stage of revitalisation works is the spatial planning of plantings selected for the revitalisation of plants and supplementing the elements of recreational development of a given object. Various ecological trends commonly used in urban landscape architecture can be used in the revitalisation of green areas. These may be the so-called green walls (facades of buildings covered with creepers and herbs), the so-called green roofs (roofs of buildings covered with low bushes and/or shrubs, as well as grasses) and the so-called buffer zones (trees, bushes and shrubs arranged in storeys bordering the object, e.g. close to roads with heavy traffic). These elements, like the selected plant species themselves, will act as pollutant filters, especially dust. Suggested design solutions are illustrated in Figures 5.1.4.1, 5.1.4.2 and 5.1.4.3.

It should be emphasised that the presented design solutions (Figures 5.1.4.1–5.1.4.3) are widely used in many cities in Western Europe (Germany, France or the Netherlands) and can also be adapted to the revitalization of degraded green areas of Polish cities. However, in each case it is necessary to take into account the specificity of the area, especially the spatial context and natural conditions of the area, the purpose, scope and arrangement of the proposed revitalisation, as well as the compliance of the planned actions with the applicable law.

Fig. 5.1.4.1. Example of a multi-storey buffer zone in J. Poniatowski Park in Łódź.

Fig. 5.1.4.2. Example of a green wall on the facade of a school building, Angers, France.

Fig. 5.1.4.3. Example of a green roof on a service building along the TGV line, Paris, France.

Photographs 5.1.4.1.–5.1.4.3 taken in the years 2011–2013 by A. Długoński (Disclaimer: copying of illustration only with author's permission).

PRACTICAL PART

Materials, media and apparatus

1. Metal spatulas (or spades) for taking soil samples from the depth of 5–20 cm, Egner's sticks for taking soil samples from the depth of 20, 30, 60 and 90 cm.
2. Soil microbial culture media as in Section 1.2.
3. Sieve with 1.5 mm mesh diameter.
4. Magnetic stirrer.
5. Laboratory pH-metre.
6. Portable chlorophyll meter for non-invasive measurement of chlorophyll content in leaves and needles of coniferous plants with a wide measuring range (50–600 mg/m²).
7. Set of reagents and apparatus for heavy metal analysis as in Section 5.7.
8. Set of reagents and apparatus for PAH determination as in Section 1.2.
9. Topographical/basic map or satellite image of the site (see Section 2.6).
10. Camera, parameters: 1) SLR camera (min. 10 Mpx), standard lens: 35–55 mm (and/or more), 2) digital camera (min. 10 Mpx) with zoom function and macro mode.

Aim of the exercise
Characteristics of degraded areas of urban green area and design of their revitalisation together with selection of plants and other components of ecological infrastructure of the area.

Procdedure

1. Soil samples collection.
   Collect soil samples with a metal spatula from a depth of 5–20 cm or more (using an Egner's stick). After air-drying, sift through a sieve with a mesh diameter of 1.5 mm, and then use for the analyses described below.

2. Determination of soil pH according to PN-ISO 10390:1997. Transfer 5.0 ml of soil dried at room temperature into a 50 ml beaker (air-dry sample) and pour 25 ml of distilled water. Mix the contents using a magnetic stirrer for 5 min and leave it to stand for 15 min. Measure the pH of the suspension with a pH-metre. Repeat the measurement after 30 min and 1 h.

   Carry out the same analysis, replacing distilled water with 1 M KCl. Take into account three parallel soil samples (n = 3).

3. Determination of the number of microorganisms capable of growing on culture media. Perform according to the methodology described in Section 1.2.

4. Analysis of soil metagenome. Perform according to the methodology described in Section 5.2.

5. PAHs analysis. Perform according to the methodology presented in Section 1.2.

6. Determination of heavy metals in the soil. Perform according to the methodology discussed in Section 5.7.

7. Determination of chlorophyll content in leaves. Follow the instructions provided by the chlorophyll meter manufacturer.

8. Simplified land inventory. Perform according to the methodology discussed in Section 5.1.2.:

   a) Analysis of aerial and satellite images: compare the management and vegetation coverage of a selected green area included in the revitalisation plan on the basis of archival aerial and satellite images/orthophoto maps (from different periods). Use the database of photographs in the Geoportal (giving geographical coordinates of the area) or materials collected in geodesy and cartography centres of public administration offices. Follow the methodology described in Section 2.6;

   b) Selection of sensitive/characteristic sites of the green area using drones (ALS technology). Set the drone's flight parameters and measuring apparatus. Determine the compactness of the stand, crown drought, stand infected by bacterial and fungal diseases, plant shortages in groups of trees and shrubs. Perform according to the methodology presented in section 2.6;

      c) Determination of air pollution for degraded green areas. Set the drone's flight parameters and measuring equipment (standard measurement: PM 2.5 and PM 10, concentration of carbon monoxide, sulphur oxide, hydrogen sulphide, methane, nitrogen, formaldehyde). Perform according to the methodology presented in Section 2.6;

9. Determination of the extent of area degradation (based on the total analysis of previously obtained data (points 1–6). Follow the methodology discussed in Section 5.1.4.

10. Determination of the possibilities of area revitalisation:

      a) present a proposal for the selection of plant species together with a justification. Reference point – plant characteristics contained in Table 5.1.4.1;

      b) select ecological elements of land development. Reference point – examples of solutions presented in Section 5.1.4.

11. Conclusions and final statements (based on the total evaluation of the previously obtained results (point 7) and the submitted proposals for site revitalisation (point 8).

5.2. Microbiological analysis of polluted environments – Next generation sequencing

INTRODUCTION

Environmental pollution by organic compounds of anthropogenic origin is a serious global problem and microorganisms play a key role in their removal (these issues are further discussed in Sections 1.2.4 and 5.1). In order to monitor bioremediation processes, it is necessary to determine which microorganisms are responsible for the different stages of pollutants removal and what role they play in the environment. It is estimated that only 1% of the microorganisms living in the environment can be cultured under laboratory conditions. The emergence of Next Generation Sequencing (NGS) has revolutionised microbiological research, including environmental studies. Thanks to these techniques, it is possible to learn about the metagenome of a given environment, i.e., the total DNA pool of microorganisms inhabiting a given environment. Metagenomics, on the other hand, is a direct cloning, sequencing and functional analysis of genetic material isolated from different environments (e.g. soil).

Despite the differences that exist between particular technologies, the stages of the process leading to the next generation of sequencing are the same (Figure 5.2.1).

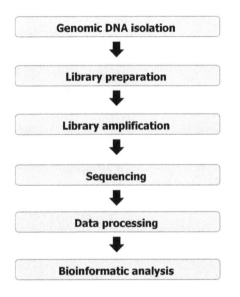

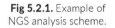
Fig 5.2.1. Example of NGS analysis scheme.

Currently, three new generation sequencing methods are available on the market: 454, PGM, Illuminia. The main advantages of these sequencing methods are high throughput and relatively low cost per sample analysis. The individual techniques are briefly described below.

454 Sequencing

The DNA library is fragmented and then short single-stranded DNA fragments are attached to the adapter (DNA fragments with a known nucleotide sequence labelled with 5' biotin). The labelled fragments are then attached to grains coated with streptavidin. DNA fragments are duplicated using emulsion PCR and then sequenced by pyrosequencing.

Ion Torrent Method

The principle of this method is similar to the 454 technique, but instead of measuring the light emission, the pH changes of the reaction mixture due to nucleotide incorporation are measured. Single-stranded DNA is introduced into the wells, covered with microbeads. There is a different DNA fragment in each of the wells, which enables mass, parallel sequencing. When subsequent complementary nucleotides are activated by the DNA polymerase, hydrogen cation is released into the reaction mixture. Specific nucleotides are added to the microplate in each cycle and the signal emitted is directly proportional to the number of nucleotides incorporated into the newly formed threads in each well. Reaction measurements are performed simultaneously in all wells with the use of a miniature pH-meter – PGM (Personal Genome Machine).

Illumina Method

The DNA library is fragmented and then the obtained single-stranded DNA fragments hybridise with adapter sequences placed on a microplate. Thus immobilised single threads of DNA are subjected to a reaction, consisting of alternating synthesis and denaturation, as a result of which the plate contains so-called clusters, built from multiple copies of a given pattern. The next step is sequencing by attaching complementary nucleotides combined with fluorescently marked reversible terminators. Attaching a given nucleotide to the thread, blocks its elongation and at the same time, a signal from the fluorescent dye is sent to the appropriate detector. On the basis of the signals recorded by the detectors, the nucleotide order in the DNA sequence can be reconstructed. Then the terminator and the dye are flushed away so that more nucleotides can be connected.

Two different approaches are most commonly used to determine the microbial population in a given environment: whole bacterial genomes sequencing or targeted sequencing.

a) sequencing of entire bacterial genomes

Shotgun sequencing provides information on the biodiversity of a given environment. In this approach, DNA is isolated from all the cells in a given sample and cut into fragments from which libraries are prepared. During sequencing, all DNA sequences present in the samples are read. The limiting factor for this technique is the subsequent, very complex analysis of the obtained metagenome sequences, which consists of assembling hundreds of thousands of randomly obtained DNA sequences into longer fragments.

b) targeted sequencing

Determination of microbiological diversity in environmental samples was until recently a challenge. Today, most research is based on targeted sequencing of the 16S rRNA region and large subunit (LSU) rRNA (bacterial markers) and ITS (Internal Transcribed Spacers) sequencing (fungal markers). The primers for the amplification of hypervariable regions are designed so that the sequences obtained are long enough to differentiate organisms. A large number of copies of rRNA genes, found in some microorganisms, can falsify the results and therefore quite often only one copy per genome is used. Such genes include e.g. rpoA (gene encoding the α subunit of RNA polymerase), rpoB (gene encoding the β subunit of RNA polymerase), recA (gene encoding a protein involved in DNA recombination) and gyrB (gene encoding the β gyrase subunit). In order to identify the bacteria present in the activated sludge, the analysis can also be directed towards specific genes encoding the proteins responsible for nitrogen metabolism, e.g. nirS (denitrification) or nifH (nitrogen fixation).

Search for genes responsible for toxic substances biodegradation

A typical metagenomic project of the contaminated environment aims to determine the number and diversity of species inhabiting the contaminated area and to determine which genes or degradation pathways are present, which species are responsible for the process and to know the changes in the microbiome over time and under different environmental conditions (e.g. changes in temperature, pH, salinity during the biodegradation process).

Metagenomic research has changed our understanding of the biodegradation processes taking place in the environment and it is now believed that the degradation of xenobiotics in the environment is most often carried out by several different species of microorganisms. The results obtained from metagenomic analyses quite often prove that one of the key stages of the degradation is carried out by the non-culturable bacteria, which are the only ones in the population to have a specific gene encoding the xenobiotic degradation pathway enzyme. Metagenomic analyses not only provide detailed information on taxonomic differentiation of the microbiome, but also enable mapping of degradation pathways. Unfortunately, metagenomics does not answer the question whether the identified genes are transcribed. Therefore, metatranscriptomics is used, which is based on mRNA analysis instead of DNA. A typical metatranscriptomic protocol is presented in Figure 5.2.2. Only those genes that are expressed are analysed in this approach.

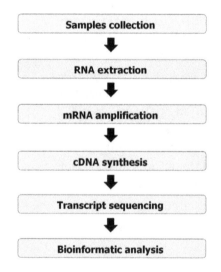

Fig. 5.2.2.
Metatranscriptomics
– main stages.

Search for molecular markers of the degradation of toxic compounds

A molecular marker of biodegradation is a fragment of genetic material (locus) that can be easily determined using molecular biology techniques. Quite often, it is a conservative fragment of a functional gene that encodes an enzyme that enables xenobiotic degradation. The analysis of the presence of these markers is usually performed using PCR technique. Examples of xenobiotics degradation markers are given in Table 5.2.1.

	Marker	Enzyme	Microorganisms
Chlorophenols	dca	Reducing dehalogenase *dcaABCT*	Firmicutes
	FcbB	4-Chlorobenzoyl-CoA dehalogenase	Bacterial consortia
	TecA	Tetrachlorobenzene dioxygenase	*Bulkholderia* sp.
Azo dyes	Azo	Azoreductase	*Bacillus subtilis*
	Lac1 MnP1 LiP2	Laccase Manganese peroxidase Lignin peroxidase	*Trametes versicolor* *Phanerochaete chrysosporium* *Phanerochaete chrysosporium*
Polycyclic aromatic hydrocarbons	nahAc, phnAc	Naphthalene dioxygenase	*Pseudomonas putida, Bulkholderia* sp.
	bcrA-bcrD	Benzoylo-CoA reductase	Denitrifying bacteria
	Lac	Laccase	Ligninolytic fungi
	nidAB	Oxygenases	*M. vanbaalenii*

Table 5.2.1. Examples of degradation markers of xenobiotics

Isolation of genomic DNA for metagenomic tests

The first stage of each metagenomic experiment is DNA isolation. Because DNA is isolated from different environments, there is no single, universal method. In the case of soil metagenomic analysis, the lysis of bacterial cells (to release DNA) is carried out directly in soil samples. The main advantage of this solution is to obtain a large amount of genetic material, well representing the diversity of the microbiome. Unfortunately, DNA isolated directly from the soil is contaminated with organic compounds (e.g. humic acids) and after the isolation is completed, it is necessary to purify the genetic material again, e.g. with commercially available DNA purification kits.

PRACTICAL PART

Materials and media

1. Materials: soil samples, Falcon tubes, glass beads.
2. Reagents: 100 mM Tris-HCl pH 8.0, 100 mM EDTA-Na pH 8.0, 100 mM phosphate buffer pH 8.0, 1.5 M NaCl, lysozyme, zymolyase, proteinase K, 20% SDS, chloroform: isoamyl alcohol mixture (24:1), 70% ethanol, isopropanol, commercial DNA purification kit.

Aim of the exercise
Metagenomic analysis of soil samples from contaminated and revitalised areas.

Procedure

1. Weigh 5 g of soil taken as in Section 5.1. from green areas located near a gas station or a road with heavy traffic, into a Falcon test tube. Take a horticultural soil sample as a reference.
2. Add 4 g of glass beads (diameter 0.6 mm) to each sample.
3. Add 5 ml of buffer (100 mM Tris-HCl pH 8.0; 100 mM EDTA-Na pH 8.0; 100 mM phosphate buffer pH 8.0; 1.5 M NaCl).
4. Homogenize samples three times for 5 min on the FastPrep-24 disintegrator (MP Biomedicals).
5. Transfer the samples to a Falcon tube (50 ml).
6. Add lysozyme (10 mg/ml), zymolyase (50 µg/ml) and incubate for 30 min at 37°C.
7. Add 50 µl proteinase K (20 mg/ml), incubate another 30 min at 37°C.
8. Add 2 ml of 20% SDS and incubate at 55°C for 18 h.
9. Centrifuge the samples (8,000 × g, 10 min), gently extract the supernatant.
10. Extract the supernatant with an equivalent volume of chloroform: isoamyl alcohol mixture (24:1). Centrifuge the sample (13,000 × g) and then transfer the top layer to a new tube.
11. Add 0.6 volume of isopropanol and place the sample in ice for 30 min.
12. Centrifuge the sample (13,000 × g, 20 min), remove the supernatant and rinse the resulting sediment with 70% ethanol solution.
13. After sample centrifugation (13,000 × g, 30 seconds) add 500 µl of deionised water.
14. Purify the obtained DNA with a commercial kit, according to the manufacturer's instructions.
15. Measure the amount of DNA on the BioDrop Duo spectrophotometer (BioDrop).
16. Sent the samples to a specialised NGS sequencing laboratory.
17. Analyse the results obtained to observe the differences in the microorganisms present in the examined soil types.

5.3. Biological wastewater treatment processes

INTRODUCTION

Sewage is wastewater that has been used, particularly for household and industrial purposes, and is not suitable for the same use again. The following can be distinguished in the commonly used division of wastwater:

- domestic wastewater – from residential buildings and public buildings;
- municipal wastewater – domestic wastewater or a mixture of domestic wastewater with industrial wastewater or rainwater or snowmelt water;
- industrial wastewater – wastewater resulting from industrial, commercial, transport or service activities.

The composition of wastewater is diverse and its physical, chemical and biological properties are determined by pollution indicators, which allow the assessment of the efficiency of the treatment plant. The content of organic compounds in wastewater is determined by total/dissolved organic carbon (TOC/DOC) and/or chemical oxygen demand (COD) and biochemical oxygen demand (BOD). An important indicator for wastewater assessment is also the amount of nitrogen and phosphorus and the amount of suspended solids. Wastewater, especially industrial wastewater, may also be burdened with toxic substances that are particularly harmful to the aquatic environment, e.g. heavy metals, pesticides, petroleum compounds, limiting microbial activity and hindering the process of wastewater treatment.

BOD (biochemical oxygen demand) – a conventional term, it is a measure of the content of organic compounds susceptible to biodegradation. BOD determines the amount of oxygen (in mg/l) needed for biological oxidation involving microorganisms, of organic compounds in the analysed sample (water, wastewater). BOD is determined under aerobic conditions, at 20°C. Usually the oxygen consumption is determined after 5 days, as the so-called Five-Day Biochemical Oxygen Demand (BOD_5).

COD (chemical oxygen demand) – specifies the amount of oxygen (in mg/l) equivalent to the amount of chemical oxidizing agent used for oxidation of organic compounds and oxidisable inorganic compounds contained in the sample (water, wastewater). COD is determined under strictly defined conditions, using strong and very strong chemical oxidants. The most commonly used oxidants are potassium permanganate (COD_{Mang}) and dichromate (COD_{Chr}).

By comparing BOD and COD values determined in samples taken from the same wastewater, it is possible to assess the susceptibility of organic compounds contained in the wastewater to biological decomposition. The BOD_5/COD_{Chr} ratio for easily biodegradable wastewater is usually 0.5–0.7, while for practically non-biodegradable wastewater, it is below 0.2.

Organic carbon, designated as TOC (Total Organic Carbon) or DOC (Dissolved Organic Carbon) – reflect the total organic compounds content in wastewater and the amount of dissolved organic matter, respectively. Determination of the organic carbon content consists of burning samples of wastewater at high temperatures, with the addition of a catalyst, into carbon dioxide, without prior filtering (TOC) or after filtering (DOC).

5.3.1. Biological methods of wastewater treatment in municipal treatment plants

Classical methods of biological wastewater treatment can be carried out under natural, partially man-made (ground treatment methods, sewage ponds) or artificial conditions, using equipment specially constructed for this purpose (activated sludge method, biological filters).

Biological wastewater treatment consists of the decomposition, mainly by microorganisms, of organic substances contained in the wastewater into inorganic compounds. Some methods also lead to the elimination of biogenic elements, nitrogen and phosphorus. Biological treatment is usually the second or third stage of wastewater treatment, which is pre-treated by mechanical and/or physicochemical methods. The two basic processes of biological wastewater treatment include treatment using microorganisms in suspended form (activated sludge) and immobilised on carriers (biological filters).

The activated sludge method is a reflection of self-purification processes taking place in surface water under natural conditions. This method uses a complex of microorganisms occurring in the form of a flocculent suspension in the chamber with simultaneous mixing and aeration (Figure 5.3.1.1). Active sludge flocs consist mainly of heterotrophic bacteria, and protozoa – ciliates and flagellate, nematodes, rotifers, and some fungi are also present. The bacteria releasing extracellular polymers play an important role in the formation of the flocs. The flocs adsorb organic substances contained in wastewater, which are then decomposed by microorganisms in cellular metabolism processes, and there is also an increase in biomass. In many technologies activated sludge is also used to remove excess nitrogen and phosphorus. The wastewater is then fed to a secondary settling tank, where the activated sludge is separated from treated wastewater. Part of the activated sludge retained in the settling tank is returned to the reactor as recirculated sludge, the remaining part is removed as excess sludge and subjected to further treatment.

Fig. 5.3.1.1. Scheme of wastewater treatment with activated sludge.

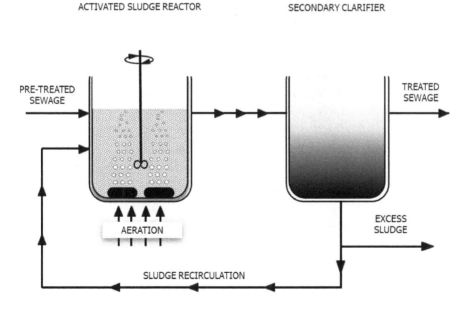

ACTIVATED SLUDGE REACTOR   SECONDARY CLARIFIER

PRE-TREATED SEWAGE

TREATED SEWAGE

AERATION

EXCESS SLUDGE

SLUDGE RECIRCULATION

Biological filters are devices with fixed or mobile filling (e.g. aggregates, slag, stones, plastic). Microorganisms in the form of a biological membrane develop on the surface of the carrier. The biological membrane consists of various organisms, mainly chemoorganotrophic bacteria, as well as fungi, protozoa, rotifers, nematodes, oligochaete, insect larvae. In conventional biological filters with solid filling, the filter is sprinkled with wastewater from above and aerated from below. The organic substances contained in the wastewater are adsorbed on the biological membrane and then decomposed by the microorganisms contained in it. The biological filters with fixed filling are divided into trickling and rinsing, and the difference in their functioning lies in different values of substrate and hydraulic load. Trickling biological filters allow for a high degree of purification. There are also disc filters, in which the biological membrane develops on rotating discs, periodically submerged in wastewater.

Anaerobic wastewater treatment methods are mainly used to process wastewater with a very high content of organic compounds and solids, e.g. to dispose of wastewater sludge, wastewater from the agri-food industry, animal husbandry waste. The basis of this process is the decomposition of organic compounds without oxygen, by means of fermentation and subsequent methanogenesis (methane fermentation) with the participation of methanogenic archaea and with the production of biogas – a mixture of methane and carbon dioxide. It is a multistage process in which different, closely interacting groups of microorganisms take part. The first phase, hydrolytic, leads to hydrolysis of high molecular compounds to monomers. In the second phase, acidogenic, the bacteria ferment the monomers to alcohols and organic acids. Carbon dioxide and molecular hydrogen are also released. In the next, acetogenic phase, most of the fermentation products are converted to acetic acid. In the last phase, methanogens produce methane both by synthesis from hydrogen and carbon dioxide and by decomposition of acetic acid.

5.3.2. Municipal-industrial wastewater treatment

Industrial wastewater, due to the presence of toxic substances, as well as the high content of compounds that are difficult to biodegrade, is usually disposed of, using physicochemical methods (e.g. wet combustion, Supercritical Water Oxidation method (SCWO)), which are not discussed in this book. However, for some industrial wastewater, e.g. dyeing wastewater (wastewater containing large quantities of dyes used in the textile industry), can be treated together with domestic wastewater after pre-treatment (Figure 5.3.2).

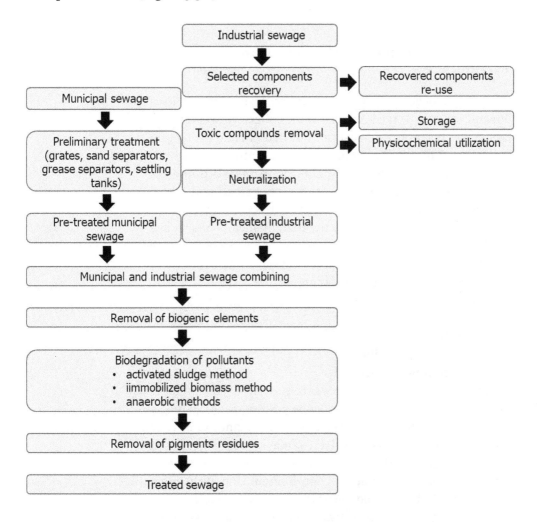

Pre-treatment usually involves the removal, by extraction, of compounds inhibiting the growth of microorganisms and their separate disposal, or further use (if they can be useful in other areas of the economy) and obtaining a pH of about 7.0. As

Fig. 5.3.2. Scheme showing common treatment of household and industrial wastewater.

industrial wastewater after pre-treatment is still more difficult to biodegrade than domestic wastewater, it is usually combined in a ratio of 20:80. Despite its low BOD_5 and COD values as well as its phosphorus content, dyeing wastewater may still contain traces of dyes, which are removed using strong oxidising agents or using microbial enzymes of the oxidoreductase class (see Section 5.11).

5.3.3. Wastewater treatment in scattered areas – small infrastructure

Elimination of domestic wastewater discharged from outflows located in a dispersed area without collective wastewater systems is one of the most serious environmental protection problems. The optimal solution, taking into account both ecological and economic aspects, is to install domestic wastewater treatment plants. After proper treatment, water should have a second class of purity and, according to Polish and EU standards, can be distributed by means of seepage drainage to the ground.

There are several types of domestic wastewater treatment plants, of which the most popular are:
- with seepage drainage;
- with a sand filter;
- soil and plants;
- with a biological filter;
- with activated sludge.

The oldest and so far, most frequently used domestic treatment plant in Poland is the one with seepage drainage. However, it occupies a relatively large area and, in case of improper operation (supervision), may pose a significant epidemiological and ecological threat. In some countries, such as France, it is completely prohibited. One of the most environmentally safe and occupying relatively small area is the treatment plant with activated sludge. Regardless of the detailed design of such plants, the processes taking place in their tanks are completely isolated from the environment and both the permeability of the soil and the groundwater level are not important for their functioning (Figure 5.3.3).

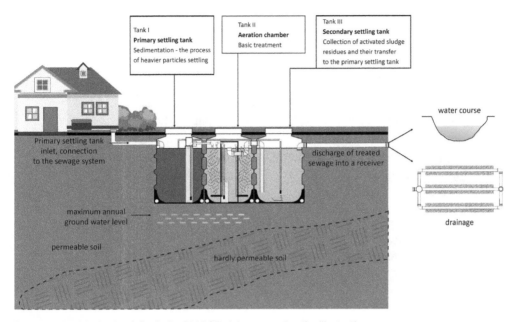

Developed on the basis of: Paniczko 2013 (disclaimer: copying the illustration only with the author's consent).

Fig. 5.3.3. Scheme of a household wastewater treatment plant with activated sludge.

Raw wastewater flows by gravity through an inlet pipe into the primary sedimentation tank (tank I), where solids (sand, paper, part of the recirculated sludge) are retained. The pre-treated wastewater flows into the aeration chamber (tank II), at the bottom of which there are diffusers to which air is pumped with a compressor. In the chamber, the wastewater is mixed with activated sludge and aerated. The process of nitrification and the growth of activated sludge takes place. Excess sludge is moved to the septic tank, from where it is periodically removed and disposed of. In the secondary sedimentation tank (tank III), as a result of sedimentation, the sludge is separated from the treated wastewater, which is gravitationally discharged to an external surface receiver, watercourse or soil through a drainage or absorbent well. The sludge remaining in tank III, together with part of the treated wastewater, is recirculated (using a pump) from the secondary sedimentation tank to the primary sedimentation tank.

PRACTICAL PART

Materials and media

1. Materials: activated sludge from the aeration chamber, raw wastewater, treated wastewater, tissue paper filters, commercial COD kits.
2. Apparatus: Imhoff funnel, thermal oven, filtration kit, spectrophotometer, OxiTop kit, heater, pH-meter, conductivity meter.

Aim of the exercise
Determination of selected parameters of activated sludge, raw wastewater and treated wastewater from group and household wastewater treatment plants.

Procedure

1. Characteristics of activated sludge:
   a) carry out microscopic observations of activated sludge taken from the household and group wastewater treatment plant, performing live preparations and Gram stained preparations;
   b) determine the sediment volume index according to the formula:

   I_o = volume of sediment (ml)/ dry residue (g)

   Pour 1,000 ml of tested samples into the Imhoff funnels and read the sediment volume after 30 min. Take 20 ml of the test samples, filter through previously weighed tissue paper filters and dry at 105°C to constant weight. Calculate the dry residue in 1,000 ml.
2. Determination of suspended solids amount in sewage. Place samples of raw and treated wastewater (1,000 ml) in the Imhoff funnel. The volume of settled suspensions should be read at 5, 10, 20, 30, 45, 60 and 120 min after filling the funnel. Graphically depict the kinetics of sediments settling.

3. COD determination in samples of raw and treated wastewater.
   Perform the analysis with the use of commercial sets for COD determination according to the manufacturer's instructions.
4. BOD_5 determination in samples of raw and treated wastewater.
   Perform the analysis with the use of OxiTop cylinders according to the manufacturer's instructions. Determine the BOD_5 value after 5 days of incubation at 20°C.
5. Determination of pH and conductivity of raw and treated wastewater samples.
6. Compare the results obtained from the analysis of samples taken in a group and household wastewater treatment plant.

5.4. Waste composting

5.4.1. Waste composting in municipal composting plants

Composting is a process of humification and mineralisation of organic matter carried out by microorganisms under aerobic conditions, at appropriate temperature and humidity. Composting is mainly done with plant biomass, e.g. wood shavings, sawdust, bark, plant mass from urban green areas, residues from vegetable processing, waste from the herbal and textile industry, peat. This process may also include sludge from biological wastewater treatment, organic and organic-mineral silt from the treatment of water reservoirs, post-consumer substrates from mushroom farms, etc. The classic method is composting in trapezoidal prisms. In the case of larger compost prisms, forced aeration, as well as prism mixing are often applied. The total composting time for the classic method is 6 to 7 months. Dynamic technologies are also used, in which the first stage takes place in rotary drums and then the partially transformed compost mass is piled up in the prism. In this method, the composting time is shortened and the process efficiency is increased.

Matured compost is mainly used in agriculture and horticulture to fertilise and improve soil structure, to produce substrates for non-soil plant cultivation, and to restore contaminated areas.

Four phases can be distinguished in the process of organic waste composting: mesophilic (A), thermophilic (B), cooling (C) and maturation (D) (Figure 5.4.1).

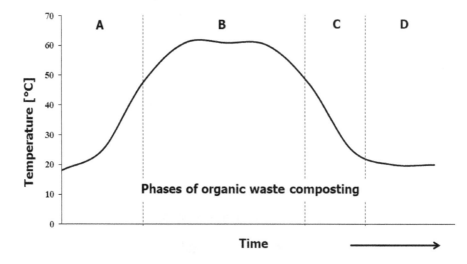

In phase one (A) it is mainly bacteria (including actinomycetes) and mesophilic fungi, whose development leads to a rapid increase in temperature. These conditions reduce the activity of mesophilic microorganisms, while the number of thermophilic microorganisms (mainly actinomycetes, fungi and bacteria *Bacillus*), characteristic of the thermophilic phase (B), increases. Under the influence of high temperature, pathogenic microorganisms and weed seeds are also eliminated (hygienisation). In the third cooling phase (C), there is a significant drop in the temperature of the compost mass and a renewed growth of mesophilic microorganisms, mainly fungi and actinomycetes, whose spores survive the extreme conditions of the thermophilic phase. The fourth phase of the composting process, maturation (D) is characterised by a slowdown in the decomposition of the compounds, a decrease in the activity of the microorganisms and a gradual drop in temperature to ambient levels.

Fig. 5.4.1. Phases of organic waste composting (explanations in the text).

5.4.2. Local use of waste from urban green areas

Every year, large amounts of bio-waste are extracted from urban green areas (parks, botanical gardens, urban forests), as well as home gardens, which can also be used on site to produce compost, and then to grow and maintain plants. Such a solution is also supported by economic reasons, allowing a significant reduction in the cost of green area maintenance. Park composters, like garden composters, can be made from impregnated wood, plastic and metal mesh (resistant to biological corrosion), usually 100 to 200 cm wide and 100 to 150 cm high (Figure 5.4.2).

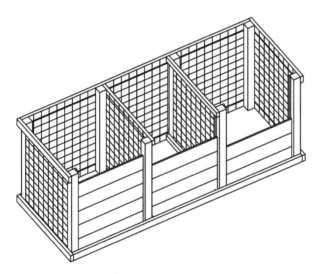

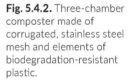
Fig. 5.4.2. Three-chamber composter made of corrugated, stainless steel mesh and elements of biodegradation-resistant plastic.

Composting, especially of larger amounts of plant waste, can also be carried out in the form of free-standing prisms with triangular or trapezoidal cross-section, located within the economic base of the facility (buffer zone or economic green area).

The composting site should be located on the side of an area with limited sunshine, sheltered from the wind and thus conducive to the non-drying of waste. The substrate of free-standing prisms and composters must be permeable. For this purpose, a 20 cm drainage layer is laid on the bottom, which may include, e.g. branches from cutting plants, wood chips, sawdust (Figures 5.4.3, 5.4.4).

Explanations: 1 – waste plant mass prepared for composting; 2, 3, 4 – green waste compost prisms covered with soil; 5, 6 – mature compost.
The photograph was taken and developed in 2019 by A. Długoński (disclaimer: copying the illustration only with the author's consent).

Fig. 5.4.3. Composting of green waste in the back yard of the Botanical Garden in Łódź.

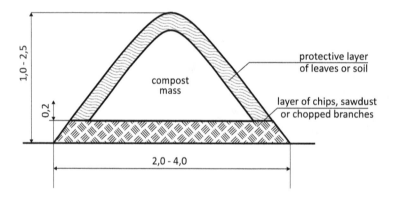

Fig. 5.4.4. Cross-section through the compost prism of green waste (dimensions in metres).

The compost mass placed on the drainage layer should not contain more than 30–50% of grasses due to the possibility of anaerobic zone formation, which limits the composting rate and promotes the secretion of odours. For this purpose, the compost mass is supplemented with waste with high initial porosity (e.g. shredded branches, plant stems). The whole is covered with a protective layer, e.g. from fallen leaves or soil. Such an arrangement of individual elements favours the proper course of organic matter decomposition with the participation of soil microorganisms, as well as invertebrates, e.g. earthworms getting into the soil prism, and obtaining compost with the desired composition and structure. During composting, the weight reduction reaches 70%. The composting time of green waste depends mainly on its composition and the availability of oxygen (material porosity) and usually lasts from 6 months to 1 year. Mature compost does not spill or stick after squeezing. It looks and smells like humus soil.

PRACTICAL PART

Materials and media

1. Media: as in Section 1.2.
2. Materials: samples of the compost mass from different phases of composting and finished compost taken from local and municipal composting plants.
3. Apparatus and set of reagents for the determination of heavy metals, as in Section 5.7.

Aim of the exercise
Quantitative and qualitative analysis of microorganisms present in the compost mass at individual composting stages and in the finished compost. Separation of strains capable of synthesising biologically active compounds. Determination of the content of heavy metals in the finished compost.

Procedure

1. Take samples of the composted mass according to PN-Z15011–1: "Compost from municipal waste. Sampling".
2. Quantitative and qualitative microbiological analysis of the samples and the evaluation of the usefulness of the isolated strains for the synthesis of biologically active compounds should be performed as in Section 1.2.
3. Determination of heavy metals in the samples should be performed according to the methodology given in Section 5.7.
4. Compare the results showing the heavy metal content of the tested composts with the standards in force in the adequate countries.

5.5. Use of municipal green waste for energy production in local biogas and incineration plants and the synthesis of fungal laccases

5.5.1. Energy production

The leaves, which fall naturally in autumn, decompose under the influence of microorganisms and the energy released during the degradation is dispersed. The energy value of leaves of trees most often planted in urban green areas, as well as along roadways, ranges from 12–17.9 kJ/g of dry matter (Table 5.5.1). Therefore, the leaves can be successfully used for energy production in local biogas plants and incinerators. This trend is particularly evident in western countries (e.g. Germany, France, the Netherlands and the Scandinavian countries), which have developed standards for municipal management and rational use of renewable energy sources. For example, in Scandinavia, many municipalities or even medium-sized companies have their own biogas or incineration plant using their own bio-waste, including municipal green waste, to produce electricity.

Tree species	Calorific value of the leaves [kJ/g-1]
Quercus petraea (Matt.) Liebl.	17.6
Quercus rubra L.	17.6
Carpinus betulus L.	12.6
Aesculus hippocastanum L.	17.9
Acer saccharinum L.	13.0
Acer platanoides L.	14.3
Platanus acerifolia	16.4
Populus alba L.	14.3
Populus nigra L.	13.3
Salix × sepulcralis 'Chrysocoma'	16.4
Tilia × euchlora K. Koch	12.0
Mean value	15.3

Table 5.5.1. Caloric values of selected tree species.

A biogas plant is a complex of installations, including fermentation chambers, used to produce gas from biomass by methane fermentation. Biogas, containing mainly methane (50–75%) and CO_2 (25–50%) and small amounts of nitrogen

(0–10%), hydrogen sulphide (0–3%), hydrogen (0–1%) and oxygen (0–0.05%), is mainly used to produce electricity and heat. According to WARMER (Eurostat) and the OECD Central Statistical Office, urban bio-waste accounts for nearly 30% of all municipal waste. European Union directives and the amended law on waste management (forcing selective collection of household bio-waste) promote an increase in its use. The possibilities of using methane fermentation for utilisation of sewage sludge, sewage from the agri-food industry, farm waste from animal husbandry, as well as microbiological aspects of methane fermentation have already been presented in more detail in Section 5.3.1.

From the areas of urban greenery mentioned at the beginning of this section, e.g. the park with an area of 2 to 10 ha, it is possible to obtain annually from 190 to 720 tonnes of bio-waste (cut grass, fallen leaves and other plant residues), which would allow the local biogas plant to produce the energy necessary to illuminate the park (a network of park lanterns in LED technology), one service building (e.g. a restaurant or a park café) or other small park infrastructure devices, thus making the area in question, self-sufficient and independent from external power sources.

Since naturally fallen leaves, mowed grass, as well as a number of other plant waste are available in excess, but only at certain times of the year, they are ensiled, which ensures an even supply of these raw materials (as a substrate for methane fermentation) throughout the year. The biogas yield of freshly cut grass and silage is similar in terms of dry organic matter content (Table 5.5.1.2).

Grass	DM %	ODM % dm	Biogas yield m³/t ODM	Methane content% vol.
After mowing	11–12	83–92	550–680	55–65
Silage	25–50	70–95	550–620	54–55

Explanations: DM – dry matter; ODM – organic dry matter.

Table 5.5.1.2.
Suitability of grass for biogas production by methane fermentation

Research conducted on the disposal of naturally fallen leaves, collected from fruit tree plantations and parks with ornamental tree plantations, also indicates the suitability of their use for biogas production. Depending on the tree species, from 417 to 453 m³ of biogas with a methane content above 50%, can be obtained from one tonne of wet mass of fallen leaves through the process of methane fermentation.

5.5.2. Use of urban green waste for biosynthesis of fungal enzymes on the example of laccases

Waste from urban green areas (hay, naturally fallen leaves) are actively decomposed by numerous species of microorganisms (bacteria and fungi), which synthesise enzymes of various activities. In this way, indirectly, this waste is a valuable raw material for the production of microbial enzymes, including laccases. Laccases are enzymes responsible for lignin degradation, and are also able to decolorize dyes, which is of great practical importance in sewage treatment, especially in the textile industry. Laccase synthesis in fungi depends, among other factors, on the presence of phenolic structure compounds in the growth environment. Naturally fallen leaves contain much more phenolic substances than green leaves. Extracts from fallen leaves are thus a valuable and, at the same time, cheap component of fungal substrates used to grow laccase-producing fungi. The characteristics of laccases and the possibilities of their application in practice are presented in more detail in Section 5.11.

PRACTICAL PART

Materials and tools

1. Simplified nature inventory of green area stand, see inventory table (Table 5.1.2.1, Section 5.1.2).
2. Basic map or topographic map or satellite image of the area.
3. Laboratory scale.
4. Camera.

Aim of the exercise
Estimation of the calorific value of leaves and their suitability as a substrate for biogas production. Determination of the usefulness of cut grass (hay) and fallen leaves of individual tree species for production of laccase by microfungi.

Calculation of weight and calorific value of fallen leaves

The starting point for the calculation is dendrological data showing that from one 100-year-old adult deciduous tree (e.g. beech) about 800,000 naturally fallen leaves can be obtained in late autumn. Knowing, after weighing several leaves, the average weight of a single leaf, it is possible to estimate the weight of fallen leaves of trees growing in the analysed area, and then their energy value.

Calculation of the dry weight of the leaves of a single deciduous tree

$$B = mL \times aL$$

where:
B – dry weight of leaves of one deciduous tree [g];
mL – average dry weight [g] of at least 15 fallen leaves collected randomly from under the tree and dried at room temperature to constant weight;
aL – the estimated number of leaves on the tree*.

* It is assumed that on one adult 100-year-old tree there are on average 800, 000 leaves (Hereźniak 2013), in case of young trees (up to 2 m high) 1/4 of this value should be assumed.

Calculation of the calorific value (energy) of the leaves of a single tree

$$Wk = SL \times aL \times b$$

where:
Wk – calorific value of leaves from one deciduous tree [kJ];
SL – dry weight of one leaf [g];
aL – number of leaves on the tree;
b – calorific value of leaves for a given tree species [kJ/g]**.

** When assessing the calorific value of leaves of other tree species the given average value (15.3 kJ/g) should be taken.

Procedure

1. Carry out simplified tree inventory, identify species, mark on the map of the analysed green area, provide their number, put the results in a table (see Table 5.1.2.1 in Section 5.1.2).

2. Collect grass from the first and last mowing of lawns (late March/April and mid-October, respectively) and possibly intermediate mowing and leaves from under particular tree species (mid-October), dry in a room with room temperature to constant mass (relative humidity of the air in the room should be about 60%).

3. Pour hay (20 g) and dry leaves of individual tree species (25 g) with deionised water to 1 l and bring to a boil with constant stirring. Filter the extract obtained through membrane filters with 0.22 μm pores and bring the pH to 6.8. Autoclave at 117°C for 20 min and make up to the final 2 and 1 % w/v with sterilised solutions of glucose and neopeptone, respectively.

4. Prepared hay and leaf extracts, inoculate with fungal strains, e.g. *Nectriella pironii*, able to synthesise laccase and culture, and determine the laccase activity according to the methodology given in Section 5.11.

5. Summarise the results obtained in the table below.

No.	Type of extract	Laccase activity [U/l]
1.	Hay – first mowing	
2.	Hay – subsequent mowing(s)	
3.	Hay – last mowing	
4.	Leaves – tree species 1	
5.	Leaves – tree species 2	
6.	Leaves – tree species 3	

6. Weigh 15 randomly selected dry leaves of a given species, calculate the average dry weight of one leaf and, using the formulae given above, calculate the calorific value of leaves of individual tree species. Summarise the results in the table below.

No.	Tree species name	Number of trees per species (I_D)	Dry weight of leaves of a single tree (B)	Calorific value of leaves of a single tree (Wk)	Dry weight of leaves of trees of a given species $(I_D \times B)$	Calorific value of tree leaves of a given species $(I_D \times Wk)$
1.						
2.						
3.						
In total						

7. Evaluate together the possibility and method of management of waste (grass and leaves) from the analysed urban green area.

5.6. Biodegradation of toxic xenobiotics

INTRODUCTION

Xenobiotics are chemical compounds, most often produced by humans, which are not natural metabolites of organisms, including microorganisms. Most microorganisms do not have enzymatic systems that could transport them into the cell and/ or transform them. In the case of substances with a higher molecular weight, their decomposition is additionally hindered by the lack of suitable hydrolases decomposing polymers to monomers. These are often toxic substances, interfering with the basic physiological processes of both microorganisms and higher organisms. Particularly noteworthy are xenobiotics belonging to the group EDCs (Endocrine Disrupting Compounds), which disrupt the proper functioning of the endocrine system in humans and animals. They are also referred to as hormonomimetics, xenoestrogens or hormonal modulators. The book adopts the term EDCs, commonly used in the scientific literature.

These compounds include numerous phenols (pentachlorophenol, bisphenol A, para-nonylphenol), pesticides (alachlor, endosulfan, DDT), organic tin compounds (tributyltin, octyltin, phenyltin), pharmaceuticals ibuprofen, naproxene, carbamazepine), heavy metals (Cd, Ni, Pb), synthetic estrogens (mestranol, 17α-ethinylestradiol), as well as natural estrogens synthesised in the mammalian body and being intermediate products during the biodegradation of phytosterols (estrone, 17β-estradiol).

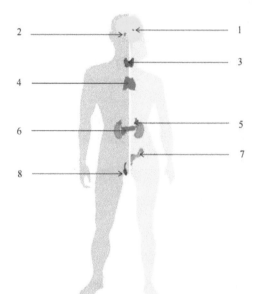

Fig. 5.6. Location of endocrine glands producing hormones in humans (man on the left, woman on the right).

Explanations:
1 – pineal gland;
2 – pituitary gland;
3 – thyroid gland;
4 – thymus;
5 – adrenal gland;
6 – pancreas;
7 – ovary; 8 – testicle.

The mechanism of EDC action usually consists of binding to receptors in target tissues (cells) and acting in a similar way as natural hormones or blocking their action. They can also interfere with hormone production or its displacement in the body.

The transformation of these compounds by microorganisms is often of a detoxification nature, as shown in Figures 8.1.6 and 8.1.17. More detailed characteristics of individual xenobiotics classified as EDCs and possibilities of their elimination with the participation of microorganisms are presented below.

5.6.1. Bisphenol A

INTRODUCTION

Bisphenol A (BPA) is one of the most frequently manufactured chemicals and is used in industry as a substrate for the production of plastics. BPA was first synthesised by Russian chemist Alexander Dianin in 1891. In 2015, global consumption of BPA was estimated at 7.7 million tonnes and is estimated to reach 10.6 million tonnes in 2022.

BPA is mainly used for the production of polycarbonates and epoxy resins, which are characterised by high stability, good mechanical properties and low moisture absorption, so they are widely used in industry. Epoxy resins are used, among others, as protective liners for pipelines, food cans and beverage cartons. They are also used to make dental fillings, as well as CDs, DVDs, electronic equipment, car parts, water bottles and food containers. Bisphenol is also an additive to thermal paper, used, among others, in cash registers and payment terminals.

BPA can be transferred to food and beverages stored in containers in which it is used. Its release is favoured by low pH, elevated temperature, repeated use of the containers and the chemical properties of the stored substances. It has been shown that epoxy resin coatings, which protect cans from corrosion, can cause bisphenol contamination of the food stored in them. The release of this compound is particularly high during sterilisation of cans when heated to high temperatures.

Due to the wide use of BPA in the manufacture of everyday products, over 90% of the population is exposed to this compound. In addition, BPA is a low molecular weight compound, easily soluble in fats, so it easily penetrates the skin and accumulates in adipose tissue. This compound was also found in urine, pregnant women's serum, amniotic fluid, breast milk, umbilical cord blood and other body fluids. BPA is classified as an endocrine disrupting compound (EDC). The endocrine disrupting action of BPA may have a significant impact on fertility. Studies also indicate a link between exposure to BPA and miscarriages, as well as the development of certain cancers, such as breast, uterine and ovarian cancer. Other disorders associated with exposure to BPA include numerous metabolic diseases such as insulin resistance, obesity and type II diabetes as well as thyroid disorders.

Due to the harmful effects of BPA on people and the environment even at very low concentrations, effective and efficient methods for the degradation of this compound are sought. Advanced oxidation processes such as the Fenton reaction, ozonation, photocatalytic oxidation and ultrasonic oxidation are effective

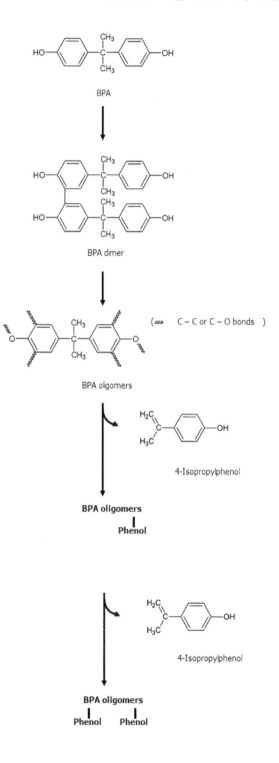

BPA

BPA dimer

(ᴧᴧ   C – C or C – O bonds)

BPA oligomers

4-Isopropylphenol

BPA oligomers
|
Phenol

4-Isopropylphenol

BPA oligomers
| |
Phenol   Phenol

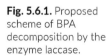

Fig. 5.6.1. Proposed scheme of BPA decomposition by the enzyme laccase.

methods to eliminate BPA from sewage. However, due to their disadvantages, such as being time consuming, high costs, and the formation of more toxic degradation products than the parent compound, the use of these methods is limited. The process of adsorption on active carbon can be used to remove BPA from sewage, however, this method is characterised by low efficiency and the need to regenerate active carbon.

Other methods of removing BPA from the environment include biodegradation with microorganisms. For many years, research has been conducted to isolate microorganisms that effectively degrade BPA. Bacterial strains capable of degrading BPA have been isolated from various environments, e.g. soil, river sediment, sea water and even food samples. They included both Gram-negative strains from the genera *Sphingomonas, Pseudomonas, Achromobacter, Novosphingobium, Nitrosomonas, Serratia, Bordetella, Alcaligenes, Klebsiella* as well as Gram-positive strains from the genera *Streptomyces* and *Bacillus*. Numerous studies have shown high effectiveness in removing this type of contaminants by ligninolytic filamentous fungi and their enzymes, such as peroxidases and laccases. The conversion of bisphenol with laccases leads to the formation of BPA oligomers as a result of successive oxidation and condensation reactions. Phenol molecules can be attached to the formed oligomers or the oligomers can split into 4-isopropenylphenol. An example of a pathway for BPA conversion using laccase is shown in Figure 5.6.1.

PRACTICAL PART

Materials and media

1. Microorganisms: *Myrothecium roridum*.
2. Media: ZT slants, Czapek-Dox medium, WHI3 medium.
3. Reagents: Bisphenol A.
4. Apparatus: ball homogeniser, vacuum evaporator, liquid chromatograph Agilent 1200 and mass spectrometer QTRAP 3200 (Sciex).

Aim of the exercise
Determination of the ability of selected filamentous fungi to degrade bisphenol A.

Procedure

1. Preparation of cultures containing BPA:
   a) prepare 48 h inoculum of *M. roridum* in WHI3 medium;
   b) prepare a BPA standard solution at a concentration of 10 mg/ml in ethanol;
   c) add 2 ml of *M. roridum* fungus inoculum to 18 ml of Czapek-Dox medium containing BPA at 10 mg/l, 50 mg/l and 100 mg/l;
   d) additionally, prepare biotic controls (containing only the tested microorganism and medium) and abiotic controls (containing only the medium and BPA in the selected concentration). Incubate the samples for 48 h at 28°C.
2. Extraction of BPA and quantitative chromatographic analysis of the extracts:
   a) prepare the salts for extraction by the Quechers method:
      - $MgSO_4$ – 2 g,
      - NaCl – 0.5 g,
      - $C_6H_5Na_3O_7$ – 0.5 g,
      - $C_6H_6Na_2O_7 \times 1.5\ H_2O$ – 0.25 g;
   b) homogenise the cultures using glass balls. Take 5 ml of homogenate and transfer to previously weighed salts, add 5 ml of acetonitrile and extract for 5 min. Pour off the acetonitrile and filter through a filter paper;
   c) prepare standard curves for quantitative calculations of tested compounds at concentrations of 0.5; 2; 4; 6; 8; 10 µg/ml in 2% ACN (BPA);
   d) standard curves of the indicated substances and appropriately dissolved test sample extracts (in a mixture of methanol : water at a ratio of 70:30) should be analysed using LC-MS/MS. Use the Agilent 1200 liquid chromatograph with AB Sciex QTRAP 3200 mass sensor. Separate the compounds using

a Kinetex C18 column (50 mm × 2.1 mm, 5 µm; Phenomenex, USA) at 40°C, solvents: water (channel A) and acetonitrile (B) with 5 mM ammonium formate, and then perform the determination using QTRAP 3200 (Sciex) spectrometer equipped with ESI-type ion source in negative polarisation mode.

Analysis of results

On the basis of the chromatographic analyses performed, determine the specific abilities of the tested filamentous fungus to degrade BPA.

5.6.2. Organotin compounds

INTRODUCTION

Organic tin compounds have been known for more than 150 years, but they were used more widely only since the 1940s, mainly as a stabilising agent in the production of polyvinyl chloride (PVC). It is estimated that the annual production of organic tin compounds worldwide is about 50 thousand tonnes, of which about 20% is used as biocides.

These substances are characterised by the occurrence of bonds of the following types: R_4Sn, R_3SnX, R_2SnX_2, $RSnX_3$. The organic substituents (R) could be methyl, phenyl, butyl, propyl or pentyl groups. Similarly, the anion (X), refers to a halogen atom, hydroxyl, carboxyl or thiol group. The attached functional groups R determine the biological properties, including toxicity of the compound, while the substituent X has an effect on volatility and solubility.

$RSnX_3$ type compounds are added to PVC as stabilisers, while R_2SnX_2 are used as catalysts in the production of PVC, polyurethane foams, silicones, and also as PVC stabilisers. R_3SnX type compounds, e.g. tributyltin (TBT) are characterised by high biological activity. TBT is used as a PVC stabiliser, a component of anti-fouling paints that cover ship hulls and underwater constructions, and it can also serve as a preservative for wood, fabric, paper and leather. R4Sn type substances are used as intermediate substrates in the synthesis of other organotin compounds.

Tin is a metal with low toxicity to living organisms. The bonding of organic groups to the tin atom results in compounds with new physicochemical properties are formed, which, depending on the type of substituent, show toxic effects of varying intensity. Toxicity of organic tin compounds decreases according to the scheme: R3SnX > R2SnX2 > RSnX3. Substances of type R4SnX are compounds of low toxicity.

The widespread use of organic tin compounds has contributed to environmental pollution by these biocides. This particularly concerns the aquatic environment. TBT and its derivatives have been detected in port waters on all continents.

The presence of TBT was also registered in human liver (Poland – 11 ng/g wet weight, Japan 96–360 ng/g wet weight). Larger amounts of this biocide in Japanese people are most likely a consequence of higher consumption of seafood. In addition, the source of these xenobiotics is a water supply network made of PVC pipes.

Organic tin compounds primarily affect the biological membranes. Thanks to their lipophilicity, they can penetrate cytoplasmic membranes, violating their integrity and increasing their permeability. Organotin compounds also have a negative effect on intracellular processes and organelles. TBT damages mitochondrial membranes, causing these to swell and rupture, and they also inhibit ion exchange and ATP synthesis.

Mono- and dialkylated octyltin derivatives are mainly used as stabilisers in PVC production and are intended to prevent deformation and discoloration of the final product. Dioctyltin is also used as a catalyst in chemical synthesis. Significantly, octyltin derivatives were initially considered to be of low toxicity, but later studies have shown that they have an immunosuppressive and immunotoxic effect and can cause a variety of anomalies in mammalian foetuses.

The most toxic organotin compounds, such as TBT or triphenyltin (TPT), belong to the group of so-called hormonal modulators, i.e., substances that damage the functioning of the hormonal system. The high toxicity of these compounds has led to a total ban on their production and use, which has been in force since 1 July 2010. However, due to their high stability and resistance to decay, TBT and TPT are still detected in the environment.

The phenomenon of "imposex", induced by TBT, plays a very important role in the environment, and is characterised by the appearance of male sexual traits in female individuals and the reduction of reproductive capacity of the population. This is related to the interference of TBT in the processes of steroidogenesis by blocking the biotransformation of male sex hormones. Cases of inhibition of the production of female sex hormones at the stage of testosterone conversion were observed, among others, in the *Ruditapes decussata* clam.

Microorganisms play a very important role in the dealkylation process in the natural environment. As a result of progressive separation of organic groups, compounds less toxic to the environment are formed, with shorter half-lives. It has been shown that *Cunninghamella elegans*, a fungus used for hydroxylation of corticosteroids, can degrade TBT (10 mg/l) to DBT and MBT with a yield of 70%.

Enzymatic systems containing cytochrome P- 450 (cyt P-450) are often involved in the transformation of xenobiotics. Also enzymes containing this hemoprotein are involved in biological degradation of organic tin compounds.

In fish, cyt P-450 is involved in the first stage of TBT dealkylation, resulting in α-, β-, γ- and δ-hydroxydibutyltin. In the next stage, they are combined with sugars and sulphides to highly polar forms, excreted from the body. The transformation and removal of xenobiotics in mammals has a similar course.

PRACTICAL PART

Materials and media

1. Microorganisms: strains of *Cunninghamella echinulata, Metarhizium robertsii.*
2. Media: synthetic with 4% glucose.
3. Reagents: hexane, tropolone, 3 M solution of methylmagnesium bromide in ether.

4. Apparatus: gas chromatograph Agilent GC7890 with mass spectrometer MS5975C.

Aim of the exercise

Comparison of selected fungal strains in terms of biodegradation efficiency of organotin compounds.

Procedure

1. Rinse the 5-day universal slants with the tested bacterial strain with 4 ml of synthetic medium. Then transfer the bacterial suspension to 30 ml of liquid medium.
2. Carry out the culture for 48 h on a reciprocal shaker at 28°C.
3. Add 1 ml of the inoculum to flasks containing 19 ml of synthetic medium with 4% glucose.
4. Add dibutyltin standard solution (20 mg/l) to the flasks.
5. Carry out the culture for 168 h on a reciprocal shaker at 28°C.
6. After the end of the incubation process, extract the organic tin compounds. Acidify the culture flasks with concentrated HCl to pH 2–3. Then extract twice with 0.03% hexane solution of tropolone (20 ml). Dry the organic phase with anhydrous sodium sulphate, evaporate on a vacuum evaporator and use for derivatisation.
7. Derivatisation of organic tin compounds.

   Add 2 ml of hexane, 50 µl TTBT and then 0.5 ml of Grignard's reagent (3 M solution of methylmagnesium bromide in ether) to the dried extract. Leave the mixture at room temperature for 20 min, then stop the reaction by adding 2 ml of 20% aqueous ammonium chloride solution. Centrifuge the sample (30 min, 500 × g) and then take 200 µl from the top layer for chromatography.

   Use a gas chromatograph with a mass spectrometric detector (GC-MS, Agilent GC 7890 and MS 5975C) equipped with a HP 5 MS column (30 m × 0.25 mm × 0.25 µm) operating in the temperature range 80–300°C, with a helium flow of 1.2 ml/min.

8. For selected cultures, separate the culture medium from the mycelium (centrifugation 2,000 × g, 10 min) and perform toxicity tests using Artoxkit M for the supernatant obtained.

Analysis of results

Compare the ability to degrade organotin compounds by selected microorganisms and assess the toxicity of the tested culture systems.

5.6.3. Use of microorganisms to eliminate pesticides

INTRODUCTION

Pesticides are a very numerous and diverse group of chemical substances introduced into the environment to prevent or limit the development of various types of pests. Among pesticides, insecticides (to control insects), herbicides (to control weeds) and fungicides (to control fungi) are most commonly used.

The widespread use of pesticides in the human economy and the progressive contamination of the natural environment, including the soil, has triggered mechanisms in microorganisms to protect themselves from the adverse effects of pesticides. Until now, many mechanisms involved in the process of degradation or detoxification of pesticides in microorganisms leading to complete decomposition of the compound or transformation of its structure have been recognised.

Pesticide biodegradation can take place under aerobic or anaerobic conditions. The most important processes of xenobiotics decomposition include oxidation reactions. Enzymatic systems containing cytochrome P-450, responsible for the introduction of the oxygen atom (formation of a hydroxy group) into the substrate molecule, are often involved in the process of biological disposal of pesticides. In the case of xenobiotics containing an aromatic ring, the hydroxylation of the substrate is often associated with the ring cleavage in the ortho or meta position.

In fungi, as in mammals, pesticides transformed into hydroxyl derivatives are additionally conjugated with compounds of natural origin such as glucuronic acid, which facilitates their excretion. The examples of pesticide conversion processes involving microorganisms are presented below.

Transformation of phenoxyacetic acid derivatives

Phenoxyacetic herbicides are most commonly used in the form of 2,4-dichlorophenoxyacetic acid (2,4-D) derivatives. Degradation of 2,4-D by *Arthrobacter, Wautersia* bacteria and *Umbelopsis* fungi has been described and starts with the removal of the phenoxyacetic residue, resulting in 2,4-Dichlorophenol (Figure 5.4.2). In the subsequent stages, the ring is split and after the subsequent transformations, acetyl-CoA is produced.

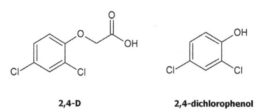

Fig. 5.6.3.1 Structural formulae of 2,4-D and 2,4-dichlorophenol.

2,4-D   **2,4-dichlorophenol**

Triazines

Triazines are compounds that consist of a heterocyclic ring with 3 nitrogen atoms. Among the compounds containing the triazine ring, the herbicide atrazine deserves special attention. Bacteria of the genera *Agrobacterium, Wautersia* and *Pseudomonas,* characterised by the ability to mineralise this pesticide, were isolated from soil contaminated with atrazine. The initial stages of atrazine metabolism are presented in Figure 5.6.3.

Fig. 5.6.3.2 Transformation of atrazine by *Pseudomonas* sp.

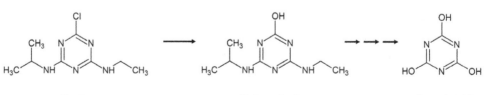

Atrazine   **Hydroxyatrazine**   **Cyanuric acid**

PRACTICAL PART

Materials and media

1. Microorganisms: microfungi strains.
2. Media: synthetic with 4% glucose.
3. Standards of selected pesticides.
4. Apparatus: liquid chromatograph Agilent 1200 and mass spectrometer QTRAP 3200 (Sciex).

Aim of the exercise
Comparison of the ability of selected microfungi strains to degrade pesticides

Procedure

1. Prepare solutions of pesticides (alachlor, atrazine, 2,4-D) at 5 mg/ml in ethyl alcohol.
2. Prepare a standard curve for quantitative calculations of tested pesticides at concentrations of 1, 5, 10, 50, 100, 500, 1,000 ng/ml in 2% ACN.
3. Setting up of culture and control samples:
   a) inoculate spores (1 ml of suspension) of a filamentous fungal strain on a mineral substrate (1% Glc) (19 ml in V flasks = 100 ml) with the addition of pesticides at a concentration of 10, 25, 50 mg/l;
   b) growth control – cultivation on mineral substrate (19 ml) of spores of filamentous fungus strain (1 ml of suspension);
   c) abiotic control – mineral substrate (20 ml) with 10 mg/l pesticides added.
4. Incubation of test and control samples 7 days at 28°C, with shaking at 140 rpm.
5. Perform micro and macroscopic observations of the obtained cultures.
6. Evaluation of degradation capacity:
   a) once the incubation is complete, homogenise the medium and mycelium;

b) filter and centrifuge the samples;

c) collect the supernatant, dilute appropriately and transfer 200 µl to HPLC plates;

d) perform a quantitative analysis with the Agilent 1200 liquid chromatograph with the AB Sciex QTRAP 3200 mass spectrometer with an ESI-type ion source. Separate the pesticides using a Kinetex C18 column (50 mm × 2.1 mm, 5 µm; Phenomenex, USA) at 37°C, flow rate 0.5 ml/min, solvents: water (channel A) and methanol (B) with 5 mM ammonium formate, and then perform the determination in ion transitions for alachlor (270–238) (positive polarisation), atrazine (216–174) (positive polarisation) and 2,4-D (219–161) (negative polarisation).

7. For supernatants obtained, perform selected toxicity tests (according to Section 5.7).

Analysis of results

On the basis of the data obtained, characterise the degradation ability of microorganisms.

5.6.4. Nonylphenol

Alkylphenols (AP) are a particularly dangerous group of anthropogenic pollutants, as they are not only toxic but also disturb the proper functioning of the endocrine system in humans and animals. Typical representatives of alkylphenols are nonylphenols (NPs), the amount of which in soil, bottom sediment and surface water, despite the prohibitions, is still at an unacceptable level.

NPs are xenoestrogens that are commonly used in the manufacturing of many household products. Para-nonylphenols (4-nonylphenols) are particularly valuable raw material used in the production processes for these products and occur in the form of dozens of isomers differing in the degree of branching of the alkyl fragment of the particle. Commonly used tNP (technical nonylphenol) is a mixture of 30 to even 80–100 isomers and is used as an intermediate product for the production of more complex compounds used in various industries. These xenobiotics, released into the environment, pose a threat to the health and life of many organisms, which is mainly related to their endocrine properties. It is known that their presence in food reduces fertility, causes body deformities and increases animal mortality. In the human body, NPs cause ejaculate volume decrease, premature puberty and thyroid dysfunction.

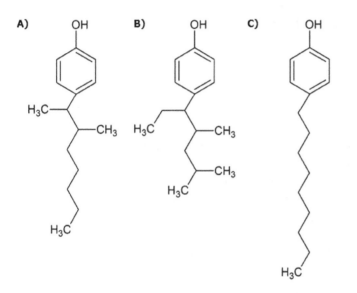

Fig. 5.6.4.1. Chemical structure of selected para- nonylphenol isomers. Explanations:

(A) – 4-(1,2- dimethyl-heptyl)phenol;
(B) – 4-(1- ethyl-2,4-dimethylpentyl)phenol;
(C) – 4-n- nonylphenol (n – not branched).

In the natural environment, NPs are characterised by relatively high resistance to degradation processes and are subject to bioaccumulation. One of the most important ways of NPs elimination from the environment is decomposition involving microorganisms. Biodegradation of these compounds is most often a multistage process and usually takes place with the participation of consortia of microorganisms, which often have a synergic effect. The pathways of microbiological decomposition of NPs have been described in great detail for some of the filamentous fungi (*Metarhizium robertsii, Aspergillus versicolor, Candida aquaetextoris*) and for bacteria belonging to the order Sphingomonadales. However, there are microorganisms that are not only capable of converting NPs to less toxic derivatives, but also of mineralisation, i.e., decomposition of xenobiotics to simple inorganic compounds (CO_2 and H_2O), both under aerobic and anaerobic conditions. In order to demonstrate the ability of the microorganism to mineralise NPs, it is necessary to use ^{14}C labelled compounds, e.g. 4-*n*-NP [ring-U-^{14}C].

Microbiological mineralisation of 4-*n*-NP is described in Section 2.2, while chromatographic analysis of decomposition products using GC-MS is described in Section 6.2.

5.6.5. Simultaneous elimination of organic and inorganic pollutants based on the example of alachlor and zinc

INTRODUCTION

Alachlor is a chloroacetanilide herbicide widely used worldwide to protect corn, cotton, soybean, rice and rapeseed crops from annual mono- and dicotyledonous weeds. The mechanism of action of this group of herbicides is based on inhibition of the VLCFA (very long chain fatty acids) synthesis pathways. Moreover, chloroacetanilides adversely affect the production of certain amino acids (leucine, isoleucine, valine), which in turn leads to disorders of protein biosynthesis.

Despite its effectiveness in destroying weeds and its relatively low toxicity and persistence in the environment, the use of alachlor was banned in European Union countries in the first decade of the 21st century. However, it is still used worldwide, among others in the USA, China and India. The reason for the withdrawal of alachlor from use in Europe was the demonstration of the induction of tumours of a cancerous nature in an animal model. Therefore, it was included in the group of 2B carcinogens and substances as potentially carcinogenic for humans. In addition, it is a hormonal modulator with estrogenic properties, i.e., it can lead to disorders of the hormonal system in animals and humans. It is characterised by relatively good solubility in water (242 mg/l), so it easily penetrates into groundwater and migrates into the aqueous environment. Alachlor is decomposed by soil microorganisms, but is completely mineralised only to a slight degree. Alachlor derivatives, some of them highly toxic, were detected in environmental samples, which additionally contributed to the decision to ban this compound in the European

Union. However, some other chloroacetanilides, such as S-metolachlor, metazachlor and dimetochlor, are still permitted for use elsewhere in Europe.

Landfills, where a mixture of organic toxicants and metal compounds often occurs, are especially dangerous for the environment. Heavy metals have a negative effect on the degradation of toxic organic xenobiotics. They inhibit enzymatic activity of microorganisms involved in their decomposition. Elevated content of heavy metals in agricultural soils may be caused by excessive fertilisation, use of organometalic fungicides such as Maneb (Mn), Zineb (Zn) or inorganic fungicides (Cu).

In special cases, the presence of metallic ions does not limit, but may even stimulate the decomposition of xenobiotics. This is due to an increased adsorption of toxic organic pollutants into the cells of microorganisms. Some heavy metals may exhibit stronger electrostatic attraction to cell shells than phenolic organic pollutants and then the cell surface becomes less hydrophilic, which promotes the adsorption of hydrophobic organic compounds. In addition, metals acting as microelements, e.g. Mn, Zn, Cu, can be cofactors of degradation enzymes and thus stimulate the breakdown of xenobiotics.

Xenobiotics, on the other hand, can increase metal uptake by influencing the metabolic processes that condition the cell wall structure and, consequently, lead to the formation of more functional groups with negative charge. The simultaneous action of xenobiotics and metals may also cause a loss of cell wall integrity and thus increase the surface area involved in metal uptake.

PRACTICAL PART

Materials and media

1. Microorganism: strain of *Metarhizium marquandii* fungus (previously *Paecilomyces marquandii*).
2. Medium: Sabouraud.
3. Apparatus: liquid chromatograph coupled with QTRAP 3200 mass sensor (ESI ion source, electrospray ionisation), atomic absorption spectrometer, confocal microscope.

4. Reagents: alachlor, zinc acetate, concentrated nitric acid, propidine iodide at 0.05 mg/ml, phosphate buffer (PBS) at pH 7.0, ethyl acetate, anhydrous sodium sulphate, methanol.

Aim of the exercise

Evaluation of simultaneous action of zinc and alachlor on the examined strain of the fungus and its ability to degrade herbicide and bind zinc in biomass.

Procedure

1. Establish *M. marquandii* cultures in 100 ml flasks:
   a) with addition of zinc acetate in concentrations of 2.5, 5, 7.5 mM Zn^{2+};
   b) with addition of alachlor (50 mg/l);
   c) with addition of alachlor (50 mg/l) and 2.5, 5, 7.5 mM Zn^{2+}
   d) biotic controls without alachlor and Zn;
   e) abiotic controls, containing medium and alachlor (50 mg/l).
2. After 24, 72, 120 and 168 h, for cultures carried out in the presence of Zn as well as Zn and alachlor, filter through a paper filter, rinse abundantly with deionised water, determine dry matter, mineralise with concentrated nitric acid at 140°C and determine the metal content of the mycelium in mg/g of dry matter – as in Section 5.7. Analysis conditions: oxidising flame: acetylene-air; gap width 1.0 nm; lamp current 5 mA; wavelength 213.9 nm.
3. Parallel cultures with alachlor and alachlor and Zn; homogenise and extract twice with ethyl acetate (collect upper organic layer).
4. Dehydrate the samples with anhydrous sodium sulphate and evaporate in a vacuum evaporator.
5. Then dissolve in 2 ml methanol (to HPLC) and dilute with methanol to obtain concentrations in the range of the standard curve.
6. Determine herbicide residues in cultures and abiotic control using HPLC-MS/MS. The content of alachlor

in the cultures should be expressed in % relative to the abiotic control. Separate the pesticides using the liquid chromatograph Agilent 1200 and the Kinetex C18 column (50 mm × 2.1 mm, 5 µm; Phenomenex, USA) at 37°C, the solvents: water (channel A) and methanol (B) with 5 mM ammonium formate, and then perform the determination using a QTRAP 3200 (Sciex) spectrometer and an ESI-type ion source. Use MRM mode in positive polarisation to monitor the transition for alachlor (270–238).

7. Take 1 ml of culture in the presence of alachlor, Zn, alachlor and Zn, as well as the control system (without metal and alachlor), centrifuge (4,000 × g, 5 min), rinse with 1 ml of phosphate buffer (PBS) and centrifuge again under the conditions as above.

8. Suspend precipitate with 1 ml of PBS buffer with 1.5 µl of propidine iodide at 0.05 mg/ml.

9. Incubate 5 min in the dark.

10. Centrifuge the samples three times, rinsing with 1 ml of PBS solution.

11. Suspend mycelium rinsed from the dye in 0.5 ml PBS.

12. Perform live microscopic preparations and, with the use of a confocal microscope, observe whether the fluorescent dye has penetrated into the hyphae and has stained them red.

13. Take a series of photos and determine the percentage of the area of red-coloured hyphae under the influence of propidine iodide in relation to the entire surface of mycelium. Red colouring of the hyphae under the influence of propidine iodide indicates increased permeability of the fungal membrane. For more information on confocal microscopy see section 2.1.

Analysis of results

On the basis of the results obtained, assess the effect of herbicide on the binding of Zn in the logarithmic and stationary growth phase and the effect of increasing Zn concentrations on the effectiveness of alachlor elimination. Compare (in relation to the control system) the effect of separate and combined action of Zn and alachlor on *M. marquandii* membranes.

5.6.6. Heterocyclic compounds

INTRODUCTION

N-heterocyclic aromatic hydrocarbons such as pyridine, quinoline or carbazole are compounds that have at least one carbon atom in the ring replaced by a nitrogen atom. They are components of coal tar, creosote and oil shale.

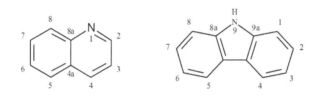

Fig. 5.6.6. Structural formulae of selected N-heterocyclic compounds.

Quinoline   **Carbazole**

Quinoline and carbazole are used in many industries, including the production of dyes, pesticides, medicines and as food additives. The most popular derivatives of quinoline are fluoroquinolone antibiotics (danofloxacin, ciprofloxacin or norfloxacin) which are active against Gram-positive and Gram-negative bacteria, anaerobic microorganisms and *Mycobacterium*. These chemotherapeutics are characterised by high bioavailability after oral administration, good tissue penetration and long half-life. Carbazole derivatives are also highly biologically active, e.g. carprofen has anti-inflammatory properties, ondansetron has antiemetic activity, and carvedilol is used to treat hypertension.

These compounds, compared to their homocyclic analogues, are characterised by higher polarity, better water solubility and higher toxicity. N-heterocyclic aromatic hydrocarbons are difficult to degrade, which combined with their high mobility makes them a common environmental pollutant. In addition, quinoline, carbazole and their derivatives are ecotoxic and can have a negative impact on organisms representing different trophic levels. Studies have also shown that these compounds have mutagenic, genotoxic and embryotoxic properties.

Elimination of N-heterocyclic aromatic hydrocarbons from the environment can be carried out by traditional physicochemical or biological methods, involving the use of living organisms, mainly plants and microorganisms. The process of biodegradation of these compounds occurs under both, aerobic and anaerobic conditions, and its efficiency depends on, among others: individual enzymatic capabilities of microorganisms, availability of oxygen, substrate concentration, temperature, pH of the environment and the presence of other pollutants.

The bacterial degradation of quinoline and carbazole is based on a series of hydroxylation reactions that lead to the breaking of the aromatic ring. Quinoline can be eliminated by four pathways: 7,8-dihydroxyquinoline, 5,6-dihydroxyquinoline, anthranilate and 8-hydroxycoumarin. The first two start with the degradation of the benzene ring, while in the others, the pyridine moiety is initially eliminated.

In the case of carbazole, the degradation process can take place through oxidation of the ring in positions C3 and C4, monohydroxylation in positions C1, C2 or C3 and oxidation of the ring in positions C1 and C9a.

Bacteria of the genera *Pseudomonas, Rhodococcus, Burkhorderia, Sphingomonas, Bacillus* or *Arthrobacter* are the most frequently described microorganisms with the ability to eliminate quinoline and carbazole. In the case of the filamentous fungi, this ability has been demonstrated for the genera *Cunninghamella, Pleurotus* and *Aspergillus*.

PRACTICAL PART

Materials and media

1. Microorganisms: fungal strains (*Cunninghamella echinulata, Curvularia lunta, Trichoderma citrinoviride*).
2. Media: ZT slants, Sabouraud (4% glc), mineral media according to Lobos (2% glc).
3. Reagents: quinoline and carbazole.
4. Apparatus: ball homogeniser, vacuum evaporator, gas chromatograph Agilent GC7890 with mass spectrometer MS5975C, liquid chromatograph Agilent 1200 and mass spectrometer QTRAP 3200 (Sciex).

Aim of the exercise
Determination of the ability of selected fungi to degrade quinoline and carbazole.

Procedure

1. Preparation of cultures containing selected N-heterocyclic compounds:
   a) add 1 ml of inoculum of selected microscopic fungi to 9 ml of Sabouraud or Lobos medium containing quinoline or carbazole at 25, 50 and 100 mg/l;
   b) additionally, prepare biotic controls (containing only the tested microorganism and medium) and abiotic controls (containing only the medium and xenobiotic at the indicated concentration).
   Incubate for 48 h at 28°C.
2. Determination of quinoline and carbazole content and their hydroxylated derivatives:
   a) after the incubation is complete, transfer the samples to 50 ml Falcon tubes containing glass beads for homogenisation and disintegrate for 5 min. Then transfer the entire homogenisate to a separator and extract 2 times with ethyl acetate (10 ml). Dehydrate the extracts with anhydrous sodium sulphate, filter and then evaporate to dryness on a vacuum evaporator (40°C);
   b) Quantitative chromatographic analysis of the extracts: Prepare standard curves of quinoline and its derivatives in the concentration range of 50 to 200 ng/ml; carbazole and its hydroxylated derivatives in the concentration range of 10 to 100 µg/ml, in 1 ml.
   Carbazole determination is performed using a gas chromatograph with attached mass spectrometer (Agilent GC 7890 and MS 5975C), HP 5 MS column (30 m × 0.25 mm × 0.25 µm) using a temperature program ranging from 80 to 300°C and helium flow at 1 ml/min.
   Use the liquid chromatograph Agilent 1200 to determine quinoline. Separate N-heterocyclic compounds using a Kinetex C18 column (50 mm × 2.1 mm, 5 µm; Phenomenex, USA) at 37°C, solvents: water (channel A)

and methanol (B) with 5 mM ammonium formate and a flow rate of 0.6 ml/min, and then perform the determination using a QTRAP 3200 (Sciex) spectrometer equipped with an ESI-type ion source operating in positive polarity mode.

Analysis of results

On the basis of the performed chromatographic analyses, determine the ability of the examined microorganisms to degrade N-heterocyclic compounds

5.6.7. Dyes

INTRODUCTION

Dyes are colour substances which upon contact with materials bind permanently to them, giving them a specific colour. Chemically, they are compounds that selectively absorb electromagnetic radiation in the visible range of 400–700 nm. The dye molecule consists of a chromophore (atoms, ions, chemical groups in which the transition of an electron from the base state to the excited state occurs under the influence of light), which gives the dye molecule its colour and usually occurs in the form of a bond $-N{=}N-$, $-NO_2$, $-CC-$, $-C{=}N-$, $-C{=}O-$ or a quinone ring. In addition, the dye molecule contains auxochromes (electron donating groups of atoms) responsible for the intensity of the colour. Some examples of auxochromes are, $-NH3$, $-OH$, $-SO_3H$ and $-COOH$ groups.

On the basis of the chemical structure and the type of chromophore present in the molecule, the following dyes can be distinguished: anthraquinone, azo, phthalein, indigo, nitro, nitroso, triphenylmethane and others (Table 5.6.7).

In addition, dyes can be classified using other criteria, i.e.: origin (there is a distinction between natural dyes, identical to natural, synthetic organic, synthetic inorganic dyes), or the dyeing method used and the type of forces responsible for this process. According to this criterion, the following dyes are distinguished: direct, vat, acid, reactive, alkaline and suspension dyes.

357

Table 5.6.7. Division of dyes according to the type of chromophore groups

Dyes	Chromophore group	Example of dye
Anthraquinone		Remazol brilliant blue R
Azo	–N=N–	Reactive black B
Phthalein		Fluorescein
Indigo		Indigotine
Nitro	–NO$_2$	Tartrazine
Nitroso	–NO	Pigment green B

Dyes are most commonly used in the paper, cosmetic and textile industries, which in total consume about 50–60% of the total world production of dyes, which is over 7×10^5 tonnes per year. They are also used in the production of paints, plastics and varnishes. Synthetic dyes are also used to monitor the condition of groundwater, to determine the specific surface area of activated sludge and to check the effectiveness of sewage treatment processes. Dyes that are harmless to humans are used in the pharmaceutical and food industries.

The significant use of dyes in various industries, such as the textile industry, leads to their release into the environment in the form of industrial sewage. This poses a serious threat to the biological balance of the entire ecosystem, including animal and human health. The presence of dye gives the water coloration, which results in reduced photosynthesis, leads to eutrophication, interferes with the proper circulation of biogenic elements and directly or indirectly affects all levels of the trophic chain. Coloured sewage poses a potential threat to human health due to the presence of toxic colourants and their degradation products, causing skin and mucous membrane ulcers, severe respiratory tract irritation, nasal septum perforation and others. In addition, dyes reduce the effectiveness of biological sewage treatment by inhibiting the growth and activity of microorganisms, as well as reducing the biodiversity of the microorganisms that make up the flocs of activated sludge and the biological membrane of trickling filters.

Due to the large variety of dyes used in industry, there is no single, universal method of removing these compounds from sewage and contaminated water reservoirs. There are three basic methods of elimination of synthetic dyes: physical, chemical and biological. However, the high costs of using physical and chemical methods and the danger of creating harmful by-products or intermediate products have led to the search for other environmentally friendly biological methods, which use microorganisms (bacteria, filamentous fungi or algae) to eliminate dyes. Especially useful are microorganisms isolated from contaminated environments because they have developed mechanisms allowing the removal of xenobiotics. An increasingly common phenomenon is the ability to decolorize dyes using fungi. White rot fungi e.g. *Phanerochaete chrysosporium, Pleurotus ostreatus* and *Trametes versicolor* strains, are most commonly used due to the presence of enzymatic complexes capable of utilising compounds which are difficult to degrade. Valuable tools include ligninolytic enzymes characterised by low substrate specificity and strong oxidising properties. These fungi also have the ability to decompose carbon bonds in the aromatic rings found in xenobiotics such as organophosphorus compounds, polycyclic aromatic hydrocarbons, dioxins and other pollutants found in sewage. These ligninolytic enzymes include lignin peroxidase (LiP, EC 1.11.1.14), manganese-dependent peroxidase (MnP, EC 1.11.1.13) and laccase (EC 1.10.3.2). Characteristics and possible applications of ligninolytic enzymes are presented in Section 5.10.

In the bioremediation of environments contaminated with toxic dyes, strains of bacteria capable of transforming and mineralising these compounds are also used. High level of pigment elimination was observed in cultures of bacteria from genera: *Pseudomonas, Bacillus, Kurthia, Shewanella, Desulfovibrio, Staphylococcus, Alcaligenes*. The processes of industrial sewage decolouration are carried out mainly using some mixed bacterial or fungal-bacterial cultures.

Microbiological elimination of toxic synthetic dyes is a promising way to remove these compounds from sewage. For this to be possible, it is necessary to search for new strains that effectively eliminate dyes from the environment and carry out a thorough evaluation of the mechanisms responsible for decolorisation while simultaneously assessing the toxicity of the resulting and intermediate products.

PRACTICAL PART

Materials and media

1. Microorganisms: *Myrothecium roridum.*
2. Medium: Czapek-Dox.
3. Reagents: Acid Blue 113 dye solution.
4. Apparatus: centrifuge, spectrophotometer.

Aim of the exercise
Analysis of the course of industrial dyes elimination by filamentous fungi.

Procedure

1. Preparation and performance of *Myrothecium roridum* culture in liquid medium:
   a) add Acid Blue 113 to 18 ml of Czapek-Dox medium to obtain a concentration of 50 mg/l and inoculate with 2 ml of a homogenous second-step preculture (24-h-old) of *M. roridum* screening inoculum;
   b) prepare biotic cultures containing no dye and abiotic controls (without addition of the microorganism);
   c) incubate the cultures for 24, 48 and 72 h on a rotary shaker (140 rpm) at 28°C;
   d) after incubation, centrifuge the cultures (3,500 × g, 15 min, 4°C);
   e) Measure the absorbance of the post-culture liquid at a wavelength of λ = 556 nm and determine the dye decolorization using the formula:

   $$D = [(A_0 - A_t) / A_0] \times 100$$

   where:
   D – Decolorization efficiency (%);
   A_0 – abiotic control absorbance;
   A_t –post-culture liquid absorbance.
2. Dry the mycelium at 105°C to obtain a constant mass.
3. Carry out microscopic analysis of mycelium using a light microscope to determine the contribution of the biosorption mechanism to the dye decoloration process. Indicate the differences in the morphology of the mycelium cultivated in the presence of the dye and without its addition.

Analysis of results

On the basis of the results obtained, determine the degree of decolourization of Acid Blue 113 and indicate the differences in the appearance of *M. roridum* mycelium hyphae in abiotic, biotic culture and the test sample.

5.7. Microbiological elimination of heavy metals from the environment

INTRODUCTION

An increased emission of metals to the environment, recorded over the past decades, is mainly due to human industrial activity. The anthropogenic sources of pollution of the atmosphere, water and soil include combustion processes (in power plants and combined heat and power plants), road transport, mining and the metallurgical industry. The metals present in industrial sewage come mainly from galvanising plants, tanneries, plants producing artificial fertilisers, pesticides, paints and from the electrical and mining industry. On the other hand, metals present in municipal sewage come from e.g. corrosion of pipes, detergents and rainwater.

Heavy metals are metallic elements with a density of more than $4.5–5$ g/cm^3, characterised by high toxicity to living organisms. This toxicity is manifested by the inactivation of enzymes, displacement of metals that perform functions in metabolic processes, inducing conformational changes in polymers and disorders of cellular transport. The mechanism of toxicity of some metals, which easily change the degree of oxidation (Cu, Ni, Cr), consists of inducing oxidative stress. As a result, the structure and function of membranes is disturbed by peroxidation of proteins and membrane lipids. Oxidative damage may impair the function of other structural proteins and nucleic acids. Heavy metals such as cadmium, nickel, chromium, manganese and beryllium have mutagenic and carcinogenic properties. In addition, some metals (e.g. Hg, Cd, Pb) belong to the group of Endocrine Disrupting Compounds (EDCs) and can disrupt the endocrine balance of humans and animals. At the same time, ions of some heavy metals, e.g. Cu, Zn, Mn, Co, are essential trace elements, necessary for proper functioning of cells and organisms. Both, their excess and deficiency can be harmful to humans. However, metals such as Cd, Pb, Cd, Ag and Au do not perform any function in cellular metabolism.

The toxic effects of the metal or its bioavailability are determined not only by its amount in the environment, but also by the form in which it occurs (soluble: ionic, complex or insoluble). The pH of the environment has the greatest influence on solubility, and thus also on the toxicity of metals. In an acidic environment, metals in the form of cations (most metals) have higher solubility and are therefore more bioavailable as microelements at low concentrations or become toxic when the metal concentration is high.

Intensive research is being carried out to develop biotechnological methods of metal removal and recovery from sewage, that are alternative to traditional physicochemical methods (ion exchange reactions, redox, precipitation). Biotechnological methods are based on the use of microbial interactions with heavy metals. Heavy metals are present in the environment in both soluble and hardly soluble form. Microorganisms, taking part in the cycling of metals in nature, can contribute both to immobilisation of metals (sorption, accumulation) and dissolution of metal compounds (leaching). Biotransformation processes of metals consist of changing the degree of oxidation, which in turn affects their solubility, bioavailability and toxicity of metals and other elements occurring in the form of inorganic compounds.

Biosorption is of the greatest practical importance for the recovery and removal of soluble metal salts. It consists of binding metals by external shells of the microorganism. The mechanism of biosorption is complex. It involves ion exchange reactions, binding between metal cations and anionic groups present in cell shells, as well as complexation and precipitation reactions. Moreover, physical interactions (van der Waals forces) may be responsible for biosorption. Biosorption is a passive process which does not require metabolic activity of cells. This creates the possibility of using microbial waste biomass from other processes, which significantly reduces costs. The second advantage of biosorption is the fact that it is a fast process. More than 90% of the metal is bound in the biomass in a few min.

The effectiveness of metal biosorption with the participation of microorganisms is greatly influenced by the construction of external shells, and more specifically – the number of active negative-charged sites. In Gram-positive bacteria there is a thick layer of peptidoglycan in the wall, with numerous glutamic acid carboxylic groups. In addition, phosphate groups of teichoic acids take part in metal bonding. The results of numerous experimental studies show that usually Gram-positive bacteria are better biosorbents than Gram-negative bacteria, in which the peptidoglycan layer is much thinner.

Chitin and chitosan, polymers that are part of the cell wall of filamentous fungi, are responsible for the effective binding of metals in this group of microorganisms. Negative-charged active groups in the microbial biomass are quickly saturated with metals. A single process using biomass leads to a fast but not too high metal accumulation per amount of biomass. It is of great practical importance that biosorption is a reversible process, i.e., the metal bound in the biomass can be desorbed through desorption and, depending on its economic viability, more valuable metals (Cu, Zn, Ni) can be recovered or toxic metals such as Cd, Pb, Hg can be safely removed. Three types of desorbents are used for desorption: acidic, alkaline and complexing compounds. Diluted solutions of mineral acids are used as acidic desorbents. They are very effective in leaching metals, but their use requires periodic regeneration of biomass. Alkaline desorbents: carbonates and hydrogen carbonates and complexing compounds, e.g. EDTA, do not have a destructive effect on the biomass and are used successfully in many cases, e.g. for desorption of uranium. Practical removal or recovery of metals with the use of biosorption is based on repeated sorption and desorption cycles, combined with biomass regeneration and metal recovery from leachate. Biosorption is most likely to be used in sewage treatment in the case of large volumes of sewage containing a metal charge of less than 100 mg/l, when the use of traditional physicochemical methods is too expensive, as well as for the pre-treatment of industrial sewage with high metal content directed to biological sewage treatment plants, using excessive activated sludge from this plant.

In addition to the process of passive biosorption, the removal of metals from sewage can use microbial interactions related to their metabolic activity (biotransformation, accumulation). Some microorganisms are capable of producing metabolites that react with metals outside the cell: organic acids, amino acids and polysaccharides and siderophores. These metabolites can complex metals or cause their precipitation by changing the environment. Biotransformation of metals consists of oxidation, reduction or methylation. For practical reasons, metal reduction reactions leading to metallic precipitation are promising, e.g. removal of mercury from sewage.

Intracellular metal accumulation is often perceived as the second stage of metal uptake, following biosorption. This leads to a slower but higher metal accumulation than biosorption. Metal is taken inside the cell, with the involvement of transport systems, and is bound in the biomass in a permanent way – it can only be recovered by breaking the cell structure.

The metals, which belong to non-renewable raw materials, are obtained from ores of different composition, metal content and deposit location. Bioleaching of metals is based on the ability of microorganisms to dissolve metals found in hard to dissolve ores, concentrates or industrial waste. Leaching with the participation of bacteria comes down to the process of converting sulphides into soluble metal sulphates, with the resulting sulfuric acid and iron sulphate. Heterotrophic leaching of metals by filamentous fungi can lead to leaching of metals from non-sulphide minerals by organic acids and complexing and chelating compounds.

As the world's reserves of rich metal deposits are increasingly depleted, it is becoming more and more important to obtain economically viable metals from poor, scattered ores or residues from rich deposits. This is made possible by biohydrometallurgical processes. Biohydrometallurgy is based on the ability of a group of acidophilic, chemo-autotrophic bacteria, mainly of the genus *Acidithiobacillus*, to oxidise inorganic sulphur and/or iron compounds, thanks to which, the metals present in sulphide and oxide ores are dissolved. In a nutshell, the role of microorganisms comes down to catalysing the reaction of sulfuric acid and iron sulphate formation, and the strong oxidants formed by biological means dissolve metals chemically. Thanks to biological extraction of the leaching solution, it is economically viable to leach Cu from ores containing less than 0.5% of this metal. About 20% of global Cu mining (Chile, USA, China, Canada, Mexico, Spain, Portugal, Australia) comes from bioleaching. Biohydrometallurgical methods have also found practical application in obtaining uranium and gold. It is also possible to bioleach other metals such as Zn, Ni, Pb, As, Sb, Bi, V and Mo. In the case of these metals, these are both attempts to recover and remove them from industrial waste or sewage sludge.

Siderophores are low-molecular compounds (200–2,000 Da) of a non-protein and non-porphyrin character, produced especially in iron deficiency conditions. Iron (Fe) is a microelement necessary for the proper growth and development of all living organisms, because it acts as a catalyst for enzymatic processes, oxygen metabolism, electron transport and also as a catalyst in the synthesis of DNA and RNA. Due to the low bioavailability of iron in the environment, microorganisms and higher organisms have developed mechanisms that enable the binding of Fe ions, among others, by producing siderophores. Siderophores enable the acquisition of Fe from insoluble sources, which facilitates survival in environments with low levels of this element.

Due to the diverse chemical structure of siderophores, they are divided into 3 classes: hydroxamic acids, catecholamines and carboxylates. The basic role of siderophores is to bind Fe^{3+} ions, but they can also form complexes with bivalent ions: Cd^{2+}, Cu^{2+}, Ni^{2+}, Pb^{2+}, Zn^{2+}, trivalent ions: Mn^{3+}, Co^{3+}, Al^{3+} and actinide ions: Th^{4+}, U^{4+}, Pu^{4+}, thus increasing the availability of these elements for living organisms. Siderophores of microbiological and plant origin (phytosiderophores) are also extremely effective in increasing the mobility and solubility of metals such as Zn, Pb, Cu, Cd, Ni, as well as actinides such as Th, U and Pu.

Chelation of heavy metals by siderophores allows the use of these compounds in environmental bioremediation. It has been shown that siderophores produced by *Pseudomonas aeruginosa* can complex e.g. Cr^{2+}, Co^{2+}, Cu^{2+}, Hg^{2+}, Al^{3+}, Ni^{2+}, Pb^{2+}, Zn^{2+}. Siderophores produced by *Agrobacterium radiobacter* participate in the elimination of As from areas contaminated with this element. In the case of phytosiderophores, it was found that they are effective in binding Fe, Cu, Zn, Ni and Cd. Siderophores can also increase the rate of degradation of petroleum hydrocarbons, facilitating the uptake of Fe ions by microorganisms, e.g. the petrobactin produced by *Marinobacter hydrocarbonoclasticus* or the ochrobactin produced by *Vibrio* sp. Siderophores, by complexing heavy metal ions, reduce both their toxicity in the environment and the level of oxidative stress in the cells of microorganisms and plants living in metal contaminated environments.

Moreover, siderophores of microbiological origin support the growth of plants by providing them with Fe^{3+} ions. It has been shown that bacteria such as *Pseudomonas* sp., *Bacillus* sp., as well as fungi such as *Trichoderma* sp., *Aspergillus niger*, *Penicillium citrinum*, which produce siderophores, stimulate plant growth. These chelators are also used in biological plant protection by reducing the availability of Fe for phytopathogens, e.g. *Fusarium oxysporum* and *Gaeumannomyces graminis*.

Popular method of quantitative siderophore analysis uses the complex of Chromazurol S (CAS) dye and hexadecyltrimethylammonium bromide (HTDMA) as an indicator. The CAS-HTDMA complex, which binds to iron ions, forms a dark blue solution. The principle of this method is to transfer Fe3+ ions from the dye complex to siderophores, accompanied by a change of colour from blue to orange. Detection of siderophores is performed on a substrate poor in Fe ions.

PRACTICAL PART

Materials and media

1. Microorganisms: selected filamentous fungi, e.g. *Metarhizium marquandii*, *Trichoderma* sp., which are characterised by high tolerance to heavy metal compounds, ability to bind them and/or produce siderophores.
2. Media: ZT slants, Sabouraud, mineral with glucose (2%) without Fe.
3. Reagents: $CuSO_4$, copper standard solution in 2% HNO_3, 0.1 M HCl, 2 mM aqueous solution of Chromazurol S (CAS), 5 mM aqueous solution of hexadecyltrimethylammonium bromide (HTDMA), 1 mM aqueous solution of $FeCl_3$ ($FeCl_3 \times 6 H_2O$, in 10 mM HCl). Immediately prior to use, mix 50 ml of 2 mM CAS solution, 40 ml of 5 mM HTDMA solution and 10 ml of 1 mM $FeCl_3 \times 6 H_2O$ solution to obtain a mixture for siderophore determination (CAS mix).
4. Apparatus: atomic absorption spectrometer, thermal mineraliser, centrifuge, spectrophotometer.

Aim of the exercise

Comparison of tolerance of selected strains of microfungi to Cu and the binding of Cu during growth and starvation. Familiarisation with the method of preparation of mycelium samples for analysis (mineralisation) and quantitative determination of heavy metals by atomic absorption spectrometry (AAS). Quantitative determination of siderophores in the cultures of filamentous fungi.

Procedure

1. Determination of strain tolerance to different concentrations of $CuSO_4$ in the range: 0–10 mM:
   a) rinse the spores from the 7-day-old culture on ZT slants with 7 ml of Sabouraud medium into a conical flask with 15 ml of this medium. Carry out the culture on a rotary shaker (160 rpm) at 28°C for 24 h. Transfer the pre-culture to a conical flask with a new portion of medium (10–20% inoculum depending on the strain) and incubate for the next 24 h under conditions as above;
   b) add the appropriate volume of Sabouraud medium, a metal solution with a starting concentration of 100 mM (with an automatic pipette) and a homogeneous 24 h culture as an inoculum (10% v/v) to 100 ml flasks to give a final volume of 20 ml;
   c) incubate for 96 h on a rotary shaker at 28°C;
   d) describe the growth of fungal strains. Filter the cultures through tissue paper filters and rinse the mycelium with deionised water. Determine the dry mass of the mycelium and determine the growth of the tested strains in the presence of different concentrations of metal, and in relation to the control system without the addition of metal.
2. Evaluation of the ability of selected fungal strains to absorb metals through live and thermally inactivated biomass:
   a) multiply the mycelium on Sabouraud medium under the conditions given in point 1a;
   b) filter 15 ml of the culture, rinsing twice with deionised water in order to wash the biomass from the medium;
   c) half a portion of mycelium should be thermally inactivated for 15 min at 100°C.;

d) transfer live and dead biomass to 50 ml flasks and suspend in 1 mM aqueous $CuSO_4$ solution;

e) incubate the samples for 24 h on a shaker;

f) then filter through a tissue paper filter, rinsing twice with deionised water, dry at 105°C to constant mass and determine dry mass;

g) transfer the dried biomass to mineralisation test tubes;

h) using an apron, gloves and protective glasses, add concentrated nitric acid under the hood (according to the ratio of 10 ml of acid to 0.5 g of biomass) and incubate in the mineraliser 4–5 h at room temperature, 1 h at 100°C, and then another 4–5 h at 140°C;

i) after cooling down, transfer the contents of the mineralisation tubes by volume into polypropylene bottles;

j) Determine the Cu content on the atomic absorption spectrometer (Spectra 240 FS, Agilent) using the flame technique. Analysis conditions: oxidising flame: acetylene-air; standard curve range up to 4.0 µg/ml; gap 0.5 nm, lamp current 4 mA, wavelength 324.7 nm. Before determination, take into account the range of the standard curve (if necessary, dilute it additionally);

k) determine the metal binding in mycelium in mg/g of dry matter. Compare the binding of Cu by live and dead biomass.

3. Testing the ability of selected filamentous fungi to remove heavy metals during culture on medium:

a) establish 100 ml of culture in 300 ml flasks with copper sulphate (at the concentration chosen from the results of the tolerance test) and 100 ml of control culture under the conditions given in point 1a;

b) after 4–5 days, filter the cultures carried out in the presence of metals and control cultures through tissue paper filters and wash twice with deionised water; divide into several portions;

c) take 100 mg of moist biomass, transfer to the weighed weighing bottle, determine the dry mass, mineralise and determine the metal content of the entire mycelium;

d) suspend 100 mg of moist biomass in 20 ml of 0.1 M HCl and incubate for 30 min at 28°C on a rotary shaker (160 rpm); then filter and rinse twice with deionised water, transfer the biomass from the filter to a weighing bottle and dry to constant weight at 105°C.;

e) mineralise and determine the concentration of copper ions in the biomass;

f) determine the concentration of metal permanently bound to the mycelium (i.e., not desorbed with hydrochloric acid).

4. Determination of siderophore activity:

a) rinse the spores from 7-day cultures of selected fungal strains on ZT slants with 7 ml of mineral medium with added glucose but without iron and filter through a funnel with sterile glass wool into a 10 ml conical flask with this medium;

b) culture on a rotary shaker (160 rpm) at 28°C for 24 h;

c) transfer the pre-culture to a new 20 ml conical flask with medium and incubate for another 24 h;

d) inoculate 2 ml of homogeneous pre-culture to 18 ml of medium (control system) and 18 ml of medium with added metal at a concentration selected from previous results;

e) culture for 7 days on a rotary shaker (160 rpm) at 28°C;

f) take 1 ml of the culture from a 24 h and 168 h cultures, into an Eppendorf tube and centrifuge (10,000 × g, 10 min);

g) add 0.5 ml of CAS mix solution to 0.5 ml of post-culture supernatant and incubate for 15 min;

h) then measure the absorbance $\lambda = 630$ nm and calculate the siderophore content according to the formula:

% siderophore content = $[(Ar - As)/Ar] \times 100$

where:

A_R – absorbance of the reference sample (CAS mix solution);

As – absorbance of the test sample.

Analysis of results

Based on the results obtained and the literature, assess the tolerance of the tested strains to copper compounds. Determine whether the metal is better bound under starvation conditions (after washing off the substrate) or during cultivation. Assess also, the sorption capacity of living and dead biomass and check what kind of binding dominates (reversible, i.e. metal is desorbed from mycelium with 0.1 M HCl, or permanent, i.e. metal is not washed out from biomass). Moreover, compare the ability of different strains to produce siderophores and check whether the addition of copper to the culture stimulates the production of these chelators.

5.8. Detoxification processes of polluted environments. Ecotoxicological tests

INTRODUCTION

Ecotoxicology is the study of the effects of toxic substances or their mixtures on organisms, populations or entire ecosystems by analyzing various life parameters of organisms under natural conditions. The purpose of ecotoxicological studies is disclosure and prediction of the effects of pollution in the context of all environmental factors.

Ecotoxicology is a multidisciplinary field that integrates two disciplines: toxicology and ecology, covering many sub-fields. Therefore, traditional, physicochemical analyses of toxicity assessment are complemented by ecotoxicological studies based on the use of biotests, in which the level of contamination is assessed on the basis of the specific, overall reaction of indicator organisms (bioindicators) to all biologically active components contained in the test sample. Such tests therefore provide comprehensive information on the occurrence, pathways, accumulation, impact or distribution of toxic substances, and thus enable the assessment of the real degree of risk.

Ecotoxicological tests using the biological response of organisms to measure environmental impacts of pollutants.

Apart from providing a more direct measure of toxicity than physicochemical methods, they also have many other benefits, such as:

- sensitivity of bioindicators to a wide range of pollutants;
- repeatability of results;
- small scale;
- effectiveness (cost);
- wide availability of indicator organisms and test materials;
- generally do not require expensive and specialised equipment;
- standardised procedures available, which can be carried out by any person without any specialist qualifications;
- low time and labour input;
- ability to store indicator organisms without loss of sensitivity and to use them at any time ("on request");
- more effective monitoring of places with difficult to define contamination.

Biotests are used to assess the risk of toxicity of tested systems, in which the object of analysis may be the whole organism or a single organ of the organism, cell culture or enzymatic reaction. Harmful effects of substances or environmental samples are assessed in these tests on the basis of measurable biological end-effects, which respond to exposure to all active components of the test sample. The release of the bioindication response under test therefore depends on the dose or concentration of the substance to which the test object is exposed and this relationship, defined as 'dose-response', forms the basis for ecotoxicological studies. Bioindicator methods using indicator organisms are based on the analysis of the symptoms of living and lethal bioindicators, which react to the presence of toxic substances at the cellular, subcellular or molecular level. In standard toxicity tests, indicator organisms give a typical bioindicator response in the form of physiological (e.g. inhibition of food intake, population size), enzymatic (e.g. inhibition of bioluminescence) or behavioural (e.g. interactions between organisms) reactions.

Ecotoxicity studies use organisms belonging to different ecosystems and representing all links in the food chain (Table

5.8.1). Due to the different sensitivity range of individual bioindicators, so-called test batteries are used in the analyses, i.e., sets containing mostly 4–5 indicator organisms belonging to different trophic levels. The use of "test batteries" allows comprehensive assessment of the quality of the sample as well as to identify the point of interruption of the food chain under the influence of toxic substances. The role of biological indicators is played by organisms with specific characteristics that allow assessment of the risk resulting from contamination with toxic compounds. The desired characteristics of bioindicators are primarily as follows:

- high sensitivity and quick response to environmental changes;
- easy to culture and maintain under laboratory conditions;
- measurable and accurate to determine, bioindication response in a short time;
- specific sensitivity and a specific tolerance range for a particular chemical or group of pollutants;
- significant role (representative species) for a given ecosystem.

The toxic properties of a test substance or environmental matrix are determined by the measurement of such parameters as: acute toxicity, chronic toxicity, bioaccumulation, mutagenicity and persistence. Ecotoxicological tests are usually based on the determination of acute toxicity (so-called short-term), observed up to 96 h after exposure to the test sample, and chronic toxicity (so-called chronic), determined over a longer period of time and usually after repeated exposure of bioindicators to a harmful compound. The result of the determinations is usually given: 1) for acute toxicity testing: LD (lethal dose), LC (lethal concentration), EC (effective concentration) or IC (inhibitory concentration), causing the tested response in a specific number of indicator organisms in the tested population (usually 50%); in determining the chronic toxicity value: NOEC – the highest concentration of the toxic substance causing the indicator response and LOEC – the lowest concentration of the toxic substance at which no adverse changes are observed in the test organisms.

373

Areas where ecotoxicological studies using biotests have found application, include:

- toxicity assessment of rivers, lakes, coastal waters, rainwater;
- determination of toxicity of untreated sewage;
- monitoring of treated industrial and municipal sewage;
- toxicity assessment of bottom sediments;
- testing the fate and effects of certain substances in the environment;
- analysis of the impact of agricultural and industrial activities on the quality of surface and ground water;
- determination of toxicity of defined chemical compounds such as pesticides, antibiotics, polychlorinated biphenyls, dyes, preservatives, etc.;
- assessment of heavy metals toxicity and mobility;
- soil monitoring;
- drinking water intakes monitoring.

The tests based on bioindication have also been used in analyses of bioremediation processes of contaminated environments, including the assessment of environmental detoxification during the microbiological decomposition of xenobiotics. Biological degradation processes can produce intermediate products of higher toxicity than the parent compound. Therefore, it is important to determine the toxic impact and risk not only of the parent compound itself, but also of the intermediate products of its degradation. The use of ecotoxicological tests results from the necessity of continuous monitoring of contaminated areas and estimating the degree of existing danger to the functioning of entire ecosystems.

Table 5.8.1. Examples of organisms used in ecotoxicological tests

Organisms	Test criterion (indicative response)	Species
Producers		
Vascular plants	Root growth inhibition	Lactuca sativa
	Number of leaves, biomass	Lemna minor
Algae	Growth inhibition	Phaeodactylum tricornutum
	Chlorophyll concentration	Chlamydomonas reinhardtii
	Growth inhibition	Raphidocelis subcapitata

Organisms	Test criterion (indicative response)	Species
Consumers		
Rotifers	Mortality	*Brachionus plicatilis*
Crustaceans	Mortality	*Artemia franciscana*
	Mortality, motility	*Daphnia magna*
Insects	Length of specimens, number of specimens	*Hexagenia bilineata*
Annelids	Motility	*Hirudo medicinalis*
Fish	Length of specimens, liver mass	*Lepomis macrochirus*
Destruents		
Bacteria	Bioluminescence measurement	*Aliivibrio fischeri*
	Growth inhibition	*Pseudomonas putida*
Yeast	Growth inhibition	*Candida boidinii*

PRACTICAL PART

Materials and media

1. Organisms: microfungus strain *A. versicolor* IM 2161, bacterial strain *P. putida*, daphnia (*D. magna*) cysts (MicroBioTests).
2. Commercial Daphtoxkit F magna (MicroBioTests) kit for crustacean testing (ISO 6341 compliant).
3. Materials, media and reagents contained in ISO 10712 for *P. putida* determination.
4. Media: ZT slants, Sabouraud liquid medium.
5. Reagents: 4-nonylphenol.
6. Apparatus: incubator with lighting, turbidity metre, vacuum filtering set, rotary shaker.
7. Other materials: sterile cellulose acetate filters with 0.45 μm pore diameter, serological pipettes, Pasteur pipettes, Erlenmeyer flask (100 ml), Thoma counting chamber.

Aim of the exercise
Evaluation of toxicity changes in the culture of Aspergillus versicolor with 4-nonylphenol based on toxicological tests with Daphnia magna and Pseudomonas putida.

375

Procedure

1. Preparation of the inoculum.
   Rinse the spores from a 7–10 day culture of *A. versicolor* IM 2161 on ZT slants with 6 ml of Sabouraud medium and then calculate the number of conidia in the obtained suspension using the Thoma chamber. Prepare culture in Erlenmeyer flasks (100 ml) in a total volume of 25 ml of Sabouraud medium with an initial spore density of 5×10^7/ml. Incubate for 24 h on a rotary shaker (160 rpm) at 28°C.

2. *A. versicolor* culture in the presence of 4-nonylphenol.
   Inculcate 28 ml of Sabouraud medium with 2 ml of the obtained 24 h fungal inoculum and then add 4-nonylphenol substrate until its final concentration in culture is 25 mg/ml. Prepare xenobiotic solution earlier in 96% ethanol. Similarly perform abiotic controls and cultures without 4-nonylphenol (biotic tests). Incubate the cultures up to 72 h on a rotary shaker (160 rpm) at 28°C.

3. Obtaining post-culture filtrates.
   Carry out a sterile filter of the test culture systems (biotic, abiotic and xenobiotic controls) after 24, 48 and 72 h of incubation through 0.45 μm pore diameter filters using a vacuum filter kit.

4. Toxicity tests.
   The tests should be conducted in accordance with the methodology described, i.e., ISO 10712 for testing with *P. putida* and the procedure provided by the manufacturer for analysis with *D. magna* (Daphtoxkit F magna test). Prior to testing, prepare a series of dilutions of the previously obtained filtrates from each time point of culture (preferably geometrically advanced) in dedicated media. In experiments with crustaceans, the range of dilutions should be chosen so as to cause mortality of the test population in the range from 0 to 100%.

Analysis of results

After incubation of bacteria and daphnia with individual dilutions of cultured filtrates, observe: 1) mortality of *D. magna* and 2) growth inhibition of *P. putida* based on measurements of culture turbidity. Taking into account the results obtained for the control samples, the results of xenobiotic culture should be presented in the form of LC50 values for *D. magna* and EC50 values for *P. putida* determined by graphical interpolation on a linear and logarithmic scale and then transformed into TU (toxic units) according to the formula: TU = [1/LC50 or EC50] × 100. Based on TU values, evaluate toxicity changes during *A. versicolor* strain incubation with 4-nonylphenol and give potential reasons for detoxification or increase in toxicity of the tested culture systems.

5.9. Use of industrial waste in microbiological biotechnology

INTRODUCTION

One of the reasons for the progressing degradation of the environment is the increase in the production of waste (both municipal and from various industries). This results from the implementation of a linear model of economy, which assumes taking raw materials from the environment and is accompanied by the production of products of low durability and waste introduced back into the environment. Since the application of such a model has led to significant pollution of the natural environment, the principles of the new model of economy are now being introduced, which is referred to as the circular economy (CE). The above model assumes the production of products with the longest possible durability and the use of the resulting waste (especially bio-waste) as raw materials for the production of new material goods. The European Commission announced in 2015 the Communication Closing the loop – an EU action plan for the circular economy. It recommends actions to the Member States to reduce bio-waste storage and to implement selective waste collection systems. In connexion with the planned implementation of a closed-loop economy, intensive research has been started on the development and implementation of innovative, energy-efficient technologies and new consumption styles of goods and services. An important part of this model is the bio-economy, enabling reuse of organic waste originating mainly from agriculture, food processing industry, forestry, fishing, households and sewage treatment plants. One of the ways of managing the above-mentioned wastes is to use them as components of traditional microbiological media. In Poland, waste or by-products used as raw materials include molasses, whey, oil cake, waste yeast, draff, fruit marc, glycerol, potato fruit juice and potato pulp, as well as lignin-cellulose waste. The characteristics of selected wastes and directions of their current and potential use are presented below.

Glycerol is obtained during the production of diesel fuels, such as Biodiesel or BIOESTER 100. Diesel fuels are produced from various oily raw materials, such as rapeseed, soybean or oil palm. The directions of glycerol development as a component of microbiological media include production of:

- 1,3-propanediol using *Klebsiella pneumoniae, Clostridium butyricum, Citrobacter freundii* or *Enterobacter agglomerans* cultures;
- succinic acid by strains of *Basfia succiniciproducens* bacteria or *Yarrowia lipolytica* yeast;
- dihydroxyacetone by acetic acid bacteria, e.g. *Gluconobacter oxydans* strains;
- feed yeast biomass.

Potato fruit juice is waste produced during potato processing. 600 tonnes of fruit juice is generated during the processing of 1,000 tonnes of potatoes. The above waste after de-proteination is still a rich source of nutrients and contains, among others, protein compounds (about 1.4%), reducing sugars (about 0.5–1.0%), biotin, organic acids (mainly oxalic, citric and malic acid), as well as microelements (including K, Mg, Ca, P, Na and Zn ions). Potato fruit juice is characterised by very high COD and BOD_5 values (about 30,000 mg O_2 and about 22,000 mg O_2, respectively). Further use of deproteinated potato fruit juice in biotechnological processes allows not only a reduction of COD and BOD_5 values by as much as 60–90%, but also results in products of high market value.

Draff is a waste obtained after wort preparation. It contains non-solidified starch, non-hydrolysed protein and husks of seed from which the malt was prepared. The draff is a very perishable waste due to its high water content and compounds that decompose quickly by fermentation. The main direction of utilising this waste is the production of feed silage. When concentrated whey is added to the draff and lactic acid bacteria are cultured (e.g. *Lactobacillus delbrueckii* subsp. *bulgaricus*), the pH is quickly lowered and the waste stabilises. Often the yeast biomass is also added to fixed draff.

Whey is obtained during the production of cheese, cottage cheese, casein (and other products containing milk proteins). There are two types of whey. Rennet whey (sweet) is produced during the secretion of milk proteins using the enzyme – rennet (complex of enzymes with chymosin as a cake component). Acidic (sour) whey is obtained by coagulation of milk proteins in an acidic pH (most often caused by the presence of lactic acid). The individual components of whey (constituting from 4.5 to 7.3% of dry matter) include: 0.02–0.4% fat, 0.4–1.1% protein and 4–5% lactose. It is estimated that cheese production in the European Union is about 9 million tonnes per year, which generates whey production of about 50 million m³. In Poland, annual whey production reaches 2–3 million m³. Whey poses a very serious threat to the environment which results from a very high level of BOD_5, generated by a high content of biodegradable organic compounds (mainly lactose). If the whey entered water reservoirs, it could possibly destroy the aquatic ecosystem. Important factors that favour the use of whey in biotechnological processes include: the content of B vitamins (B_2, B_5, B_{12}), microelements (mainly easily assimilated Ca and P ions), lactose and lactic acid. Nowadays, the use of technologies based on membrane processes, such as reverse osmosis and ultrafiltration, allows separation of the components of whey and produce e.g. whey powders, whey protein concentrates and lactose preparations from this waste. Lactose is a good source of energy for many lactic acid bacteria (mainly from the genera: *Lactobacillus, Lactococcus* and *Leuconostoc*), participating in the conversion of the above-mentioned disaccharide to lactic acid during lactic acid fermentation. In addition, lactose hydrolysis products (glucose and galactose) are more easily assimilated by a larger number of microorganisms, which improve opportunities of whey application in biotechnological processes. Currently developed methods of whey management include production of protein dietary supplements, products included in functional food (e.g. high-protein supplements for athletes), animal feed, microbial biomass, organic acids, amino acids, enzymes and dyes.

Oil cake is a by-product obtained after oil extraction from seeds (in some cases fruits) of oil plants. The most important is sunflower, rapeseed, sesame, soya, coconut, palm and mustard cake. The oil cake produced during pressing of oil from cotton and olive seeds is also obtained on a large scale. Rape and sunflower seeds are among the main sources of vegetable oils in the world. The largest producers of food oils (and thus oil cake) are the European Union, China, Canada, the United States, India, Brazil, Mexico and Russia. In Poland, rapeseed oil is the main product. Detailed and constantly updated statistical data describing the level of production of various products and waste from the agri-food industry in particular countries are published on the IndexMundi pages in the agriculture section (https://www.indexmundi.com/agriculture). Oil cakes are divided into two groups: 1) edible (with protein content above 15%) and 2) inedible. The chemical composition of oil cakes are variable and depend on many factors (Table 5.9.1).

Table 5.9.1.
Composition of rapeseed oil cake*

Component	Content [%]
Moisture content	7.2–9.8
Ash	6.1–11.7
Protein	33.9–36.0
Lignin-cellulose compounds (fire): cellulose, hemicellulose, lignin	11.2–13.9
Microelements (Ca, P, Fe, Mn, Zn, K, Mg)	0.3–1.3
Lipids	1.7–3.8

* The content of individual components depends on, among others, the variety of rape, cultivation conditions (climate, soil type), weather conditions, oil extraction methods used and the method of storage.

So far, the methods developed allow oil cakes to be used as:
- feed additives for farmed animals;
- components of microbiological media used for cultivation of microorganisms (submerged and solid state fermentation);
- substrates for the production of edible fungi (Table 5.9.2);
- soil fertiliser – generally, inedible cakes are used for increasing the content of organic nitrogen compounds available for plants, as a source of microelements and positively influencing the soil structure;
- energy source (as fuel and substrate for biogas production);
- starting material for chemical production of polyols.

Table 5.9.2. Directions of biotechnological management of oil cakes

Products	Microorganism
Enzymes: proteases lipases phytases α- amylases glucocoamylases inulinases L- glutaminase mannanase xylanase cellulases and hemicellulases	*Bacillus* sp., *Penicillium* sp., *Aspergillus* sp. *Penicillium* sp., *Rhizomucor* sp., *Aspergillus* sp. *Aspergillus* sp., *Rhizopus* sp. *A. oryzae* *A. niger Staphylococcus* sp. *Zygosaccharomyces rouxii* *A. niger* *Trichoderma reesei, Trametes versicolor*
Antibiotics: cefamycin C bacitracin	*Streptomyces clavuligerus* *Bacillus licheniformis*
Biopesticides: δ-endotoxins	*Bacillus thuringiensis*
Biosurfactants: iturin	*Bacillus* sp.
Clavulanic acid (β-lactamase inhibitor)	*Streptomyces clavuligerus*
Lactic acid	*Lactobacillus casei*
Polyunsaturated fatty acids	*Mortierella alpina*
Biomass of edible fungi	*Pleurotus ostreatus, P. sajorcaju*

PRACTICAL PART

Materials and media

1. Microorganisms: *S. cerevisiae, Myrothecium roridum* (or other micromycetes capable of overproduction of extracellular laccase), strains of *Bacillus* (IM 13, I'-1a, KP7, DSM 3257 or other bacterial strains producing surfactants, so-called biosurfactants, whose characteristics are given in Section 4.1.2).

2. Media: Sabouraud, LB, wort (6°Blg); and media prepared from whey, rapeseed oil cake, carrot peels supplemented with 0.2% tryptone.

3. Reagents: 20% aqueous solution of tryptone, Loeffler Methylene Blue, Lugol's iodine, ABTS (diammonium 2,2'-azobis(3-ethylbenzothiazoline-6-sulfonate), kerosene, 5% aqueous solution of sodium dodecyl sulphate (SDS), synthetic dyes, e.g. Acid Orange 7 (AO7) and Indigo Carmine (IC).

4. Apparatus: spectrophotometer, polypropylene Petri dishes, tubes (with a 20 ml scale) and caps for tubes.

Aim of the exercise

Comparison of yeast biomass and biological surfactants production in microbial cultures with the use of standard microbial media and by-products of the agri-food industry. Assessment of the suitability of the above-mentioned media for the production of laccase.

Procedure

1. Describe selected features (colour, clarity, smell, pH-value) of the media: Sabouraud (control, defined composition), wort (6°Blg) and media prepared from whey, oil cake and carrot peels.

2. Compare the production of *S. cerevisiae* yeast biomass in flask cultures*, using the media mentioned in point 1 above, using for the evaluation:
   a) growth intensity – OD measurement;
   b) cell survival – Loeffler Methylene Blue staining;
   c) glycogen content in cells – staining with Lugol's iodine.
   (*The performance and course of the examination of yeast multiplied in the bioreactor in a batch and continuous system is described in Section 4.1.1).

3. Compare the production of biomass and production of biosurfactants (description of the tests is presented in the theoretical part of Section 4.1.2) in cultures of different *Bacillus* strains (e.g. IM 13, I'-1a, KP7, DSM 3257) carried out in LB medium and in carrot peel medium:
   a) determine OD at 610 nm for the given medium as a blank sample;
   b) determine emulsifying activity against kerosene;
   c) evaluate the surface activity of biosurfactants using the modified drop collapsing test.

4. Determine the laccase activity according to the Eggert et al. (1996) method described in Section 5.11 in the post-culture liquids obtained from the cultivation of selected *Myrothecium* strains in Sabouraud (control, defined composition), whey, and oil cake media.

5. Compare the ability of the waste cake (obtained after the preparation of the cake medium) to absorb different

textile dyes: take 1 ml from flasks containing 50 mg/l of dye (control) and dye solution and addition of 4 g/l of rapeseed cake (test sample) and centrifuge (3,500 × g, 15 min); measure the absorbance at the wavelength appropriate for the dye (AO7 λ = 485 nm, IC λ = 611 nm) using deionised water as a blank test; calculate the percentage of dye decoloration according to the method described in Section 5.6.7.

Analysis of results

On the basis of the obtained results, determine the suitability of the rapeseed oil cake and media prepared from by-products of the agri-food industry to conduct the analyzed biotechnological processes

5.10. Biodeterioration caused by fungi

INTRODUCTION

Microorganisms are the cause of destruction of many construction materials, both of natural origin and those based on plastics or inorganic materials. Harmful influence of microorganisms, as well as other organisms, on materials valuable from the point of view of human activity (including construction materials) is referred as biodeterioration. In case of construction materials, the term includes:

- typical biodegradation processes, during which microorganisms use compounds contained in construction materials (e.g. wood, plasterboard, floor coverings with natural fibres) as a source of carbon and energy;
- initiation of biological corrosion, during which microorganisms develop on the surface of the construction material, e.g., glass, concrete, mortar, brick (Figure 5.10.1), using superficial organic pollutants as a source of carbon and energy and producing metabolic products (usually low-molecular organic acids) that cause deterioration of the construction material.

Wood and essential wooden parts of buildings (e.g. roof truss, load-bearing beams, boards) or elements of green area developments (see Section 5.1) within parks (e.g. pergolas, benches, wooden bridges) are natural construction materials that are most often and most easily degraded by microorganisms. If the water content of the wood exceeds 18–20% and the wood is not properly impregnated with biocides, the development of moulds capable of breaking down cellulose and/or lignin, which are the main components of the wood, is inevitable. This process is referred as wood rot in construction and woodworking. During decomposition, quantitative and qualitative changes in the chemical composition of the wood take place, which leads to the decomposition of the cell walls of the wood (disappearance of lignin and cellulose). As a consequence, the strength of wood as a construction material decreases. Decomposition is accelerated with constant contact between wood and soil.

Therefore, when assembling structures in open areas (e.g. park pergolas), elements limiting contact of soil microbiota with wood are used (Figure 5.10.2 A, B).

Fig. 5.10.1. Initiation of biological corrosion of the wall caused by fungi:

visible mycelium (→) on the surface of the bricks.

Fig. 5.10.2.
A. Galvanized garden anchor (driven into the ground base of the pole); B. Pergola base with a grounded anchor with a fixed beam impregnated with an antifungal agent with green dye (as an indicator of impregnation).

There are three main types of wood rot:

1. Brown rot (destructive, fibre red) – fungi decompose cellulose and hemicellulose, leaving red-brown lignin. The disappearance of cellulose causes the wood to darken, shrink, and then crack and disintegrate into characteristic prismatic fragments (Figure 8.2.4 C). In the final phase of decomposition, the wood usually takes the form of a powdery or fibrous mass. The brown rot is caused by: *Serpula lacrymans, Coniophora puteana, Fibroporia vaillantii*. These species cause 90% of the rot of wooden construction materials in buildings.

2. White rot (Figure 8.2.5) – fungi decompose all components of wood. The affected wood becomes relatively (to healthy wood) lighter, as more cellulose remains in the final stage when both components are evenly decomposed. This is due to the higher cellulose content in the wood compared to lignin. White rot is caused by: *Trametes versicolor, Neolentinus lepideus, Gloeophyllum sepiarium*.

3. Soft rot – fungi decompose lignin, leaving the cellulose. This rot is characterised by the formation of cavities and pockets unevenly distributed in the wood, filled with white powdered cellulose. White cavity rot is caused by *Porodaedalea pini, Cylindrobasidium laeve, Heterobasidion annosum* (Figure 8.2.4 A, B).

In addition, fungi may also cause the grey rot of wood, giving it a greyish colour in a state of strong moisture, which adversely affects its appearance and lowers the class of wood, but does not significantly affect its strength as a structural material (Figure 8.2.6). This type of rot is caused by fungal species mainly belonging to Ascomycota, e.g. *Chaetomium globosum*, *Ophiostoma* spp. (see also Section 5.11).

In order to prevent the destructive effect of ligninolytic microorganisms, wood is impregnated with preparations containing biocides. Effective agents are considered to be those which, after wood contamination with standard lignolytic strains and incubation under laboratory conditions (optimal for the growth of the microorganisms used), do not allow weight reduction of the wood by more than 3% within 16 weeks. Other construction materials such as plasterboardas, roofing felt, floor coverings, mortars in continuous contact with organic matter (e.g. in sewers) are also impregnated with biocides. Most often these are heterocyclic compounds containing sulphur and/or nitrogen atoms in their composition.

When repairing concrete surfaces, mortars containing various additives that limit the formation of biofilms and the destructive influence of microorganisms, are often used, e.g. lithium silicate compounds, nanopolymers of polyester resins. Lithium silicate compounds, by reacting with free $Ca(OH)_2$, form hydrated silicates and calcium aluminosilicates and provide a permanent bond between all the components in the concrete, while the resins make the system more flexible and reduce the formation of stresses, which reduces the formation of microcracks and the deposition of microorganisms on the concrete surface.

It is worth noticing that an important factor determining the development of microorganisms, both ligninolytic and non-degrading cellulose or lignin, is water availability. Excessive air humidity caused by improper ventilation of the rooms as well as damp walls, e.g. due to faulty water drainage from the roof of the building, can lead to the development of moulds using even traces of simple organic compounds present in dust or other pollutants (Figure 5.10.3).

Fig. 5.10.3. Mouldy wall (plasterboard) https://fixly.pl/ blog/ naprawy/jak-sie- pozbyc-grzyba-ze- sciany/ (access: 7.07.2020).

Conditions promoting the development of microorganisms on wallpaper:
1) indoor air humidity – above 60.5%;
2) water content:
   a) in wallpaper – above 10%,
   b) in concrete substrate – above 4%,
   c) in gypsum substrate – above 3%.

The number of colony-forming units (cfu) of mould, indoors and outdoors is usually similar and usually does not exceed 200 cfu/m³ of air. In the autumn period, when mould develops intensively on the dead remains of the plants, as well as in permanently damp apartments where the relative air humidity is above 60%, this number can be a few or even several times higher. This poses a serious threat to the housing infrastructure (especially wood) and also to human health. Humans intake 12–15 cubic metres of air through breathing, in a 24 h period, and this could potentially contain pollutants, including mould spores, which can cause, directly or indirectly, the following problems:
- allergies – manifested by conjunctivitis, allergic cold and cough, bronchial asthma, alveolitis;
- mycotoxicosis – in the form of acute or chronic poisoning caused by ingestion of mycotoxins with food or by inhalation;

- mycoses – especially in immunocompromised people, caused by fungal spores, which due to their small size can penetrate into the lungs and cause infections leading to various forms of mycoses of the respiratory system (e.g. lung mycosis).

The most common types of non-lignolytic fungi isolated in damp apartments, which have harmful effects on human health, are: *Cladosporium, Alternaria, Penicillium, Aspergillus, Sporobolomyces, Fusarium* and *Geotrichum*.

The air in the buildings should not contain pathogenic, allergenic and toxigenic fungi, especially *Aspergillus fumigatus, A. flavus, Stachybotrys chartarum*. The presence of more than 50 cfu of one fungal species in 1 m³ of air is a signal of serious fungal danger in the room. The presence of up to 200 cfu in 1 m³ of air is allowed, if it is a mixture of several species of fungi.

PRACTICAL PART

A. Microbiological analysis of wood rot

Materials and media

1. Standard strains of wood rot: *Serpula himantioides* – brown wood rot fungus; *Phanerodontia chrysosporium* (*Phanerochaete chrysosporium*) and *Trametes versicolor* – white wood rot fungi; *Chaetomium globusum* – soft wood rot fungus.
2. Strains isolated from wood infected by fungi.
3. Materials and reagents: wood blocks seasoned for 3 months (treated with biocides and not containing biocides) of comparable weight and size, glass jars, lignin plugs, sterile garden soil.

Aim of the exercise
Evaluation of the ability of fungi isolated from the wood surface to decompose it, and resistance to biocides.

Procedure

1. Dry wooden blocks in a dryer at 105°C for 24 h and weigh them on an electronic scales with an accuracy of 0.001 g after 10 min and then store them for 24 h in a vessel with sterile distilled water.
2. Immediately before the test, cover the blocks (impregnated with fungicides and unimpregnated) with a suspension of spores (standard and isolate strains) in NaCl (1 ml/block).
3. Introduce a portion of sterile moist horticultural soil with 60% moisture content by weight into sterile glass jars with a volume of about 1 l, to cover the bottom of the jar.
4. Then place on the soil, wooden blocks of known weight infected with fungi.
5. Cover the blocks with a layer of sterile, moist soil to a height of about 5 cm.
6. Cover the jars with a plug. Incubate the samples for 4, 8 and 16 weeks at room temperature. During the incubation period, control soil moisture, do not allow samples to dry out.
7. Dry and weigh the wood blocks in the same way as in point 1.
8. On the basis of the formula, determine the percentage loss of wood mass compared to wood mass before inoculation, reflecting the degree of wood decomposition by the examined fungi:

$$U = \frac{(M_0 - M_1)}{M_0} \times 100\%$$

where:
U – percentage loss of sample weight,
M_0 – mass of the block before incubation,
M_1 – mass of the block after a fungal infection.

Analysis of results

Based on the results obtained, assess the ability of isolated fungal strains to degrade the wood, taking into account the incubation time and type of rot, as well as resistance to the analysed antifungal preparations. The reference system is made up of the reference strains of wood decay.

B. Initiation of biological corrosion of construction materials by fungi

Materials and media

1. Media: plates with ZT and Czapek-Dox media.
2. Materials and reagents: template with a 10 cm × 10 cm hole, glass pipettes, automatic pipettes, sterile test tubes, sterile 0.85% NaCl, heater (temperature 28°C), rotary shaker, light microscope.
3. Standard toxigenic strains

Aim of the exercise

Evaluation of the degree of fungal infestation of inorganic construction materials, determination of the ability of isolated fungal strains to produce mycotoxins and determination of taxonomic affiliation of the most dangerous isolates.

Procedure

1. The moulded surface, limited by a template of the above-mentioned sizes, should be washed off with a sterile swab. Transfer the swab to a conical flask containing 100 ml of saline and incubate for 30 min on a shaker at 180 rpm.
2. After performing a series of dilutions, inoculate 0.1 ml (from appropriate dilutions) on plates with ZT and Czapek-Dox media.
3. After 7 days of incubation at 28°C, count the fungal colonies. Calculate the total number of fungi [cfu/100 cm²]. The results should be presented in the table.

Number of colonies on Czapek-DOX medium	Number of colonies on ZT medium	Total fungi count cfu/100 cm²

4. Compare macroscopic morphologies of colonies of isolated fungi and standard toxigenic strains and select strains for chromatographic analysis for mycotoxin production capacity.

5. Check for the presence of mycotoxins in cultures of the selected strains according to the procedure described in Section 5.14.

6. Describe the macroscopic appearance and make microscopic preparations of isolated fungi. Determine the generic origin of fungi from construction materials (Section 3.2).

7. Determine the species affiliation of isolates according to the methodology given in Section 3.2.3.

5.11. Characteristics and use of ligninolytic enzymes produced by fungi in environmental protection, industry and medicine

INTRODUCTION

Ligninolytic fungi are a taxonomically heterogeneous group with a unique ability to depolymerise and mineralise lignin. Lignin is a natural, (as opposed to cellulose), hard-degradable polymer whose degradation is only possible through the action of extracellular, non-specific enzymes, which include laccases, peroxidases and oxidases. The activity of these enzymes, their catalytic and biochemical characteristics, and thus the possibilities of their potential use, vary according to the species and even the strain of microorganism by which they are synthesised, as well as the culture conditions. A popular group of microorganisms capable of producing ligninolytic enzymes are the so-called white rot fungi (WRF), which belong to Basidiomycota. The activity of laccases, peroxidases and oxidases has also been found in many species of Ascomycota and in bacteria.

Laccases are the best studied and described, and at the same time the most numerous group among ligninolytic enzymes. These enzymes catalyse oxidation reactions of organic and inorganic compounds with simultaneous reduction of molecular oxygen to water. There are three groups of reactions catalysed by laccases: coupling reactions, oxidation of simple phenolic compounds occurring without the use of a mediator and substrate oxidation requiring the presence of a mediator. Due to the high redox potential and the large particle size of the substrates, non-phenolic substrates cannot be directly oxidised by laccases. In this case, chemical mediators are used, i.e., low molecular weight compounds that are oxidised by enzymes with the formation of highly reactive radical cations, which then oxidise the non-phenolic substrates. The most commonly used synthetic mediators are ABTS (2,2'-Azino-bis(3-ethylbenzothiazoline-6-sulfonic acid) diammonium salt), hydroxyanthranilic acid (HAA), hydroxybenzotriazole (HBT) and N-Hydroxyphthalimide (NHPI). Natural mediators, e.g. vanillin and syringaldehyde, also have similar effects. In fungi, laccases are involved in morphogenesis, lignin degradation and defensive reactions to stress. These enzymes are produced by saprophytic and mycorrhizal fungi and take part in the circulation of organic matter in the soil, participating in the degradation of mulch polymers or the formation of humic compounds. In the group of fungi capable of laccase biosynthesis, the most numerous are the fungi of white rot of wood, which include strains from the species *Trametes versicolor, Pleurotus ostreatus, Phanerochaete chrysosporium, Phlebia radiate, Cerrena unicolor*. The synthesis of these enzymes has also been described in fungi belonging to the Ascomycotus genera *Myrothecium, Aspergillus, Neurospora* and *Trichoderma*. Bacterial laccases have been identified in cultures of strains including *Azospirillum lipoferum, E. coli, B. subtilis* and several species of *Streptomyces* sp. Bacterial laccases are more active and stable at high temperatures, alkaline pH and high concentrations of chlorine and copper ions.

Lignin peroxidase (LiP) was first described in *Phanerochaete chrysosporium* fungus in 1983. Since then, the activity of this enzyme has been demonstrated in cultures of *Trametes, Bjerkandera, Phlebia* and some bacteria, e.g. *Acinetobacter calcoaceticus* and *Streptomyces vidosporus*. LiPs are hemoproteins active at low pH, dependent on H_2O_2, with high redox potential. The substrate specificity of LiPs is low. They exhibit the ability to oxidise phenolic and non-phenolic compounds which have a similar structure to lignin, as well as inorganic compounds. In the catalytic cycle of lignin peroxidase, the iron ion is oxidised from Fe^{3+} to Fe^{4+} with hydrogen peroxide, followed by a reduction of the LiP molecule with simultaneous oxidation of the substrate molecule. The catalytic cycle ends when the LiP molecule is reduced to its native form again.

Manganese peroxidase (MnP) is produced by almost all the fungi of the white rot of the Basidiomycota division. This enzyme has a catalytic cycle similar to other peroxidases with dual electron oxidation, but MnP oxidises Mn^{2+} with the formation of highly reactive Mn^{3+} ions, capable of penetrating the cell wall and oxidising phenolic substrates. Mn^{3+} ions combine with organic acids generated by fungi to form, for example, oxalates or malates, which can act as low-molecular redox mediators capable of oxidising organic substrates.

As mentioned earlier (Section 5.5), phenolic compounds stimulate the production of laccases by fungi. Therefore, waste from urban green areas, especially naturally fallen leaves in the autumn, containing significant amounts of these compounds, is a cheap and easily accessible raw material for the production of laccases. This also applies to trees commonly found in our climatic zone, such as *Betula pendula* Roth (silver birch), *Aesculus hippocastanum* L. (chestnut) or *Salix* sp. L. (willow). It should also be borne in mind that plant pathogens that cause white rot in the wood and are the cause of numerous types of damage can be valuable producers of laccases for industrial, medical and environmental applications (Figure 5.11.1).

Of all the enzymes included in this group, laccases are particularly interesting for potential industrial applications. These enzymes have the ability to oxidise many substrates, are extracellular and quite stable over a wide range of pH and temperature. Laccases are used in many areas of the food industry, such as:
- baking – to improve bread structure, taste and durability of cakes;
- fruit and vegetable processing;
- winemaking and brewing – to eliminate undesirable phenolic derivatives that cause darkening and turbidity of fruit juices, beers and wines.

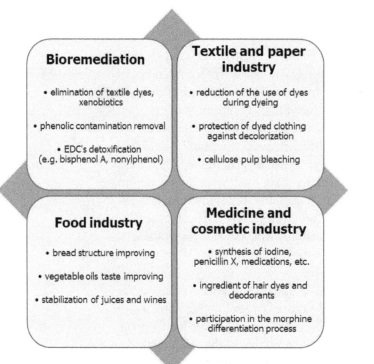

Fig. **5.11.1.** Examples of the application of laccases

Bioremediation

• elimination of textile dyes, xenobiotics

• phenolic contamination removal

• EDC's detoxification (e.g. bisphenol A, nonylphenol)

Textile and paper industry

• reduction of the use of dyes during dyeing

• protection of dyed clothing against decolorization

• cellulose pulp bleaching

Food industry

• bread structure improving

• vegetable oils taste improving

• stabilization of juices and wines

Medicine and cosmetic industry

• synthesis of iodine, penicillin X, medications, etc.

• ingredient of hair dyes and deodorants

• participation in the morphine differentiation process

Laccases and peroxidases can also be used:
- as biocatalysts in reactions, whose products are substances with antimicrobial, anticancer, disinfectant or antioxidant properties;
- for the construction of biological, enzymatic and immunochemical tests (e.g. in detecting the presence of morphine);
- in the cosmetics industry – for the production of dyes used in non-toxic hair dyes and skin lightening preparations;
- in the textile and paper industry – to modify cellulose fibres, to remove dyes from sewage, in fabrics decoloration processes;
- for the elimination of a wide range of toxic compounds such as: phenolic compounds, chlorophenols, cyclic aromatic hydrocarbons, alkenes; for the decoloration of industrial dyes.

PRACTICAL PART

Materials and media

1. Microorganisms: reference strains of *Trametes versicolor, Phanerochaete chrysosporium* fungi and strains isolated from wood infected with fungi (white rot of wood) collected in the city park.
2. Kirk-Farell liquid medium.
3. Apparatus: centrifuge, spectrophotometer.
4. Reagents: Mc Ilvaine buffer pH 4.5; 10 mM ABTS.

Aim of the exercise
Determination of the activity of laccase produced by fungi causing white rot in wood.

Procedure

1. Evaluation of the laccase production by *Trametes versicolor, Phanerochaete chrysosporium* and isolates from wood infected with fungi (Section 8.2) during growth in liquid medium:

   a) Pour 7 ml of Kirk-Farell medium into 1 slant of the tested fungus and insert into flasks containing 20 ml of this medium. Incubate on a shaker (120 rpm) at 28°C;

   b) after 7 and 14 days determine the activity of laccase, manganese-dependent peroxidase and lignin peroxidase in post-culture supernatant. For this purpose, take 1 ml of the post-culture liquid and transfer to 1.5 ml Eppendorf tubes.
   Centrifuge the samples at 3,500 × g, 10 min, 4°C. The supernatant obtained is a test sample.
   - Determine the laccase according to the modified Eggert et al (1996) method. Prepare the following systems:

	Test sample [µl]	Blind sample [µl]
Mc Ilvaine buffer pH 4.5	740	750
Post-culture liquid	10	0
10 mM ABTS	250	250

- Measure the absorbance for 1 min at $\lambda = 420$ nm.
- Calculate the laccase activity from the formula:

$$A_L = (\Delta Abs \times V_{total} \times 10^6)/(\varepsilon \times \Delta t \times V_{sample})$$

where:
A_L – laccase activity (U/l);
ΔAbs – absorbance increase during 1 min at $\lambda = 420$ nm;
V_{total} – total sample volume (ml);
ε – molar absorbance coefficient ($36{,}000$ M^{-1} cm^{-1})
Δt – measurement time (min);
V_{sample} – test sample volume (ml).

- Determine the activity of manganese-dependent peroxidase according to the modified Wariishi et al (1992) method. Prepare the following systems:

	Test sample [µl]	Blind sample [µl]
Malonate – sodium buffer pH 4.5	600	750
Post-culture liquid	150	0
20 mM 2,6-DMP	150	150
20 mM MnSO$_4$ × H$_2$O	150	150
4 mM H$_2$O$_2$	100	100

- Incubate 30 min in the dark. Measure the absorbance for 2 min at $\lambda = 470$ nm.
- Calculate the manganese-dependent peroxidase activity from the formula:

$$A_{MnP} = (\Delta Abs \times V_{total} \times 10^6)/(\varepsilon \times \Delta t \times V_{sample})$$

where:
A_{MnP} – MnP activity (U/l);
ΔAbs – absorbance increase during 2 min at $\lambda = 470$ nm;
V_{total} – total sample volume (ml)
ε – molar absorbance coefficient ($14{,}800$ M^{-1} cm^{-1});
Δt – measurement time (min);
V_{sample} – test sample volume (ml).

- Determine the activity of lignin peroxidase according to the modified Tien and Kirk (1983) method. Prepare the following systems:

	Test sample [µl]	Blind sample [µl]
Sodium tartrate buffer pH 4.5	200	500
Post-culture liquid	300	0
Veratryl alcohol	300	300
4 mM H_2O_2	100	100

- Incubate 30 min in the dark. Measure the absorbance for 2 min at $\lambda = 310$ nm.
- Calculate the lignin peroxidase activity from the formula:

$$A_{LP} = (\Delta Abs \times V_{total} \times 10^6)/(\varepsilon \times \Delta t \times V_{sample})$$

where:
A_{LP} – LP activity (U/l);
ΔAbs – absorbance increase during 2 min at $\lambda = 310$ nm;
V_{total} – total sample volume (ml)
ε – molar absorbance coefficient (9,300 M^{-1} cm^{-1});
Δt – measurement time (min);
V_{sample} – test sample volume (ml)

2. Application of *Tramtes versicolor* laccase for elimination of azo dyes:
   a) prepare aqueous solutions of selected dyes at concentrations of 10, 25 and 50 mg/l;
   b) introduce commercial *Trametes versicolor* laccase (Sigma-Aldrich) into the solutions so that its activity in the solution is 1 U/ml. Additionally, prepare an enzyme free control system.
   - Incubate the samples on a rotary shaker (120 rpm), at room temperature.
   - Measure the absorbance of the solutions after 0.5, 1, 2 and 4 h incubation at the appropriate wavelength for the dye. Determine the degree of

decoloration according to the method described in Section 5.6.7.

- Prepare a diagram showing the decolourization of the dye in different concentrations during incubation.

Analysis of results

Determine and compare the production of selected ligninolytic enzymes in standard strains and isolates from wood infected with fungi causing white wood rot. Assess the ability of *Trametes versicolor* laccase to eliminate azo dyes.

5.12. Determination of antimicrobial properties of macromolecules (dendrimers) and newly synthesised silver compounds

INTRODUCTION

Growing interest in nanomaterials characterised by spatial dimensions in the range of 1–100 nm results from the change of basic properties of substances after their transition to nanometric level. These specific physicochemical characteristics include higher specific surface area and higher surface to volume ratio of nanomaterials, lower melting point, lower ionisation potential while reducing the size of these molecules, higher chemical reactivity, higher mechanical strength while maintaining high flexibility, as well as magnetic, photocatalytic, photoconductive and photoemission properties. All this makes nanomaterials widely used in food, cosmetic, textile industry, electronics, as well as in environmental protection and medicine.

Dendrimers, which are classified as nanoparticles, are regular, repeatedly branched molecules with strictly defined molar mass and three-dimensional structure. Their name originates from the Greek word dendron, meaning tree, and was first used in 1985 in relation to strongly branched spherical polymers synthesised by Tomalia.

Three different topological parts can be distinguished in the structure of the dendrimer: core, dendrons and functional groups. The core is located in the central part of the dendrimer, and repeated arm building units, i.e., dendrons, at the end of which there are appropriately selected functional groups, radially depart from the core. The strictly defined three-dimensional structure and the strongly branched structure of the dendrimer make the core and the spaces around it a perfect place to encapsulate various molecules and chemicals. The outer part of the polymer, which is responsible for contact with the environment, allows control of the properties of the molecule by making appropriate modifications on its surface.

Dendrimers can be synthesised using two methods. The first, referred to as divergent synthesis, involves attaching successive layers of dendrimer from the core towards the outside. A different approach is represented by the second method of obtaining dendrimers, so-called convergent synthesis. This technique assumes first of all obtaining extreme dendrimer fragments and then attaching them to the central core.

Due to their chemical structure, dendrimers can be divided into several types, of which polyamidoamine dendrimers (PAMAM), poly(propylene imine) dendrimers (PPI), polyether dendrimers, carbosilane dendrimers, phosphorus dendrimers and peptide dendrimers are the most popular.

The presence of free spaces inside the dendrimers and the possibility to connect different functional groups on their surface made them widely used in nanotechnology and medicine. The research has shown that dendrimers can be used as medicines transporting agents. Such systems can be obtained by encapsulation of substances inside the molecule or its covalent binding to the dendrimer surface. The incorporation of the drug into a suitable carrier such as a dendrimer, can reduce the toxicity of the drug, improve its solubility and stability in an aqueous environment and facilitate its uptake by cells. Such complexes are also characterised by longer duration of release of the active substance. The introduction of modifications to the dendrimer surface also enables direct delivery of the drug to specific cells, e.g. cancer cells, thus reducing its toxicity to healthy tissues and limiting the side effects of the applied therapy. Examples of drugs whose combination with dendrimers has significantly improved their therapeutic properties include ibuprofen, methotrexate, niclosamide, cisplatin, 5-fluorouracyl and doxorubicin.

In recent years, more and more attention has also been paid to the possibility of using dendrimers as antimicrobial agents. The mechanism of the biocidal action of dendrimers may be different, and results from their specific structure and the properties of functional groups present on the surface of the molecule. It has been shown that dendrimers interact with negatively charged bacterial membrane, increasing its permeability, which can lead to lysis of the microorganism. These molecules may also limit microbial adhesion or block the key receptors for infection. Antimicrobial activity has been demonstrated, among others, for poly(propylene imine) dendrimers, carbosilane dendrimers, carbosilane modified with ammonium group or nitric oxide modified PAMAM dendrimers.

Another group of nanomaterials with high antibacterial potential are metal and metal oxide nanoparticles. Metallic elements such as Al, Au, Cu, Fe , Mn, Ti, Zn or Ag, as well as their oxides are most commonly used for the synthesis of antimicrobial nanoparticles. They are mainly synthesised from metallic elements classified as heavy metals with a density > 5 g/ml. These metals are usually transition metal elements, which means that their electron configuration in orbital d is partially filled. This property makes these metals usually more reactive, which facilitates the synthesis of nanoparticles. The simplest way to produce nanoparticles is to reduce the metal precursor (metal salt) with a strong reducing agent such as sodium borohydride. The reaction involves reduction of the metal cation to an inert state, which enables the aggregation of metal atoms and the final formation of nanoparticles. Many transition metals have important biological functions, but at higher concentrations they become toxic to living organisms. The degree of oxidation of the metals that make up the nanoparticles is 0, so they are not likely to exceed the cell membrane barrier. However, it is known that metal nanoparticles slowly release their ions, which may interact with the cell membrane, disturbing its functioning. The bactericidal activity of metal nanoparticles can be attributed to many different properties, the most important being the ability to generate reactive oxygen species. In addition, metal ions, such as Ag^+ or Hg^{2+}, can easily bind to thiol (-SH) groups found e.g. in the structure of cysteine, which can directly interfere with enzymes or damage sulphide bridges necessary to maintain the integrity of proteins, causing harmful effects on cell metabolism and physiology.

The antibacterial activity of metal nanoparticles and their oxides depends on their size. The size of nanoparticles plays a key role in their interaction with enzymes, membrane proteins and other components of the bacterial cell. Numerous studies indicate that copper and silver nanoparticles of different dispersive and phase composition have different antibacterial activity. Moreover, many studies confirm the correlation between the size of the nanoparticles and their antibacterial activity. Smaller nanoparticles have a higher surface to volume ratio, which leads to their stronger interaction with the bacterial cell, thus increasing their antimicrobial activity.

PRACTICAL PART

Materials and media

1. Microorganisms: *Staphylococcus aureus, Streptococcus pyogenes, Pseudomonas aeruginosa, Escherichia coli.*
2. Media: agar plates, Mueller-Hinton medium.
3. Reagents: poly(propylene imine) dendrimers.
4. Apparatus: spectrophotometer, microplate reader.

Aim of the exercise
Determination of the antibacterial activity of the studied nanoparticles.
Determination of the minimum inhibitory concentration (MIC) and the minimum bactericidal concentration (MBC).

Procedure

Determination of antimicrobial activity of tested substances should be performed by the microdilution method in accordance with The Clinical and Laboratory Standards Institute (CLSI), assuming spectrophotometric measurement of optical density of bacteria after incubation with tested compounds:

1. Inoculate the stocks of examined microorganisms into 20 ml of Mueller-Hinton medium and incubate 24 h at 37°C.
2. Prepare the initial solutions of the dendrimers (concentrations dependent on the weight and generation of dendrimers used) and then apply 5 μl each to the titration plate so that the final concentrations of the test samples are between 0.1 and 100 μM.
3. Dilute the bacterial suspension so that the density of the inoculum used is $3 - 7 \times 10^5$ CFU/ml.
4. Apply 195 μl of bacterial culture to the titration plate containing serial dendrimer dilutions, incubate for 24 h at 37°C.

   Additionally, prepare biotic controls (containing only the microorganism and medium) and abiotic controls (containing dendrimer and medium). All culture variants should be performed as 4 replicates.
5. Incubate the samples for 24 h at 37°C and then measure optical density at 630 nm.
6. Determination of MIC and MBC concentration:
   a) MIC – minimum concentration of dendrimer, at which no growth of the examined microorganisms is observed;
   b) MBC – to determine the minimum biocidal concentration, inoculate 100 μl of each sample in which no growth of the microorganism was observed on an agar substrate using the "smear plates" method. Incubate the media for 24 h at 37°C.

Analysis of results

Determine the antibacterial activity of the dendrimers tested and determine the values of MIC and MBC.

5.13. Entomopathogenic fungi and their use in biocontrol

INTRODUCTION

Entomopathogenic fungi are organisms parasitising on live arthropods, which in natural conditions are an important factor in the biocontrol of natural insect populations. These microorganisms do not form a single monophyletic group and so far 12 Oomycota species, 65 Chytridiomycota species, 339 Microsporidia species, 474 Zoopagomycota species, 238 Basidiomycota species and 476 Ascomycota species are described. The most common species in nature belong to Ascomycota and Zygopagomycota (Table 5.13.1).

Table 5.13.1.
Entomopathogenic fungi belonging to Asco- and Zygopagomycota

Ascomycota	Zoopagomycota (genus Entomophthorales)
Metarhizium (M. anisopliae, M. acridum, M. robertsii)	Entomophthora sp. (E. muscae)
Lecanicillium (L. lecani)	Zoophthora (Z. radicans)
Beauveria (B. bassiana, B. brongniartii)	Pandora (P. neoaphidis)
Isaria (I. fumosorosea)	Entomophaga (E. maimaiga)
Cordyceps (C. militaris)	Erynia (E.conica)
Paecilomyces (P. variotti)	Furia (F. gastropachae)

The life cycle of entomopathogenic fungi includes several stages (Figure 5.13.1):

1. Adhesion of the spore to the cuticle of insects. It is a complex process which consists of both non-specific factors (hydrophobic and electrostatic interactions) and specific factors (production of proteins by the fungus – adhesins and low-molecular hydrophobins).
2. Germination and formation of an appressorium. Spores adhering to the insect's cuticle begin to germinate and form a swollen fragment of the hypha (appressorium) which allows the fungus to attach itself to the insect's outer shell. The appressorium releases hydrolytic enzymes such as proteases, esterases, N-acetylglucosamidases, chitinases and lipases which cause enzymatic degradation of the insect's cuticle. In addition, the appressorium

403

causes a mechanical pressure on the cuticle, which facilitates the breakage of coatings and the penetration of the fungus into the insect.

3. Penetration of the arthropod cuticule. A hypha grows from the appressorium, and it penetrates the shells of the invertebrate, penetrating its body cavity (hemocoel).

4. Production of blastospores (spores formed by budding). The infectious hypha, when it reaches the hemolymph, begins to produce blastospores. There is considerable damage to the insect's body as a result of mechanical damage to internal organs and exhaustion of the nutrients contained in hemolymph. The death of the insect occurs within 3–7 days after blastospores begin to multiply.

5. Production of spores outside the body of the insect. After the death of the host, hypha are formed, which first spread inside the body, and then break through the cuticle and emerge on the outside, where they produce numerous spores to infect other insects.

Fig. 5.13.1. Scheme of the infection process of entomopathogenic fungi.

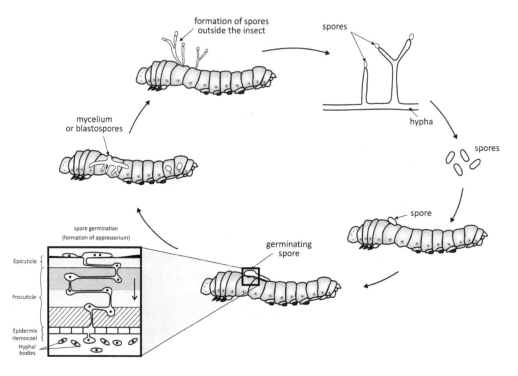

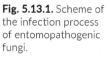

Typically, fungi belonging to Ascomycota do not show any specificity in the choice of host, but the exception is *Metarhizium acridum*, which only infects Acrididae.

Fungi belonging to the Entomophtorales carry out the entire development cycle in the body of insects, leading to their death. Entomophtorales play a very important role in reducing the number of forest and agricultural crop pests, which cause plagues of larvae, pupae and adult beetles, butterflies, hymenoptera and aphids. Of all entomopathogenic fungi, Entomophtorales species are the most specialised and adapted to attacking insects, and are also highly selective against the infected hosts. Their characteristic feature is their ability to shoot spores over long distances, which increases their infective potential. Unfortunately, due to difficulties in laboratory culturing of these microorganisms, they are not used in practice as biopesticides.

In turn, entomopathogens belonging to Ascomycota are commonly used in agriculture and organic horticulture as biological agents against arthropod populations. They are most often used to reduce the populations of mosquitoes, termites and ticks. It is estimated that several hundred different commercial preparations containing entomopathogenic fungi such as *Paecilomyces fumosoroseus*, *Lecanicilium muscarium*, *Metarhizium anisopliae* are available worldwide.

Unfortunately, the use of entomopathogenic fungi in pest control is still only an alternative to commonly used toxic chemical insecticides. It should also be mentioned that the widespread use of pesticides, as well as their persistence in the environment, has a negative impact on soil microflora, including entomopathogenic fungi that (as already mentioned) control arthropod populations under natural conditions.

PRACTICAL PART

I. Examination of the influence of chemical insecticides on the viability of entomopathogenic fungi (see Section 2.1).
II. Determination of the infectious ability of entomopathogenic fungi in relation to mealworm beetle (*Tenebrio molitor*).

405

Materials and media

1. Microorganisms: *Metarhizium robertsii* cultures on slants and plates with ZT medium.
2. Media: ZT slants and plates.
3. Reagents: sterile deionised water.
4. Materials: glass pipettes, automatic pipettes, sterile tubes, mealworm beetle larvae.
5. Apparatus: incubator (temperature 28°C).

Aim of the exercise
Determination of the infectious ability of Metarhizium robertsii in relation to mealworm beetle larvae.

Procedure

1. Examination of infectious ability of entomopathogenic fungi – immersion method:
   a) rinse *Metarhizium robertsii* slant with 5 ml of sterile deionised water;
   b) calculate the density of spores in Thoma chamber;
   c) prepare a suspension of spores of the tested fungus with a density of 1×10^6 spores per ml in sterile deionised water;
   d) apply 50 µl of the suspension to mealworm beetle larvae;
   e) count live and dead specimens after 7 and 14 days of incubation.
2. Examination of infectious ability of entomopathogenic fungi – plate method:
   a) apply 5 larvae of the mealworm beetle to the dishes with entomopathogenic fungus;
   b) after 15 min incubation, move the larvae to a new container;
   c) count live and dead specimens after 7 and 14 days of incubation.

Analysis of results

Calculate insect mortality. The obtained mortality results expressed as a percentage (for both methods) should be presented in a table. Indicate which of the applied insect control methods is more effective.

5.14. Toxigenic fungi. Search for and identification of aflatoxins

INTRODUCTION

Fungal diseases include three groups:
- mycoses;
- allergies – the allergenic factor may be spores or hyphae;
- mycotoxicosis – acute or chronic poisoning with fungal toxins following ingestion. Fungi capable of producing toxins belong to Basidiomycota and Ascomycota – microfungi, also described by the term "mould fungi".

Mycotoxins are low molecular weight fungal metabolites (Figure 5.14.1) that are toxic to animals and humans even at very low doses. The effect of mycotoxins on health depends on the type of compound, dose, frequency of intake, as well as species and age of the organism. The health risk may be a mycotoxin contamination as low as a few micrograms per kg of raw material/ food product. Most mycotoxins are hydrophobic and heat-stable compounds, and their decomposition is favoured by an alkaline environment. Most often the production of mycotoxins starts after the mycelium has completed the logarithmic growth phase. Therefore, mycotoxins are considered to be secondary metabolites. The ability to produce mycotoxins is a characteristic of the strain, not the species. In the case of species belonging to toxigenic fungi (e.g. *Aspergillus flavus*), strains incapable of producing mycotoxins, strains producing only one toxic compound and strains producing several toxic metabolites were described. The production of mycotoxins depends on the metabolic abilities of the fungus, however, environmental factors such as temperature, oxygen availability, humidity, availability of microelements (e.g. Zn, Mg, Co ions), time and the presence of other microorganisms, play a very important role. It is also possible that the same mycotoxin is produced by fungi of different genera. For example, patulin is produced by certain strains of *Aspergillus* and *Penicillium*.

Due to their particular affinity to specified organs/tissues and biological activity, individual mycotoxins are classified as nephrotoxins, hepatotoxins, neurotoxins, dermatoxins, carcinogenic compounds or hormonal modulators. Usually mycotoxins introduced into mammalian organisms cause abdominal pain, vomiting and diarrhoea. When the dose is increased or the exposure time is prolonged, liver and kidneys – the organs responsible for detoxification – are damaged.

The beginning of intensive research on toxigenic fungi and mycotoxins dates back to the 1960s. In 1960 a mass poisoning of young poultry fed with feed prepared from cereals from tropical and subtropical countries was recorded in England. The studies showed that the cause was the toxins produced by *Aspergillus flavus* strains. The four basic aflatoxins (B1, B2, G1 and G2) were obtained in crystalline form in 1964, and the continuation of the studies allowed several additional aflatoxins to be identified. In subsequent years, toxic metabolites of *Fusarium* and *Penicillium* were characterised (currently the number of described mycotoxins exceeds 400 compounds) and standards were developed, describing recommended methods of detection of toxigenic fungi and mycotoxins in raw materials and food products.

The most important international organisation dealing with mycotoxins is the Codex Alimentarius Commission (CAC). The work is carried out, among others, within the framework of the Joint Programme for Food Standards established by the Food and Agriculture Organisation (FAO) and the World Health Organisation (WHO). In Europe, the acceptable levels of selected mycotoxins in raw materials and food products are defined in regulations and recommendations of the European Commission.

Aspergillotoxicoses are diseases caused by the ingestion of toxins produced by *Aspergillus* fungi. Most often, these toxins are produced by strains belonging to species: *A. flavus, A. parasiticus, A. nomius, A. ochraceus* and *A. fumigatus*. The

408

most important mycotoxins produced by fungi of the genus *Aspergillus* include aflatoxins and ochratoxin A.

The *A. flavus* species is a cosmopolitan fungus due to its very wide geographical range, covering many climate zones. Optimal temperature of growth and aflatoxin production by *A. flavus* strains is 28–39°C. Toxigenic *A. flavus* strains are often isolated from the surface of peanuts, corn, soybean, cocoa and other plant raw materials (especially cereals) from tropical and subtropical regions. In Europe, this fungus is mainly detected on cereal grains. The basic aflatoxins are a set of four compounds described by symbols: B1, B2, G1 and G2, among which aflatoxin B1 is the most toxic. Currently, several aflatoxins are known. For example, aflatoxins M1 and M2 are hydroxylated derivatives of aflatoxins B1 and B2, respectively, which are isolated from milk from cows fed with feed containing the initial toxins.

Characteristics of aflatoxins B1, B2, G1, G2:
- compounds with a ring structure, similar to coumarin;
- soluble in water, ethanol, chloroform;
- not persistent in aqueous solutions, sensitive to light, UV radiation;
- very resistant to heating (dry to 160°C);
- glow under UV light (254–366 nm) in blue and green; the names of toxins B1, and B2 are derived from blue, and toxins G1, G2 from green;
- toxic to all animals tested so far, they also show strong mutagenic and carcinogenic properties.

Long-term exposure to low doses of aflatoxins leads to liver cancer. The reason is that these compounds (mainly aflatoxin B1) bind to the DNA of the nuclei of liver cells and interfere with DNA polymerase. Therefore, high prevalence of primary liver cancer in the population of Uganda, Nigeria, Kenya or the Philippines is associated with the widespread occurrence of aflatoxin B1 in some raw materials and foods (at levels exceeding even 1 mg of toxin/kg of food). According to the WHO, the upper acceptable concentration of aflatoxin B1 in feed should not exceed 20 μg/kg. However, for food intended for human consumption, the aflatoxin content should not exceed 0.1 μg/kg.

Toxin-forming strains of *Aspergillus ochraceus* produce ochratoxins that are nephrotoxic, neurotoxic, immunotoxic and teratogenic. The most toxic compound of this mycotoxin group is ochratoxin A. Intensive growth of mycelium is favoured by improper grain storage, mechanical damage to the seeds and their contamination with soil, as well as high environmental humidity. Ochratoxin A is also produced by some strains of *A. sulphureus, A. sclerotium, A. mellus* and *P. verrucosum*. The contamination of feed with ochratoxin A (like aflatoxins) is very dangerous for human health, due to the accumulation of this toxin in animal tissues.

The strains belonging to the species *Aspergillus fumigatus* grow in a wide range of temperatures, and the values conducive to intensive mycelium growth and mycotoxin production close to 37°C. Some strains of *A. fumigatus* are capable of synthesising various toxic metabolites. These include fumigalin, helivelic acid and alkaloids (similar to those of *Claviceps purpurea* ergot). The above compounds are mainly neurotoxins – they cause convulsions and cramps, but prolonged exposure can cause liver and kidney damage. Many strains of Aspergillus produce patulin (similar to some *Penicillium* strains). Patulin has mutagenic, carcinogenic, hepatotoxic and haemorrhagic properties. Significant concentrations of patulin (reaching up to 1 mg/g of raw material in infected areas) are detected in badly stored fruit, including apples, peaches, apricots, bananas and pineapples.

Fusariotoxicosis is poisoning with toxins of *Fusarium* fungi. The toxinic strains belong mainly to species: *F. sporotrichioides, F. culmorum, F. cerealia, F. poae* and *F. graminearum*. Fungi of the genus *Fusarium* usually show phytopathogenic properties – they cause damping-off of seedlings, stalks and fusariosis of ears and cobs. The most important mycotoxins produced by fungi of the genus *Fusarium* include: zearalenone, trichothecenes and fumonisins. Fusariotoxicoses occurred mainly during periods of hunger, in countries with temperate climates (in 1942–1945 in Japan and in some countries of the former USSR). A slow development of *Fusarium* fungi in winter was observed in cereal grains, which were not harvested in the autumn. In early spring (at a temperature close to 0°C) the growth of mycelium, spore production and production of toxic metabolites accelerated. The consumption of bread obtained from such grains caused diarrhoea and vomiting, accompanied by convulsions and imbalances (for this reason, the term "drunken bread" was created). The above toxins are characterised by very high persistence – they were detected in the grains even after 6–7 years, with concurrently negative results in microbiological tests. Zearalenone (ZEA) is a hormonal modulator and is classified as a non-steroidal mycoestrogen. Pigs are most sensitive to this toxin. Depending on the dose and exposure time, ZEA causes changes in female reproduction called estrogenic syndrome and can even cause total infertility of farm animals. Trichothecenes are very strong toxins that are classified as sesquiterpenic epoxides. There are four groups of these compounds, marked with symbols A, B, C and D. The group A includes: T-2 toxin, HT-2 toxin and F-2 toxin, while group B is represented by: fusarenon, nivalenone and deoxynivalenol (called vomitoxin). Trichothecenes of group C and D are also characterised by a strong toxic effect. Most trichothecenes have an irritating effect on the skin (which is used in biological tests), cause nervous disorders, extend the period of blood clotting (up to 8 h), increase the permeability of blood vessels, resulting in the formation of oedemas and haemorrhages and leading to a drastic decrease in the number of erythrocytes – up to 1 million/ml of blood and, consequently, to death. F-2 toxin shows oestrogenic effects – in animals, it causes miscarriages and disorders of the sexual cycle. This compound is detected in the milk of cows fed with infected feed.

Toxicoses caused by toxins of fungi of the genus *Penicillium* are quite frequent, but usually mild, and are caused by mycotoxins of strains of *P. rubrum*, *P. rugulosum*, and *P. purpurogenum*. These fungi are common in the soil and cause cereal grain infections. Poisoning of domestic animals – mainly pigs, horses and geese – has been reported. Symptoms of poisoning include salivation, disorderly movements, skin lesions as well as liver and kidney damage. *P. rubrum* strains produce rubratoxin A and B. These compounds show the ability to shine under UV light and, unlike other mycotoxins, are produced during intensive mycelium growth.

Scientific literature on mycotoxins also includes publications on the following subjects:

1) **agricultural and storage techniques to limit the development of toxigenic fungi and the production of mycotoxins;**

2) methods used to eliminate mycotoxins;

3) characteristics of rapid tests for mycotoxin detection;

4) requirements for acceptable levels of mycotoxins in feed and food products.

Fig. 5.14.1. Structural formulae of selected mycotoxins.

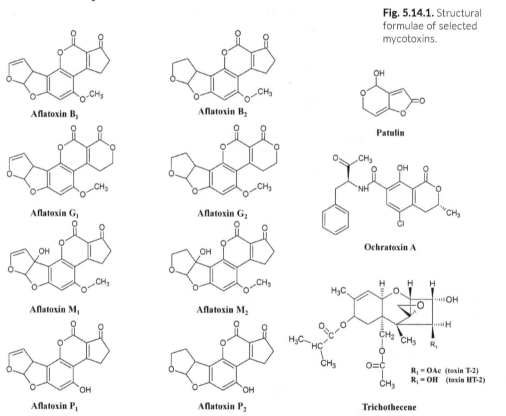

Aflatoxin B₁   Aflatoxin B₂   Patulin

Aflatoxin G₁   Aflatoxin G₂   Ochratoxin A

Aflatoxin M₁   Aflatoxin M₂

Aflatoxin P₁   Aflatoxin P₂

Trichothecene

R_1 = OAc (toxin T-2)
R_1 = OH (toxin HT-2)

PRACTICAL PART

Materials and media

1. Microorganisms: strains of filamentous fungi: *Aspergillus flavus, Penicillium rubrum and Fusarium graminearum*.
2. Media: Czapek-Dox and Czapek-Dox with 2% agar.
3. Materials and reagents: Fodder and food product samples tested for the presence of toxigenic fungi and/or aflatoxin content, reagents used for QuEChERS extraction, aflatoxin standards, mobile phase components used in aflatoxin analysis by LC-MS/MS (phase A – water with 0.2% formic acid and 0.2 mM ammonium formate, phase B – acetonitrile with 0.2% formic acid and 0.2 mM ammonium formate).
4. Apparatus: incubator for shaking cultures, centrifuge, microscope, liquid chromatograph Agilent 1200 and mass spectrometer QTRAP 3200 (Sciex).

Aim of the exercise
A. Determination of the content of fungi belonging to species A. flavus, P. rubrum and F. graminearum in feed and evaluation of the ability of isolated A. flavus strains to produce aflatoxins.
B. Qualitative and quantitative analysis of selected food products for the presence of aflatoxins.

Procedure

1. Describe the macroscopic and microscopic characteristics of the test strains of *A. flavus, P. rubrum* and *F. graminearum*.
2. Carry out a feed analysis for the presence of toxigenic fungi from the above-mentioned genera and determine the capacity of the *A. flavus* strain isolated from the feed for aflatoxin production (microbiological analysis of the feed for the presence of toxigenic fungi – 1st and 2nd stage).
3. Perform fodder (or selected food products) analysis for the presence of aflatoxins.

Microbiological feed analysis for the presence of toxigenic fungi:

1. Determination of the total content of microorganisms (including saprophytic fungi and macroscopic fungi corresponding to *A. flavus*) – (1st stage) should be carried out according to the procedure scheme presented in Figure 5.14.2.

2. Determination of the content of potentially toxigenic fungi (fungi macroscopically corresponding to *A. flavus*) – perform according to the procedure scheme presented in Figure 5.14.3.

3. Quantitative analysis of aflatoxins for samples:
   A) feed qualified for 2nd stage of the study (point 1);
   B) feed macroscopically suspected of toxigenic fungi presence (inappropriate smell, macroscopically visible mould); C) post-culture liquid from a 10-day culture of an isolated *A. flavus* strain:

   a) extract a sample of culture supernatant (obtained from a 10-day culture of the examined fungus, a water suspension of feed or beer using the QuEChERS procedure);

   b) Dilute the extract properly (at least 10 times) with the start phase of the chromatographic system or with water;

   c) perform a quantitative analysis using the LC-MS/MS technique. The analysis (sample injection volume – 10 µl) should be carried out using a gradient elution in a reversed phase arrangement on a C18 "coreshell" 100 Å column of 2.1 × 100 mm. Perform MS/MS detection in positive ionisation in MRM mode. An example of a gradient system is shown in Table 5.14.1, while basic parameters of MS/MS scanning (QTRAP 3200 Sciex) are shown in Table 5.14.2.

Fig. 5.14.2. Scheme of procedure: microbiological analysis of feed for the presence of toxigenic fungi – 1st stage of analysis.

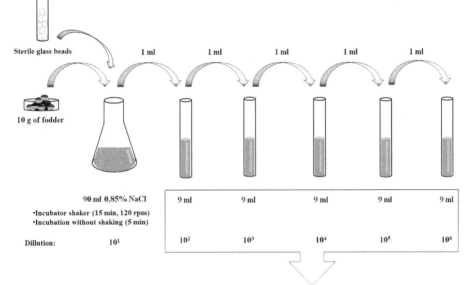

Sterile glass beads

10 g of fodder

90 ml 0,85% NaCl

•Incubator shaker (15 min, 120 rpm)
•Incubation without shaking (5 min)

Dillution: 10^1

1 ml — 1 ml — 1 ml — 1 ml — 1 ml

9 ml — 9 ml — 9 ml — 9 ml — 9 ml

10^2 — 10^3 — 10^4 — 10^5 — 10^6

Inoculation of the samples (0,1 ml) on growth medium Petri dishes

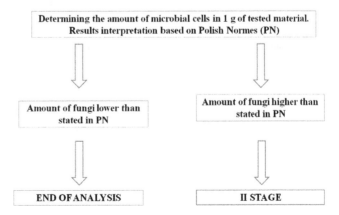

Determining the amount of microbial cells in 1 g of tested material.
Results interpretation based on Polish Normes (PN)

Amount of fungi lower than stated in PN

Amount of fungi higher than stated in PN

END OF ANALYSIS

II STAGE

415

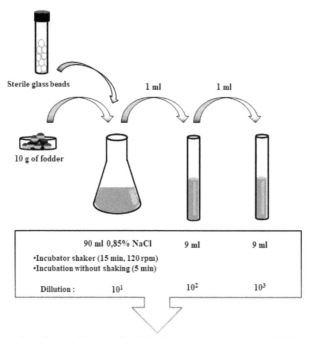

Fig. 5.14.3. Scheme of procedure: microbiological analysis of feed for the presence of toxigenic fungi – 2ⁿᵈ stage of analysis.

Sterile glass beads

10 g of fodder

1 ml   1 ml

90 ml 0,85% NaCl   9 ml   9 ml

•Incubator shaker (15 min, 120 rpm)
•Incubation without shaking (5 min)

Dillution : 10^1 10^2 10^3

Inoculation of the samples (0,1 ml) on growth medium Petri dishes

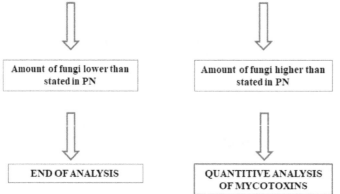

Determining the amount of fungi morphologically simillar to
toxigenic fungi in 1 g of tested material

Amount of fungi lower than stated in PN	Amount of fungi higher than stated in PN

END OF ANALYSIS	QUANTITIVE ANALYSIS OF MYCOTOXINS

Stage	Time [min]	Flow [µl/min]	A [%]	B [%]
0	0.0	500	95.0	5.0
1	0.5	500	95.0	5.0
2	8.0	500	10.0	95.0
4	12.0	500	5.0	95.0
5	12.1	750	95.0	5.0
6	14.0	750	95.0	5.0

Table 5.14.1. Example of the gradient system of the LC-MS/MS chromatographic separation during aflatoxin analysis (2nd stage of analysis)

Explanations: Phase A – water with 0.2% formic acid and 0.2 mM ammonium formate, phase B – acetonitrile with 0.2% formic acid and 0.2 mM ammonium formate.

Q1	Q2	Name	CE
313	128.1	Aflatoxin B1 1	91
313	241.1	Aflatoxin B1 2	51
315.2	259.1	Aflatoxin B2 1	43
315.2	287.1	Aflatoxin B2 2	41
329.1	243	Aflatoxin G1 1	37
329.1	200	Aflatoxin G1 2	59
330.9	245.1	Aflatoxin G2 1	45
330.9	189	Aflatoxin G2 2	59

Table 5.14.2. Chromatographic analysis LC-MS/MS of aflatoxins

Explanations: basic parameters of MS/MS scanning (QTRAP 3200 Sciex), collision energy (CE).

Analysis of results

1. Describe characteristics of macroscopic morphology of *A. flavus, P. rubrum* and *F. graminearum* strains.
2. Determine the content of toxigenic fungi in the tested feed samples on the basis of the description of macroscopic mycelium morphology of *A. flavus, P. rubrum* and *F. graminearum*.
3. Evaluate the ability of *A. flavus* strains isolated from feed to produce aflatoxins.
4. Determine the level of aflatoxins in selected food products and determine the suitability of the products tested for consumption.

Literature

Books

Bachorz, M., 2017. *Polska droga do gospodarki o obiegu zamkniętym. Opis sytuacji i rekomendacje*, http://igoz.org/wp/wp-content/uploads/2017/04/Polska_droga_do_GOZ_IGOZ.pdf (access: 19.06.2020)

Borowski, J., Fortuna-Antoszkiewicz, B., Łukaszkiewicz, J., Wiśniewski, P., 2016. *Standardy kształtowania przestrzeni miejskiej*. Polskie Towarzystwo Dendrologiczne, SGGW, Warszawa.

Chelkowski, J. (red.), 2014. *Fusarium: Mycotoxins, Taxonomy, Pathogenicity (Topics in Secondary Metabolism)*. Elsevier, Amsterdam.

Cullen, G. (red.), 2011. *The Concise Townspace*. Architectural Press, London.

Długoński, A., Szumański, M. 2019. *Idea parku ekologicznego*, w: Górski, F., Łaskarzewska, M. (red.), *Biocity*, t. 2. Fundacja Wydziału Architektury Politechniki Warszawskiej, Warszawa, s. 163–170.

Długoński, J., 2016. *Microbial Elimination of Endocrine Disrupting Compounds*, w: Długoński, J. (red.), *Microbial Biodegradation From Omics to Function and Application*. Caister Academic Press, Norfolk, s. 99–117.

Długoński J., Bernat P., Lisowska K., Paraszkiewicz K. 2008. Ksenobiotyki, w: Libudzisz, Z., Kowal, K., Żakowska, Z. (red.), *Mikrobiologia techniczna. Tom 2: Mikroorganizmy w biotechnologii, ochronie środowiska i produkcji żywności*. Wydawnictwo Naukowe PWN, Warszawa, s. 500–517.

Długoński, J. (red.), 1997. *Biotechnologia mikrobiologiczna – ćwiczenia i pracownie specjalistyczne*. Wydawnictwo Uniwersytetu Łódzkiego, Łódź.

Gawroński, S.W., Gawrońska, H., 2017. *Air phytoremediation*, w: Ansari, A., Gill, S., Gill, R., Lanza, G., Newman, L. (red.), *Phytoremediation*. Springer, Cham, s. 487-504.

Hereźniak, J., 2013. *Mocarze czasu. Pomnikowe drzewa w świecie i na ziemi łódzkiej*. Łódzkie Towarzystwo Naukowe, Łódź.

Jasińska, A., Góralczyk, A., Długoński, J., 2016. *Dye decolorization and degradation by microorganisms*, w: Długoński, J. (red.), *Microbial Biodegradation From Omics to Function and Application*. Caister Academic Press, Norfolk, s. 119–142.

Jasińska, A., Paraszkiewicz, K., Słaba, M., Długoński, J., 2014. *Microbial decolorization of triphenylmethane dyes*, w: Singh, S.N., (red.), *Microbial Degradation of Synthetic Dyes in Wastewaters. Environmental Science and Engineering*. Springer, Cham, s. 169–186.

Jędrczak, A., 2008. *Biologiczne przetwarzanie odpadów*. Wydawnictwo Naukowe PWN, Warszawa.

Kabała, C., Karczewski, A., 2017. *Metodyka analiz laboratoryjnych gleb i roślin*, http://www.up.wroc.pl/~kabala (access: 19.06.2020)

Klimiuk, E., Łebkowska, M. (red.), 2007. *Biotechnologia w ochronie środowiska*. Wydawnictwo Naukowe PWN, Warszawa.

Kronenberg, J., Bergier, T., Maliszewska, K., 2011. *Usługi ekosystemów jako warunek zrównoważonego rozwoju miast – przyroda w mieście w działaniach Fundacji Sendzimira*, w: Kosmala, M. (red.), *Miasta wracają nad wodę*. Polskie Zrzeszenie Inżynierów i Techników Sanitarnych, Toruń, s. 279–285.

Lovett, B., St. Leger, R., 2016. *Genetics and Molecular Biology of Entomopathogenic Fungi*. Academic Press (Elsevier).

Lisowska, K., Bernat, P., Paraszkiewicz, K., Długoński, J., 2008. *Zanieczyszczenia pochodzenia naturalnego*, w: Libudzisz, Z., Kowal, K., Żakowska, Z. (red.), *Mikrobiologia techniczna*. Tom 2: *Mikroorganizmy w biotechnologii, ochronie środowiska i produkcji żywności*. Wydawnictwo Naukowe PWN, Warszawa, s. 479–499.

Miksch, K., Sikora, J. (red.), 2018. *Biotechnologia ścieków*. Wydawnictwo Naukowe PWN, Warszawa.

Paniczko, S., 2013. *Budowa i zasada działania przydomowych oczyszczalni, w: Przydomowe oczyszczalnie ścieków. Poradnik*. Wydawnictwo Podlaska Stacja Przyrodnicza NAREW, Białystok, s. 9–85.

Różalska, S., Iwanicka-Nowicka, R., 2016. *Organic pollutants degradation by microorganisms: genomics, metagenomics and metatranstriptomics backgrounds*, w: Długoński, J. (red.), *Microbial Biodegradation From Omics to Function and Application*. Caister Academic Press, Norfolk, s. 1–12.

Słaba, M., Hrynkiewicz, K., Gadd, G.M., 2016. *Heavy metals elimination by microbial cells*, w: Długoński, J. (red.), *Microbial Biodegradation From Omics to Function and Application*. Caister Academic Press, Norfolk, s. 197–218.

Surmacz-Górska, J., 2018. *Biotechnologia dla gospodarki o obiegu zamkniętym*, w: Twardowski, T. (red.), *Biogospodarka, biotechnologia i nowe techniki inżynierii genetycznej. Nowoczesna biotechnologia podstawą biogospodarki*. Ośrodek Wydawnictw Naukowych ICHB PAN, Poznań, s. 56–63.

Wesołowska-Trojanowska, M., Targoński, Z., 2014. *Wykorzystanie serwatki w procesach biotechnologicznych*. Nauki Inżynierskie i Technologie 1 (12), 102–119.

Ziembińska-Buczyńska, A., 2016. *Molecular markers in biodegradation processes*, w: Długoński, J. (red.), *Microbial Biodegradation From Omics to Function and Application*. Caister Academic Press, Norfolk, s. 27–42.

Review articles

Abdullaeva, Z., 2017. *Nanomaterials in daily life. Compounds, synthesis, processing and commercialization*. Springer, Cham.

Ahmed, E., Holmström, S.J.M., 2014. *Siderophores in environmental research: Roles and applications*. Microbial Biotechnol. 7, 196–208.

Aw, K.M.S., Hue, S.M., 2017. *Mode of Infection of Metarhizium spp. Fungus and Their Potential as Biological Control Agents*. J. Fungi (Basel) 3 (2), 1–30.

Baig, T., Nayak, J., Dwivedi, V., Singh, A., Srivastava, A., Tripathi, P.K., 2015. *A review about dendrimers: Synthesis, types, characterization and applications*. Int. J. Adv. Pharm. Biol. Chem. 4, 44–59.

Banat, I.M., Satpute, S.K., Cameotra, S.S., Patil, R., Nyayanit, N.V., 2014. *Cost effective technologies and renewable substrates for biosurfactants' production*. Front. Microbiol. 12, 697.

Barabasz, W., Pikulicka, A., 2017. *Mykotoksyny – zagrożenie dla zdrowia ludzi i zwierząt. Część 2: Mykotoksyny zamaskowane – powstawanie, występowanie w żywności i paszach, metody identyfikacji i eliminacji mykotoksyn, prawodawstwo dotyczące mykotoksyn*. Journal of Health Study and Medicine 3, 109–132.

Bharathiraja, S., Suriya, J., Krishnan, M., Manivasagan, P., Kim, S.K., 2017. *Production of enzymes from agricultural wastes and their ptential industrial applications*. Adv. Food Nutr. Res. 80, 125–148.

Capriotti, A.L., Caruso, G., Cavaliere, C., Foglia, P., Samperi, R., Laganà, A., 2012. *Multiclass mycotoxin analysis in food, environmental and biological matrices with chromatography/mass spectrometry*. Mass Spectrom. Rev. 31, 466–503.

Corrales, J., Kristofco, L.A., Steele, W.B., Yates, B.S., Breed, C.S., Williams, E.S., Brooks, B.W., 2015. *Global assessment of Bisphenol A in the environment: review and analysis of its occurrence and bioaccumulation*. Dose-Response 13, 1–29.

Felczak, A., Zawadzka, K., Lisowska, K., 2014. Efficient biodegradation of quinolone. Factors determining the process. Int. Biodeterior. Biodegradation 96, 127–134.

García-Gallego, S., Franci, G., Falanga, A., Gómez, R., Folliero, V., Galdiero, S., de la Mata, F.J., Galdiero, M., 2017. *Function oriented molecular design: dendrimers as novel antimicrobials*. Molecules 22 (10), 1581–1610.

Ghosh, P., Thakur, I.S., Kaushik, A., 2017. *Bioassays for toxicological risk assessment of landfill leachate: A review*. Ecotox. Environ. Safe 141, 259–270.

Góralczyk, A., Jasińska, A., Długoński, J., 2016. *Mikroorganizmy w usuwaniu toksycznych barwników przemysłowych*. Post. Mikrobiol. 55, 424–432.

Góralczyk-Bińkowska, A., Jasińska, A., Długoński, J., 2019. *Characteristics and use of multicopper oxidases enzymes*. Advancements of Microbiology 58, 7–18.

Jacyno, M., Korkosz-Gębska, J., Krasuska, E., Milewski, J., Oniszk-Popławska, A., Trębacz, D., Wójcik, G., 2013. *Koncepcja biogazowni wykorzystującej odpady komunalne.* Rynek Energii 2 (105), 69–77.

Jalal, N., Surendranath, A.R., Pathak, J.L., Yu, S., Chung, C.Y., 2018. *Bisphenol A (BPA) the mighty and the mutagenic.* Toxicology Reports 5, 76–84.

Jansson, J.K., Hofmockel, K.S., 2018. *The soil microbiome-from metagenomics to metaphenomics.* Curr. Opin. Microbiol. 43, 162–168.

Janusz, G., Pawlik, A., Sulej, J., Świderska-Burek, U., Jarosz-Wilkołazka, A., Paszczyński, A., 2017. *Lignin degradation: Microorganisms, enzymes involved, genomes analysis and evolution.* FEMS Microbiol. Rev. 41, 941–962.

Kokkali, W., van Delft, W., 2014. *Overview of commercially available bioassays for assessing chemical toxicity in aqueous samples.* Trac-Trend Anal. Chem. 61, 133–155.

Kowalska, A., Walkiewicz, K., Kozieł, P., Muc-Wierzgoń, M., 2017. *Aflatoksyny – charakterystyka i wpływ na zdrowie człowieka.* M. Postepy Hig. Med. Dośw. 71, 315–327.

Kües, U., 2015. *Fungal enzymes for environmental management.* Curr. Opin. Biotechnol. 33, 268–278.

Li, P., Zhang, Z., Hu, X., Zhang, Q. 2013. *Advanced hyphenated chromatographic-mass spectrometry in mycotoxin determination: Current status and prospects.* Mass Spectrom. Rev. 32, 420–452.

Li, P.S., Tao, H.C., 2015. *Cell surface engineering of microorganisms towards adsorption of heavy metals.* Crit. Rev. Microb. 41, 140–149.

Liu, S.G., Zeng, G.M., Niu, Q.Y., Liu, Y., Zhou, L., Jiang, L.H., Tan, X., Xu, P., Zhang, C., Cheng, M., 2017. *Bioremediation mechanisms of combined pollution of PAHs and heavy metals by bacteria and fungi: A mini review.* Biores. Technol. 224, 25–33.

Lomascolo, A., Uzan-Boukhris, E., Sigoillot, J.C., Fine, F. 2012. *Rapeseed and sunflower meal: A review on biotechnology status and challenges.* Appl. Microbiol. Biotechnol. 95, 1105–1114.

Lovett, B., St. Leger, R.J., 2018. *Genetically engineering better fungal biopesticides.* Pest. Manag. Sci. 74 (4), 781–789.

Malaviya, P., Singh, A., 2016. *Bioremediation of chromium solutions and chromium containing wastewaters.* Crit. Rev. Microb. 42, 607–633.

Marchut-Mikołajczyk, O., Kwapisz, E., Antczak, T., 2013. *Enzymatyczna bioremediacja ksenobiotyków.* Inżynieria i Ochrona Środowiska 16, 39–55.

Marin, S., Ramos, A.J., Cano-Sancho, G., Sanchis, V., 2013. *Mycotoxins: occurrence, toxicology, and exposure assessment.* Food Chem. Toxicol. 60, 218–237.

Martinkova, L., Kotik, M., Markova, E., Homolka, L., 2016. *Biodegradation of phenolic compounds by Basidiomycota and its phenol oxidases: a review.* Chemosphere 149, 373–382.

Mazzoli, R., Giuffrida, M.G., Pessione, E., 2018. *Back to the past: "find the guilty bug-microorganisms involved in the biodeterioration of archeological and historical artifacts".* Appl. Microbiol. Biotechnol. 102 (15), 6393–6407.

Michałowicz, J., 2014. *Bisphenol A: sources, toxicity and biotransformation.* Environ. Toxicol. Pharmacol. 37, 738–758.

Mishra, A., Malik, A., 2013. *Recent advances in microbial metal bioaccumulation.* Crit. Rev. Environ. Sci. Technol. 43, 1162–1222.

Nesic, K., Ivanovic, S., Nesic, V., 2014. *Fusarial toxins: secondary metabolites of Fusarium fungi.* Rev. Environ. Contam. Toxicol. 228, 101–120.

Oleńska, E., Małek, W., 2013. *Mechanizmy oporności bakterii na metale ciężkie.* Post. Mikrobiol. 52, 363–371.

Quispe, C.A.G., Coronado, Ch.J.R., Carvalho, J.A., 2013. *Glycerol: production, consumption, prices, characterization and new trends in combustion.* Renew. Sust. Energ. Rev. 27, 475–493.

Rachoń D., 2015. *Endocrine disrupting chemicals (EDCs) and female cancer: informing the patients.* Rev. Endocr. Metab. Dis. 16, 359–364.

Ramachandran, S., Singh, S.K., Larroche, C., Soccol, C.R., Pandey, A. 2007. *Oil cakes and their biotechnological applications: a review.* Bioresour. Technol. 98, 2000–2009.

Ryan, M.P., Walsh, G. 2016. *The biotechnological potential of whey*. Rev. Environ. Sci. Biotechnol. 15, 479-498.

Sadh, K., Duhan, S., Duhan, J.S., 2018. *Agro-industrial wastes and their utilization using solid state fermentation: a review*. Bioresour. Technol. 5, 1.

Saha, M., Sarkar, S., Sarkar, B., Sharma, B.K., Bhattacharjee, S., Tribede, P., 2016. *Microbial siderophores and their potential application: a review*. Environ. Sci. Pollut. Res. 23, 3984–3999.

Salam, L.B., Ilori, M.O., Amund O., 2017. *Properties, environmental fate and biodegradation of carbazole*. 3Biotech. 7, 111.

Seachrist, D.D., Bonk, K.W., Ho, S.M., Prins, G.S, Soto, A.M., Keri, R.A., 2016. *A review of the carcinogenic potential of bisphenol A*. Reprod. Toxicol. 59, 167–182.

Shafei, A., Ramzy, M.M., Hegazy, A.I., Husseny, A.K., El-Hadary, U.G., Taha, M.M., Mosa, A.A., 2018. *The molecular mechanisms of action of the endocrine disrupting chemical bisphenol A in the development of cancer*. Gene 647, 235–243.

Slavin, Y.N., Asnis, J., Häfeli, U.O., Bach, H., 2017. *Metal nanoparticles: understanding the mechanisms behind antibacterial activity*. J. Nanobiotechnol. 15, 65–85.

Stępień, Ł., Chełkowski, J., 2010. *Fusarium head blight of wheat – pathogenic species and their mycotoxins*. World Mycotox J. 3, 107–119.

Subramanian, J., Ramesh, T., Kalaiselvam, M., 2014. *Fungal laccases – properties and applications: A Review*. Int. J. Pharm. Biol. Arch. 2, 8–16.

Udiković-Kolić, N., Scott, C., Martin-Laurent, F., 2012. *Evolution of atrazine-degrading capabilities in the environment*. Appl. Microbiol. Biotechnol. 96 (5), 1175–1189.

Umesha, S., Manukumar, H.M., Chandrasekhar, B., Shivakumara, P., Kumar, J., Raghava, S., Avinash, P., Shirin, M., Bharathi, T.R., Rajini, S.B., Nandhini, M., Rani, G.G., Shobha, M., Prakash, H.S., 2017. *Aflatoxins and food pathogens: impact of biologically active aflatoxins and their control strategies*. J. Sci. Food Agric. 97, 1698–1707.

Yagub, M.T., Sen, T.K., Afroze, S., Ang, H.M., 2014. *Dye and its removal from aqueous solution by adsorption: a review*. Adv. Colloid Interface Sci. 209, 172–182.

Yin, S., Wang, L., Kabwe, E., Chen, X., Yan, R., An, K., Zhang, L., Wu, A., 2018. *Coper bioleching in China: review and prospect*. Minerals 8, 32.

Zhang, W., Yin, K., Chen, L., 2013. *Bacteria-mediated bisphenol A degradation*. Appl. Microbiol. Biotechnol. 97, 5681–5689.

Original scientific papers

Bernat, P., Długoński, J., 2006. *Acceleration of tributyltin chloride (TBT) degradation in liquid cultures of the filamentous fungus Cunninghamella elegans*. Chemosphere 62, 3–8.

Bernat, P., Szewczyk, R., Krupiński, M., Długoński, J., 2013. *Butyltins degradation by Cunninghamella elegans and Cochliobolus lunatus co-culture*. J. Hazard. Mater. 246–247, 277–282.

Długoński, A., Szumański, M., 2016. *Atlas ekourbanistyczny zielonej infrastruktury miasta Łodzi. Tom 1a: Tereny zieleni strefy śródmiejskiej*. Łódzkie Towarzystwo Naukowe, Łódź.

Długoński, A., Szumański, M., 2016. *Use of recreational park biowaste as a renewable energy resource*. Journal of Chemistry and Environmental Engineering A, 23 (3), 265–274.

Dwivedi, U.N., Singh, P., Pandey, V.P., Kumar, A., 2011. *Structure – function relationship among bacterial, fungal and plant laccases*. J. Mol. Catal. B: Enzym. 68, 117–128.

Dziewanowska, M, Dobek, T., 2006. *Wartości cieplne liści wybranych gatunków drzew zbieranych na terenach zabudowanych*. Acta Agrophys. 8 (3), 551–558.

Eggert, C., Temp, U., Dean, J.F.D., Eriksson, K.E.L., 1996. *A fungal metabolite mediates degradation of non-phenolic lignin structures and synthetic lignin by laccase*. FEBS Letters 391, 144–148.

Fukuda, T., Uchida, H., Suzuki, M., Miyamoto, H., Morinaga, H., Nawata, H., Uwajima, T., 2004. *Transformation products of bisphenol A by a recombinant Trametes villosa laccase and their estrogenic activity.* J. Chem. Technol. Biotechnol. 79, 1212–1218.

Gutarowska, B., Celikkol-Aydin, S., Bonifay, V., Otlewska, A., Aydin, E., Oldham, A.L., Brauer, J.I., Duncan, K.E., Adamiak, J., Sunner, J.A., Beech, I.B., 2015. *Metabolomic and high-throughput sequencing analysis-modern approach for the assessment of biodeterioration of materials from historic buildings.* Front Microbiol. 6.

Hawkins, T.S., Gardiner, E.S., Comer, G.S., 2009. *Modeling the relationship between extractable chlorophyll and SPAD-502 readings for endangered plant species research.* Journal for Nature Conservation 17, 123–127.

Haase, D., Clemens, J., Wellmann, T., 2019. *Front and back yard green analysis with subpixel vegetation fractions from earth observation data in a city.* Landscape and Urban Planning 182, 44–54.

Janicki, T., Długoński, J., Krupiński, M., 2018. *Detoxification and simultaneous removal of phenolic xenobiotics and heavy metals with endocrine-disrupting activity by the non-ligninolytic fungus Umbelopsis isabellina.* J. Hazard. Mater. 360, 661–669.

Kopeć, D., Woziwoda, B., Fortysiak, J., Sławik, J., Ptak, A., Charąża, E., 2016. *The use of ALS; botanical, and soil data to monitor the environment hazards and regeneration capa city of areas devastated by highway construction.* Environ. Sci. Pollut. Res. 23, 13718–13731.

Monnet, J.-M., Bourrier, F., Dupire, S., Berger, F. 2016. *Suitability of airborne laser scanning for the assessment of forest protection effect against rockfall.* Landslides. 14, 299–310.

Nykiel-Szymańska, J., Stolarek, P., Bernat, P., 2018. *Elimination and detoxification of 2,4-D by Umbelopsis isabellina with the involvement of cytochrome P450.* Environ. Sci. Pollut. Res. Int. 25 (3), 2738–2743.

Paraszkiewicz, K., Bernat, P., Kuśmierska, A., Chojniak, J., Płaza, G., 2018. *Structural identification of lipopeptide biosurfactants produced by Bacillus subtilis strains grown on the media obtained from renewable natural resources.* J. Environ. Manage. 1, 65–70.

Parker, G.G., Harding, D.J., Berger, M.I., 2004. *A portable LIDAR system for rapid determination of forest canopy structure.* Journal of Applied Ecology. 41, 755–767.

Rajkowska, K., Koziróg, A., Otlewska, A., Piotrowska, M., Nowicka-Krawczyk, P., Brycki, B., Kunicka-Styczyńska, A., Gutarowska, B., 2016. *Quaternary ammonium biocides as antimicrobial agents protecting historical wood and brick.* Acta Biochim. Pol. 63 (1), 153–159.

Schlemmer, M., Gitelson, A., Schepers, J., Ferguson, R., Peng, Y., Shanahan, J., Rundquist, D. 2013. *Remote estimation of nitrogen and chlorophyll contents in maize at leaf and canopy levels.* International Journal of Applied Earth Observation and Geoinformation 25, 47–54.

Słaba, M., Szewczyk, R., Bernat, P., Długoński, J., 2009. *Simultaneous toxic action of zinc and alachlor resulted in enhancement of zinc uptake by the filamentous fungus Paecilomyces marquandii.* Sci. Total Environ. 407, 4127–4133.

Sławik, Ł., Niedzielko, J., Kania, A., Piórkowski, H., Kopeć, D., 2019. *Multiple flights or single flight instrument fusion of hyperspectral and ALS data? A comparison of their performance for vegetation mapping.* Remote Sens. 11 (8).

Tien, M., Kirk, T.K., 1983. *Lignin-degrading enzyme from the hymenomycete phanerochaete chrysosporium burds.* Science 221, 661–663.

Tomczyk-Żak, K., Kaczanowski, S., Drewniak, Ł., Dmoch, Ł., Sklodowska, A., Zielenkiewicz, U., 2013. *Bacteria diversity and arsenic mobilization in rock biofilm from an ancient gold and arsenic mine.* Sci. Total Environ. 461–462, 330–340.

Wariishi, H., Valli, K., Gold, M.H., 1992. *Manganese (II) oxidation by manganese peroxidase from the basidiomycetes Phanerochaete chrysosporium. Kinetic mechanism and role of chelators.* J. Biol. Chem. 267, 23688–23695.

Wellmann, T., Haase, D., Knapp, S., Salbach, Ch., Selsamd, P., Lausch, A., 2018. *Urban land use intensity assessment: The potential of spatio-temporal spectral traits with remote sensing.* Ecological Indicators 85, 190–203.

Websites

Ginalski, Z., 2011. *Substraty dla biogazowni rolniczych*, https://cdr.gov.pl/pol/OZE/substraty.pdf (access: 13.08.2019).

Gmina Korczyna, 2016. *Kompostujmy odpady zielone*, http://korczyna.pl/wiadomosc/kompostujmy-odpady-zielone (access: 13.08.2019).

Mobilny System Obserwacji i Wspomagania Analizy Powietrza SOWA, http://usm.net.pl/sowa-2 (access: 20.09.2019).

WHO Global Urban Ambient Air Pollution Database, 2016, http://www.who.int/phe/health_topics/outdoorair/databases/cities/en (access: 20.09.2019).

Związek Szkółkarzy Polskich, 2007. *Pasy zieleni między ulicami – nie tylko trawniki*, https://zszp.pl/roslina/zielen-miejska/pasy-zieleni-miedzy-ulicami-nie-tylko-trawniki (access: 13.08.2019).

http://agroenergetyka.pl/?a=article&idd=226 (access: 5.10.2019).

https://cdr.gov.pl/pol/OZE/substraty.pdf (access: 5.10.2019).

https://pl.wikipedia.org/wiki/Biogaz (access: 5.10.2019).

Conference materials

Długoński, A., 2017. *Wpływ procesu rewitalizacji terenów zielonych na zdrowie mieszkańców miast (studium przypadku centrum miasta Łodzi)*, w: *Gospodarka przestrzenna, stan obecny i wyzwania przyszłości – ujęcie interdyscyplinarne*, Konferencja naukowa, 25–26.09.2017, Uniwersytet Przyrodniczy we Wrocławiu, Wrocław.

Długoński, A., Siewiera, P., Bernat, P., Słaba, M., Długoński, J., 2017. *The significance of interdisciplinary research in the field of landscape architecture and environmental biotechnology. A case study of the green space revitalization in the city of Łódź (Poland)*. 6th Central European Congress of Life Science EUROBIOTECH, Kraków.

Góra-Drożdż, E., 2001. *Produkcja i wykorzystanie kompostów*. Materiały Konferencji Naukowej „Produkcja i wykorzystanie kompostów z terenu miasta Krakowa", http://www.zb.eco.pl/inne/kompost/2.htm (access: 20.09.2019).

Marczyński, S., 2009. *Stosowanie roślin okrywowych*. Materiały z seminarium II Wiosennej Wystawy Szkółkarskiej „Mazowiecka Zieleń – jakość i asortyment", 26–27.02.2009 r., Pęchcin k. Ciechanowa.

Standards and legal acts

Norma PN-ISO 10390:1997. Jakość gleby – oznaczanie pH.

Rozporządzenie Ministra Środowiska z dnia 1 września 2016 r. w sprawie sposobu prowadzenia oceny zanieczyszczenia powierzchni ziemi, Dz. U. z 2016 r., poz. 1395.

Ustawa z dnia 20 lipca 2017 r. Prawo wodne, Dz. U. z 2017 r., poz. 1566.

Ustawa z dnia 9 października 2015 r. o rewitalizacji, Dz. U. z 2015 r., poz. 1777.

6
Omics in microbial biotechnology

6.1. Proteomics in microbiological analysis of xenobiotics degradation

INTRODUCTION

Proteomics is a relatively complex field of science, because the number and concentration of proteins present in a cell constantly changes due to interactions with intra- and extracellular factors. The expression of proteins in cells may vary depending on the location (e.g. different types of tissues), cell cycle phase, or conditions in the surrounding environment (e.g. presence of toxic factors). In addition, proteins are polymers made up of more than 20 amino acids, which directly influences the significant diversity of possible sequences, and the function of proteins depends on the sequence, spatial structure and posttranslational modifications (PTMs). The total number, concentration, type and function of proteins present in the body during a complete life cycle is called a proteome. In proteomics, we typically try to:
- identify the proteins in the sample (qualitative proteomics);
- quantify the proteins in the sample (quantitative proteomics).

The determination of proteins is a complex procedure and despite the development of technology, it still encounters several basic problems:
- the concentration of different proteins in a sample can vary greatly – devices with a very wide and dynamic measuring range are required (wide measuring linearity range);
- identification: unique sequence of amino acids and one source of origin:
  — different genes can encode an identical sequence,
  — different organisms can synthesise identical proteins;

- splicing variants – the gene can be the source of different versions of mRNA;
- polymorphism – many genes occur in the form of alleles encoding different versions of the sequence;
- PTMs – are very diverse and significantly change the function of proteins.

6.1.1. Isolation and separation of proteins

The procedure of isolation, separation and determination of single proteins, protein fractions or the whole proteome varies depending on the organism from which the proteins are isolated (e.g. tissue vs. culture of unicellular organisms), cell structure (e.g. fungi vs. bacteria), cell growth phase /age, protein location (extra- and intracellular proteins) etc. Despite numerous detailed differences, several basic stages of proteomic analysis can be distinguished (Figure 6.1.1):

- purification of the cell mass to remove residual substrate or other cells, such as blood (in the case of tissues);
- cell homogenisation – release of proteins into solution;
- separation of cell fractions (optional);
- protein precipitation – obtaining a pure protein sample;
- dissolving the precipitate in an appropriate buffer;
- protein separation – 1-D and 2-D electrophoresis, capillary electrophoresis, chromatography (mainly ion-exchange or gel-type) and fractionation;
- digestion – peptides obtained either after separation or without separation (total proteome digesting);
- sequence analysis and PTMs – currently, mainly by mass spectrometry.

The process of cell homogenisation is carried out in different types of homogenisers, depending on the tissue – mainly mechanical methods (ball mills, presses) or using ultrasound. The process occurs at a reduced temperature (0–4°C) with the addition of cell protease inhibitors. The precipitation of proteins from the solution aims to obtain a purified protein preparation and can be carried out in two basic ways: by adding acetone or trichloroacetic acid (TCA) solution in acetone and by desalination with a concentrated solution of Na_2SO_4, $MgSO_4$

or $(NH_4)_2SO_4$. The second method is a completely reversible process and is mainly used to obtain biologically active protein preparations (e.g. enzymes).

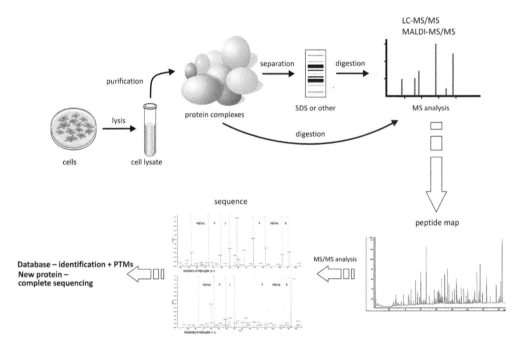

The most common technique used for protein separation is electrophoresis. Electrophoresis is the movement of dissolved ionised substance under the influence of an externally applied potential difference. A particle with an electric charge q located in an electric field E is affected by a force Fe = qE. This force causes the particle to move (uniformly accelerated motion, according to Newton's Second Law). The movement of the particle is counteracted by a viscous resistance force, which in the case of a spherical particle is described by the Stokes' formula: F = 6πrvη. The velocity of motion of a given particle is constant, if other parameters remain constant during the process, such as:

- potential difference;
- current rate;
- environment pH;
- temperature;
- solution conductivity and viscosity.

Fig. 6.1.1. General scheme of a typical analytical procedure for proteomic studies.

The movement depends on:
- particle charge;
- particle size (mass) and geometry.

Electrophoresis is performed in a thin layer of gel in the presence of appropriate buffers (electrolytes), which are designed to: transfer voltage, maintain pH and determine the charge of the analyte. In the case of proteomics, polyacrylamide gels and barbital buffers or Tris-EDTA with the addition of sodium dodecyl sulphate (SDS) – SDS-PAGE electrophoresis are used. SDS is anionic detergent. Molecules in SDS solution are negatively charged over a wide pH range and polypeptide chains bind SDS proportionally to their relative weight. Negative SDS charges destroy most complex protein structures and are strongly attracted to the anode (+) in an electric field. The molecules in the electric field move in the gel matrix at different rates, which ultimately leads to their separation.

The proteins separated on the gels are invisible, so it is necessary to use specific staining to visualise and analyse the gels (image analysis techniques). Different dyes are used, depending on the assumed analytical requirements, especially in terms of detection limits (LOD), but also taking into account the final protein sequencing method (not all dyes are suitable for MS/MS analysis) (Table 6.1.1).

Dye	LOD	MS/MS
Ponceau S	1–2 mg	++
Amido Black	1–2 mg	++
Coomassie Blue	1.5 mg	+++
India Ink	100 ng	++
Silver stain	10 ng	+
Colloidal gold	3 ng	+
Fluorescent (DIGE)	125 pg	+++

Table 6.1.1.
Dyes used for visualisation in protein electrophoresis.

Electrophoresis can be one-dimensional (1-D) or two-dimensional (2-D). 1-D electrophoresis is a standard procedure based on protein weight separation (SDS-PAGE). 2-D electrophoresis is a two-stage technique (Figure 6.1.2). In the first part, proteins are separated according to their isoelectric points using gel strips with an immobilised pH gradient (IPGE), isoelectric focusing (IEF) (Figure 6.1.3) or electrophoresis in the pH gradient. The second stage involves SDS-PAGE separation by mass.

Fig. 6.1.2.
2-D electrophoresis scheme.

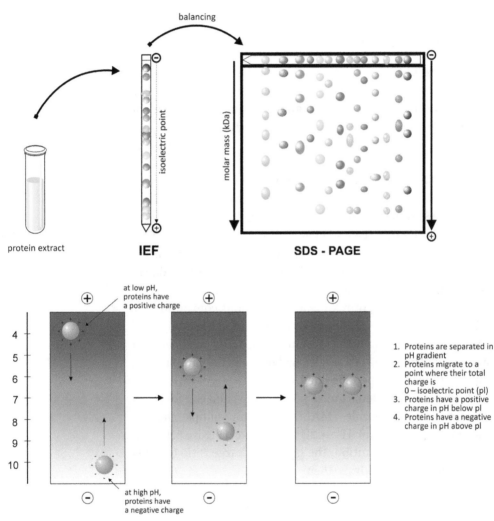

Fig. 6.1.3. Electrical focusing.

1. Proteins are separated in pH gradient
2. Proteins migrate to a point where their total charge is 0 – isoelectric point (pI)
3. Proteins have a positive charge in pH below pI
4. Proteins have a negative charge in pH above pI

Capillary electrophoresis, as a modern variant of electrophoresis, is based on the same principles as other electrophoretic techniques, but is carried out in a gel placed in a capillary (average below 1 µm). This technique is characterised by very high resolution and can be combined with detectors such as UV-VIS or MS/MS.

Both, analytical and preparative chromatographic techniques are also often used for protein separation, mainly for fractionation of protein preparations. Chromatographic techniques are discussed in Section 2.3.

6.1.2. Identification of proteins

The separated proteins or total protein are then subjected to a chemical or enzymatic digestion process to obtain a mixture of peptides undergoing MS/MS analysis. Currently, the most commonly used method is enzymatic digestion using trypsin, which catalyses the hydrolysis of peptide bonds between arginine and lysine.

The gold standard used in protein identification and/or quantitative analysis is currently mass spectrometry, using both MS and MS/MS scans to obtain information about the mass of peptides (MS Scan), their sequence, type and location of PTMs (MS^2 scan) and their quantity (MRM). Fragmentation spectra of peptides and proteins are relatively simple to interpret due to the fact that the main fragmentations occur within the peptide bonds (Figure 6.1.4), providing a series of ions conventionally named a, b, c and x, y, z. The most important in the analysis of the sequence are usually the most intense ions – y and b, formed as a result of breaking of a single bond between the carbonyl and amine groups of the peptide bond.

Peptides and proteins ionise positively (+) and can ionise as a single-charged molecules (MALDI source) or multiple-charged molecules (ESI source): $H^+ \rightarrow 2H^+ \rightarrow 3H^+$ etc. (Figure 6.1.5). If, according to the definition, the mass-to-charge ratio (m/z) is determined in MS analysis, then:

1,000 Da $H^+ \rightarrow$ 1001 m/z $\rightarrow$ isotopic spectrum every 1 amu
1,000 Da $2H^+ \rightarrow$ 501 m/z $\rightarrow$ isotopic spectrum every 0.5 amu
1,000 Da $3H^+ \rightarrow$ 334,33 m/z $\rightarrow$ isotopic spectrum every 0.33 amu etc.

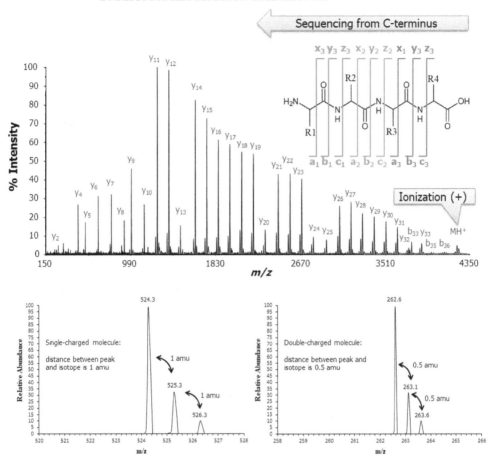

LAADEDDDDDDEEDDDDEDDDDDDFDDEEAEEKAPVKK

The MS^2 mass spectrum of a multiple-charged molecule is slightly more complex because we observe both single-charged fragmentation ions F$^+$ with larger and smaller m/z values than the multiple-charged pseudo-molecular ion [M + H]$^{n+}$ and multiple-charged fragmentation ions F^{n+} with smaller m/z values than the multiple-charged pseudo-molecular ion [M + H]$^{n+}$ (Figure 6.1.6). It should also be emphasised that while a single-charged peptide connects the proton to a specific amino acid – usually from the N-terminus, in the case of multiple-charged molecule, the second and subsequent protons do not have a specific location in the amino acid chain (Figure 6.1.6).

Fig. 6.1.4. Peptide fragmentation in MS/MS type detector according to the scheme: y2 – y1 (etc.) = AA m/z.

Fig. 6.1.5. The mass and isotopic spectrum of a single and double-charged molecule of 523.3 Da.

439

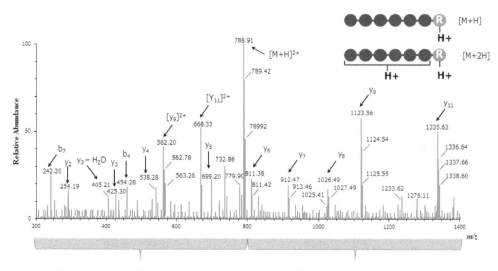

Fragmentation ions 1+ and 2+   Fragmentation ions 1+

Data (mass spectra) obtained during peptide sequencing with MS/MS technique are the basis for determining the sequence together with PTMs and identifying the protein on the basis of these data in database applications. Databases used in proteomics contain both experimentally verified amino acid sequences and known DNA and RNA sequences of organisms whose genome has been sequenced but not experimentally verified for the product of gene expression ("predicted" proteins). Currently, two basic algorithms are used to search the proteomic databases – MASCOT and Paragon. The first one is a very fast algorithm that allows the search of up to 9 possible PTMs in one query (out of more than 850 available) and returns information about the identification of a protein with the appropriate score, statistical significance of the match and the exact location of the selected PTMs. The second algorithm operates more slowly, but searches the databases for all defined post-translation modifications (over 900) and various other possible biological modifications (e.g. conversion of amino acids in a sequence) or non-biological ones, introduced e.g. as a result of a specific sample preparation. As a rule, the results obtained with the Paragon algorithm are characterised by a much higher sequence coverage compared to the MASCOT algorithm. If proteins are identified in organisms with an unknown genetic code, the proteins are identified by

Fig. 6.1.6. Mass spectrum MS^2 of peptide from ion [M + H]$^{2+}$

searching for homology in the most related organisms (e.g. DELTA-BLAST algorithm in the NIST database), which often results in finding constitutional sequences and assigning the overall function of the tested protein.

PRACTICAL PART

Materials and media

1. Universal medium.
2. Sealing gel:
   - demineralised water .... 0.87 ml
   - 0.5% (w/v) SDS .... 0.4 ml
   - 625 mM Tris-HCl, pH 6.8 .... 0.4 ml
   - acrylamide : bis-acrylamide .... 0.33 ml
   - 10% (w/v) APS .... 20 µl
   - TEMED .... 5 µl
3. Separating gel 12.5%:
   - acrylamide : bis-acrylamide .... 2.5 ml
   - 1.88 M Tris-HCl, pH 8.8 .... 1.2 ml
   - 0.5% (w/v) SDS .... 1.2 ml
   - demineralised water .... 1.1 ml
   - 10% (w/v) APS .... 30 µl
   - TEMED .... 5 µl
4. Thickening gel:
   - demineralised water .... 0.87 ml
   - 0.5% (w/v) SDS .... 0.4 ml
   - 625 mM Tris-HCl, pH 6.8 .... 0.4 ml
   - acrylamide : bis-acrylamide .... 0.33 ml
   - 10% (w/v) APS .... 10 µl
   - TEMED .... 2 µl
5. Load buffer:
   - glycerin .... 1 ml
   - 1 M Tris .... 0.5 ml
   - 10% (w/v) SDS .... 0.5 ml
   - β- mercaptoethanol .... 0.5 ml
   - 1% (w/v) bromophenol blue .... 0.5 ml

6. SSSB-thio buffer:

	Per litre	Per 10 ml	Per 20 ml	Per 30 ml
7 M urea	420 g	4.2 g	8.4 g	12.6 g
2 M thiourea	152 g	1.52 g	3.04 g	4.56 g
4% (w/v) CHAPS	40 g	0.4 g	0.8 g	1.2 g
40 mM Tris	4.84 g	48.4 mg	96.8 mg	145.2 mg
0.2% ampholytes (stock 40%)	50 ml	50 µl	100 µl	150 µl

7. Electrode buffer:
- Tris $\qquad$ 25 mM
- glycine $\qquad$ 192 mM
- SDS $\qquad$ 0.1%
- pH 8.3

8. Staining reagent:
- Coomassie Brilliant Blue R250 $\quad$ 0.5 g
- methanol $\qquad$ 225 ml
- demineralised water $\qquad$ 225 ml
- acetic acid $\qquad$ 50 ml

9. Discolouring reagent:
- demineralised water $\qquad$ 325 ml
- methanol $\qquad$ 125 ml
- acetic acid $\qquad$ 50 ml

10. Reagents for protein digestion:
- 100 mM NH_4HCO_3 in distilled water: 0.158 g/ 20 ml;
- 50 mM NH_4HCO_3 / CH_3CN (50:50 v/v);
- 10 mM DTT in 100 mM NH_4HCO_3 (freshly prepared). To obtain 1.5 ml 10 mM DTT, add 15 µl of 1 M DDT stock (storage: -20°C for one year) to 1.485 ml 100 mM NH_4HCO_3;
- 50 mM of freshly prepared solution of iodoacetamide (IAA) in 100 mM NH_4HCO_3. To obtain 1.5 ml of 50 mM IAA, add 75 µl of 0.5 M IAA stock to 1.425 ml of 100 mM NH_4HCO_3;
- freshly prepared trypsin solution. To obtain 1 ml of the solution, dissolve the contents of the tube (20 µg trypsin sequencing grade, Promega) in 200 µl dissolving buffer (Promega) and make up to 800 µl 25 mM NH_4HCO_3;
- 2% acetonitrile (ACN) / 0.1% aqueous solution of trifluoroacetic acid (TFA);

11. Other reagents:
    - 0.1 M ethanol solution of phenylmethane sulfonyl fluoride (PMSF);
    - 10% trichloroacetic acid solution (TCA) in acetone;
    - acetone (ultra pure);
    - Bardford reagent;
    - MALDI matrix – α- cyano-4-hydroxycinnamic acid (10 mg/ml) in 50% ACN with 0.1% TFA.
12. Apparatus: MALDI-TOF/TOF 5800 system (Sciex).

Aim of the exercise
To conduct the full procedure of identification of selected proteins with altered expression in the control and tested systems during the xenobiotic biodegradation process.

Procedure

1. Carry out *Pseudomonas* B219 culture on liquid medium at 37°C, shaking minimum 140 rpm, minimum 24 h. Control system – culture; test sample – culture with added xenobiotics.
2. Centrifuge the culture to obtain cell suspension. Discard the supernatant and rinse the biomass with water and centrifuge again (repeat 2–3 times).
3. Protein isolation from the cellular fraction:
   a) mechanically homogenise the biomass obtained (0.1 mm bed – bacteria) in SSSB-thio buffer with addition of 1 mM PMSF;
   b) centrifuge homogenate for 5 min at 4°C at 10,000 × g;
   c) precipitate the solution over sediment with 10% TCA in acetone for 30–45 min at 4°C;
   d) centrifuge samples for 10 min (10,000 × g) at 4°C;
   e) rinse the precipitate with a cold solution of acetone : water 90:10 (–20°C);
   f) centrifuge the samples for 5 min (10,000 × g) at 4°C (repeat 2–3 times together with point 3e above);
   g) transfer the protein precipitate into an Eppendorf Lobind tube (1.5 ml);

h) evaporate the remaining acetone from the precipitate in the furnace for about 2–5 min at 40°C or under pressure;

i) dissolve the solution in 500 µl SSSB-thio buffer (sonication, vortex from 10 to 30 min or 4°C per night);

j) determine protein content in the test and reference samples by the Bradford method (5 µl of sample + 250 µl of Bradford reagent – wait 10 min and measure the absorption, plot curves/read the protein content of the samples);

k) store the samples at −80°C.

4. Stain the protein gel with Coomasie Brillant Blue for 24 h on the horizontal cradle.

5. Discolour gel with a discolouring agent.

6. Take photos of the gel and analyse the expression of proteins in GelAnalyzer (or other equivalent). On the basis of the gel analysis, select proteins with significant relative changes of expression (at least two changes of expression).

7. In-gel protein digestion:

   a) cutting out the protein strips from the gel:

- cut the gel strips into small pieces (1 mm^2) in tubes up to 1.5 ml (Protein Lo-bind type);
- rinse with 100 µl 50 mM NH_4HCO_3/ACN (50:50 v/v), mixing within 15 min. If the gel pieces are still blue, repeat the NH_4HCO_3/ACN rinsing (the gel pieces should be finally transparent);
- add 100 ul ACN to dehydrate the gel, shake 5–10 min. The pieces of gel should shrink and turn white. Discard the supernatant;

   b) protein (Cys) reduction/alkylation:

- add 50 µl/sample of 10 mM DTT in 100 mM NH_4HCO_3;
- incubate at 56°C for 30 min. Discard the supernatant;
- add 50 µl/sample of 50 mM iodoacetamide in 100 mM NH_4HCO_3;
- incubate at room temperature for 30 min in the dark. Discard the supernatant;

- dehydrate with 100 ul ACN, vortex 5–10 min. Discard the supernatant;
- rinse gel pieces with 100 μl 100 mM NH_4HCO_3, vortex for 10–15 minutes. Discard the supernatant;
- dehydrate with 100 μl ACN, vortex 5–10 min. Discard the supernatant;
- rinse gel pieces with 100 μl 100 mM NH_4HCO_3, vortex for 10–15 minutes. Discard the supernatant;
- dehydrate with 100 ul ACN, vortex 5–10 min. Discard the supernatant;
- dry samples under pressure (5 min) or at room temperature (to dry state);

c) protein digestion in gel:
- add 10 to 30 μl (or more to cover the gel) of trypsin solution in 25 mM NH_4HCO_3 and incubate at 37°C for 24 h.

8. Peptide extraction:
   a) collect the remaining solution in the new described tubes (0.5 ml Protein Lo-Bind);
   b) add 100 μl of the following to the gel fragments (enough to cover the gel):
   - 2% ACN/0.2% TFA, vortex for 20 min;
   - collect the supernatant from point 8a in tubes;
   - 50% ACN/0.2% TFA, vortex for 15 min,
   - collect the supernatant from point 8a in tubes;
   - 90% ACN/0.2% TFA, vortex for 10 min,
   - collect the supernatant from point 8a in tubes;
   c) the supernatant collected contains peptides. Store in the fridge at 4°C;
   d) concentrate samples under pressure or evaporate to dryness;
   e) dissolve evaporated samples in 20–50 μl 2% ACN/0.1% TFA.

9. MALDI-TOF/TOF analysis:
   - mix the extracts with the matrix (CHCA) in a 1:1 ratio;
   - apply samples (1 μl) to MALDI plates (4 repetitions per sample) and wait until they are completely dry;

- analyse in the range of 800–4000 m/z MS. Scan with automatic selection of up to 20, and from 20 to 40 of the most intense signals from appropriate repetitions of the MS/MS analysis.

Analysis of results

1. Identification of proteins by sequence and determination of PTMs using ProteinPilot software with MASCOT and Paragon algorithm.
2. Determination of the function of proteins in the examined process based on data from their identification, based on literature data and/or NIST database search with delta-BLAST algorithm.
3. Formulation of conclusions and preparation of report.

6.2. Metabolomic analysis as a tool for multi-level characterisation of the biodegradation process

INTRODUCTION

Metabolomics is a science that deals with the examination and the quantitative and qualitative analysis of metabolites in cells and tissues of living organisms. Together with other departments of omics, metabolomics is an element of systems biology, dealing with the study of complex and complicated interactions that occur in biological systems.

Metabolites are products, intermediates or derivatives created in various biochemical reactions taking place in living cells. It is assumed that the subject of metabolomic studies are molecules whose molecular weight is no greater than 1,000 Da, but this type of generalisation is increasingly neglected, including all monomeric molecules regardless of their weight in metabolomic analyses. Metabolites are divided into two basic groups: the so-called primary metabolites, resulting from the basic transformations necessary for growth, development and reproduction, and secondary metabolites, not directly

involved in these processes but having other important
biological functions.

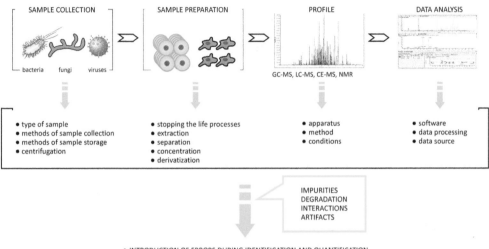

• INTRODUCTION OF ERRORS DURING IDENTIFICATION AND QUANTIFICATION
• MISINTERPRETATION OF METABOLIC PATHWAYS OR BIOMARKERS

Metabolome research requires modern high-capacity
techniques such as gas chromatography (GC), liquid
chromatography (LC) or capillary electrophoresis (CE), coupled
with mass spectrometry (MS and MS/MS) or nuclear magnetic
resonance (NMR). Analytical methods used in metabolomics
can be divided into targeted and non-targeted ones. The first
group includes analytical methods which of the total number
of molecules present in the sample, focus on user-defined
compounds – mainly MS/MS (MRM) methods. The second
group includes methods characterised by very general data
filtering assumptions (e.g. mass range 100–400 Da, intensity,
etc.), allowing a theoretical analysis of all molecules present
in the sample regardless of their structure and providing
an overall picture of the sample – MS and NMR methods.
A typical analytical procedure used in metabolomic analysis is
presented in Figure 6.2.1. Despite the use of high-tech analysis,
metabolomics is still a complex and difficult field of science due
to the following limitations and problems:

  • the metabolome analysed is a collection of numerous
    compounds, including: amino acids and their derivatives,

Fig. 6.2.1. Standard
procedures used in
metabolomic analysis.

lipids, carbohydrates, nucleotides, organic acids, hormones, signal substances, cofactors, alcohols etc.;
- great biological diversity in the concentration of metabolites from pM to mM;
- great diversity in terms of chemical structure and physicochemical properties (masses up to 1,000 Da or more, hydrophobic, hydrophilic, volatile, viscous, delicate, aromatic, aliphatic, etc.) – there is no single good sample preparation method for all types of compounds;
- some of the metabolites are so reactive and/or delicate that they do not survive the sample preparation process (a derivative or other form is produced, e.g. reduced or oxidised);
- chromatography and detection limitations – not everything can be done with the same analytical method.

Metabolomics has found many applications in various fields of science and routine analysis, such as:
- identification of the influence of external and internal factors (e.g. temperature, xenobiotic, antibiotic, age, growth phase) on metabolic pathways;
- study of the toxicity of the substance and its effect on intracellular metabolism;
- agriculture – GMO plants (overproduction or extinction of certain metabolic products);
- searching for biomarkers of diseases, environmental and other;
- personalised medicine.

The increasing knowledge and amount of data related to metabolomic research contributes to the creation of a variety of bioinformatics tools that facilitate data analysis and allow for the continuous development of science for further discoveries. A good example of such tools are available online: Metaboanalyst v4.0 (www.metaboanalyst.ca), which contains tools for processing, analysing and interpreting data, or KEGG Pathway database (www.genome.jp/kegg/pathway.html), used mainly to overlay experimental data on developed metabolic maps to visualise the studied changes in selected organisms or create new maps, e.g. for xenobiotics (Figure 6.2.2).

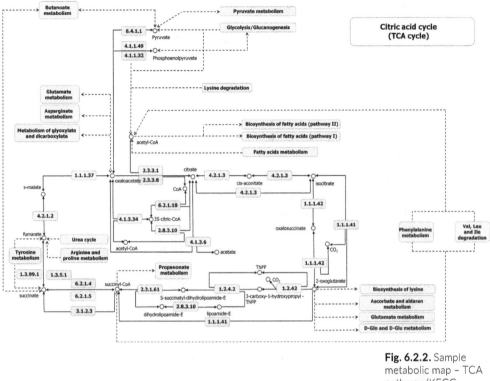

Fig. 6.2.2. Sample metabolic map – TCA pathway (KEGG Pathway database, 2018).

PRACTICAL PART

Materials and media

1. Media: mineral with 1% glucose added.
2. Reagents: 4-*n*-NP (5 mg/ml) in ethyl acetate; N, O-Bistrifluoroacetamide (BSTFA), anhydrous $Na_2(SO_4)_2$, reagents for GC-MS analysis (ultra pure): ethyl acetate, hexane; reagents for LC-MS/MS analysis (ultra pure): water, methanol, acetonitrile (ACN), formic acid, ammonium formate.
3. Apparatus: gas chromatograph Agilent GC7890 with mass spectrometer MS5975C, liquid chromatograph Agilent 1200 and mass spectrometer QTRAP 3200 (Sciex).

Aim of the exercise
*Quantitative and qualitative analysis of the 4-n-nonylphenol (4-n-NP)
biodegradation process with targeted metabolomic analysis of intracellular
extracts from Aspergillus versicolor culture.*

Procedure

1. Preparation and conduction of *A. versicolor* culture in liquid medium:
   - take 2 ml of *A. versicolor* inoculum (24 h) and add to 18 ml of mineral medium with 1% glucose.;
   - add xenobiotic to obtain the final concentration of 50 mg/l in the culture;
   - incubate the cultures are for 24, 48 and 72 h on a rotary shaker (140 rpm) at 28°C.;
   - prepare biotic cultures and abiotic controls and incubate under the same conditions as the test samples.
2. Preparation of the 4-*n*-NP standard curve:
   - prepare a 4-*n*-NP standard curve in ethyl acetate in the following xenobiotic concentration system: 1, 5, 10, 25, 50, 75, 100 µg/ml;
   - analyse samples by GC-MS method using a gas chromatograph with mass spectrometer (Agilent GC 7890 and MS 5975C) with HP 5 MS column (30 m × 0.25 mm × 0.25 µm) in the temperature range 70–300°C in SIM mode.
3. Preparation of samples for quantitative and qualitative GC-MS analysis:
   - disintegrate the cultures using ultrasonic homogeniser (3 min in system: 10 sec homogenisation – 5 sec break);
   - extract the samples three times with ethyl acetate (1:1), dehydrate with anhydrous $Na_2(SO_4)_2$, filter and evaporate on a rotary evaporator;
   - for quantitative analysis, dissolve the post-culture extract in an appropriate amount of ethyl acetate, filter, take 1 ml into the chromatographic vials;

- quantitative determination should be performed by GC-MS method in SIM mode;
- for qualitative analysis, dissolve the extract in an appropriate amount of ethyl acetate, take 50 μl of the extract into the chromatographic vials and then add 50 μl of derivatisation reagent (BSTFA);
- derivatisation should be carried out for 1 h at 55°C;
- add 100 μl of ethyl acetate to the samples after derivatisation;
- qualitative determination is performed by means of the GC-MS method using a gas chromatograph with mass spectrometer (Agilent GC 7890 and MS 5975C) on HP 5 MS column (30 m × 0.25 mm × 0.25 μm) in the temperature range 70–300°C in SCAN mode.

4. Preparation of samples for metabolomic LC-MS/MS analysis:
   - filter cultures (biotic controls and test samples);
   - freeze mycelium in −70°C (for storage for further preparation);
   - gradually thaw at 4°C;
   - weigh 100 mg of wet mycelium (3 repetitions from each sample);
   - homogenise the samples (1 mm glass bed) in 1 ml of 80% ethanol with 0.1% formic acid. Conduct the process cold (!) at least three times;
   - incubate the samples from 1 to 2 h at −20°C;
   - centrifuge the samples (4°C, 10,000 rpm, 5–10 min.);
   - take 100 μl of supernatant in new tubes and make up to 1 ml with 2% ACN solution in water (10-fold dilution).;
   - dilute the samples 1,000 and 5,000 times 2% with ACN solution in water.;
   - transfer the samples to a 96-well plate;
   - perform LC-MS/MS (MRM) analysis using liquid chromatograph Agilent 1200 and QTRAP 4500 spectrometer (Sciex).

5. Perform quantitative analysis of 4-*n*-NP loss in *A. versicolor* cultures:

- plot the standard curves on the basis of the analysed standard samples;
- collect quantitative data.

6. Perform a qualitative analysis of the 4-*n*-NP decomposition process:
- on the basis of chromatograms and analysis of mass spectra from control and test samples, search for potential metabolites and propose their structure.

7. Perform PCA analysis of samples after metabolomic targeted LC-MS/MS analysis.

8. Describe the results obtained, conclusions, and procedures carried out in a report.

Analysis of results

1. Construct graphs and/or tables with quantitative results.

2. Based on the analysis of the trend and structure of the identified metabolites, propose a 4-*n*-NP decomposition pathway.

3. Indicate and interpret key and collateral changes in the metabolic profile of the test organism resulting from the presence and biodegradation of 4-*n*-NP.

4. Prepare the report.

6.3. Application of lipidomics in the study of detoxification processes in microorganisms

INTRODUCTION

Over the last dozen or so years, thanks to the development of analytical techniques, our knowledge of metabolomics, i.e., the study of small molecules in biological samples, has been greatly expanded. One of the elements of metabolomics is lipidomics – a discipline that characterises lipid compounds in biological samples.

Many publications in the field of lipidomics describe the phenomenon of adaptation of microbial membranes in response to the influence of such factors as temperature changes, nutrient deficiencies or the presence of toxic substances in the living environment of microorganisms. Thanks to such modifications in the structure and functioning of cell membranes, the survival of microorganisms in changing environmental conditions is possible. To understand what modifications occur in the membranes, lipids involved in the formation of cell structures are identified. Until a dozen or so years ago, the most popular way to study changes in cell membrane lipids was to determine fatty acids using gas chromatography coupled to a mass spectrometer (GC-MS) or a flame ionisation detector (GC-FID). The development of molecular ionisation methods in mass spectrometers and mass analysers is the beginning of the exact identification of lipids from different classes. Special attention is paid to phospholipids – the basic building blocks of the biological membranes of cells.

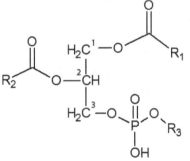

Fig. 6.3.1.
Phospholipid scheme.

R1, R2 - fatty acid
R3 - polar functional group

Phospholipids belong to the group of lipids. Due to their chemical structure, they are divided into glycerophospholipids (discussed later) and sphingophospholipids. Glycerophospholipids contain in the hydrophobic part, two chains of fatty acids attached to the glycerol residue. In their hydrophilic part there is a phosphate group associated with ethanolamine (phosphatidylethanolamine, PE), choline (phosphatidylcholine, PC), serine (phosphatidylserine, PS), inositol (phosphatidylinositol, PI) or glycerol (phosphatidyloglycerol, PG) (Figure 6.3.1). Cardiolipins (diphosphatidylglycerols) are also determined in bacterial cells and mitochondrial membranes. In fungal cells, the main fatty acids attached to the glycerol residue are 16- and 18-carbon acids. They occur in the form of saturated or unsaturated acids. The latter have one, two or three double bonds.

Phospholipid isolation

The first stage of phospholipid isolation from biological samples is their separation. The most popular technique of biomass fragmentation is homogenisation – cellular disintegration. From the biomass prepared in this way, the extraction of lipids is most often carried out using solutions of chloroform, methanol and water, according to the methods described by Folch et al. (methanol: chloroform, 1:2 v/v) or Bligh and Dyer (methanol: chloroform, 2:1 v/v). As chloroform is a very toxic compound, the laboratory also uses a phospholipids extraction method based on a mixture of Methyl tert-butyl ether (MTBE) and methanol (methanol: MTBE, 1:5.5 v/v). The organic phase is dehydrated (e.g. using anhydrous sodium sulphate) and concentrated. A small amount (0.01%) of butylated hydroxytoluene (BHT), a synthetic antioxidant that limits the rate of lipid oxidation, can be added during extraction.

Phospholipid analysis using liquid chromatography and tandem mass spectrometry (LC-MS/MS)

Separation of extracted phospholipids is performed using liquid chromatography techniques. A normal phase system (NP) was used initially, in which the stationary phase is much more polar than the mobile phase. The order of outflow of individual phospholipids was determined by the increasing polarity of individual classes (PG, PI, PE, PS and then PC). Currently, the reversed phase system is most often used (RP, in which the stationary phase is less polar than the mobile phase). In this technique, the retention time of phospholipids is determined, among others, by the length of the fatty acid residue chains and the presence of double bonds. Silica gel modified with octadecyl groups (phase C18) is most often used as a column filler.

Using the resolving power of chromatography, individual components of the mixture of compounds extracted from the biological sample are obtained. Quantitative and qualitative analysis of separated phospholipids classes is performed using mass spectrometers. Based on the results obtained using qualitative methods and analysis of phospholipids mass spectra (Figure 6.3.2), the so-called Multiple Reaction Monitoring (MRM) pairs – parent and daughter ions, are selected for quantitative determination.

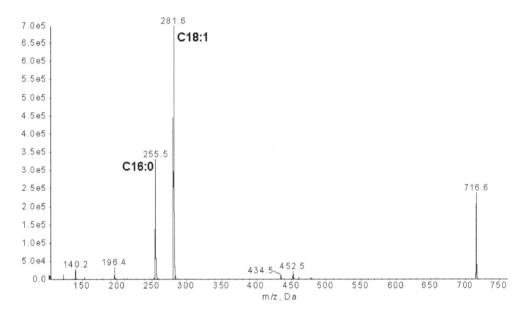

On the basis of the obtained surface fields and standards, it is possible to determine the amount of individual phospholipids in the extract. Phospholipids obtained by synthesis which do not occur naturally in microbial cells are used as internal standards. A separate standard should be prepared for each class.

When trying to identify individual phospholipids, it may be helpful to use software containing the most important information about the mass and structure of lipids, e.g. Lipid MS Predictor (LIPID MAPS). However, due to the large amount of data received and difficulties with their interpretation, the PCA can also be used.

Fig. 6.3.2. Mass spectrum of phospholipid PE 16:0/18:1. From this spectrum, the transition pair, 716.6-281.6, was used for MRM

Comparative analysis

Comparative lipidomic analysis uses data sets from samples obtained in at least two systems that differ in biological variables. Lipidomic analysis of fatty acids are most often described. Changes of fatty acids in microbial cells in response to environmental pollution have been discussed in many articles in the last 30 years. However, there is little information available on the role of phospholipid groups.

In many bacterial strains, PE and PG are the most numerous phospholipids, which accounts for 60–95% of all cell membrane phospholipids.

Organic solvents, such as benzene and toluene, destabilise the structure of the PE double layer. To counteract this, some bacteria have developed several adaptive mechanisms. A decrease in the amount of PE and an increase in PG and CL were observed in the culture of *Pseudomonas putida* S-12 with toluene. This change is likely to stabilise the membrane by lowering its fluidity, which reduces the permeability to lipophilic particles.

PC is the main phospholipid forming cell membranes in eukaryotes. However, it is estimated that probably no more than 10% of all bacteria contain PC as membrane phospholipid.

The methods described above were used, among others, to analyse the effect of the toxic compound tributyltin (TBT) on the phospholipid profile of *Cunninghamella elegans* microfungus. A decrease in the DBI (double bond index) was shown in particular classes of *C. elegans* phospholipids exposed to TBT, i.e., the predominance of lipids containing saturated fatty acids. Taking into account the fatty acids included in the phospholipids, a decrease in the amount of lipids from different classes containing gamma-linolenic acid was found. Then, the obtained results were subjected to the principal components analysis (PCA), confirming changes in the phospholipid profile of the microorganism depending on the growth phase and the presence of the biocide. Comparing the quantities of individual classes of phospholipids, an increase in the amount of the main membrane building block PC and a decrease in PE were observed in the presence of butyltin. Phospholipids provide biological membranes with adequate fluidity and permeability. The ratio of PC to PE indicates the functioning of the membrane, because PC is a lipid that stabilises the structure of the lipid bilayer, while PE prefers the hexagonal structure.

PRACTICAL PART

Materials and media

1. Microorganisms: microfungus *Umbelopsis isabellina*.
2. Media: Sabouraud.
3. Reagents: TCA trichloroacetic acid – 15% aqueous solution (H_2O deionised), BHT (7.2%) dissolved in ethanol (95%), TBA (thiobarbituric acid)-TCA = 20 mM TBA dissolved in 15% TCA.
4. Equipment: liquid chromatograph Agilent 1200 and mass spectrometer QTRAP 3200 (Sciex).

Aim of the exercise
Observation of changes in the lipid profile of a microorganism under the influence of toxic substances.

Procedure

1. Culture of *Umbelopsis isabellina* fungus strain on liquid medium:
   a) wash off the slant with 4 ml of medium, then transfer the suspension to 20 ml of the same medium, place the flask with culture on a rotary shaker (140 rpm, 28°C, 48 h).
2. Using phospholipid standards (PE 28:0, PC 28:0), optimise the parameters of the quantitative and qualitative method. Select ion transition pairs for MRM mode analysis.
3. For optimal performance of the spectrometer create a method using a liquid chromatograph.
4. For concentrations in the range 1–10 µg/ml, prepare a curve for the selected phospholipid.
5. Prepare cultures of the tested microorganism with 2,4-D at the final concentration of 25, 50, 100 mg/l and a pesticide-free control system. The inoculum should constitute 10% of the culture and the total volume of the system should be 20 ml.
6. Incubate the culture at 28°C for 120 h on a rotary shaker (140 rpm).

7. Determine the profile of phospholipids of *U. isabellina* culture:

   a) separate the biomass from the culture medium;

   b) transfer 50 mg of wet biomass into an Eppendorf tube (1.5–2 ml);

   c) add 666 µl of chloroform and 333 µl of methanol and glass beads. Break the cells in the FastPrep disintegrator for 1 min;

   d) centrifuge the samples (9,200 rpm for 1 min). Transfer the solvent into a new Eppendorf tube;

   e) add 0.2 ml deionised H_2O, vortex, centrifuge (500 × g for 5 min), transfer the lower layer to a new tube and then evaporate at 40°C.;

   f) centrifuge the samples (2,000 × g, 5 min), dissolve in 1 ml of methanol, filter and prepare for LC-MS/MS analysis;

   g) analysis on liquid chromatograph HPLC Agilent Technologies® 1200, coupled with a tandem mass spectrometer Sciex® 3200 QTRAP MS/MS equipped with an ESI source (for more information on creating chromatographic methods see section 2.3). Use the Kinetex C18 column (50 mm × 2.1 mm, 5 µm; Phenomenex, USA, temperature 40°C) and the mobile phase: water (channel A) and methanol (B) with the addition of 5 mM ammonium formate and a flow rate of 0.5 ml/min..

8. Based on the IDA methods (suitable precursor, e.g. 153 m/z, 168, 241, 196, 227) and NL 87, select MRM pairs for quantitative determinations.

9. Determine the phospholipid profile of 5-day-old *U. isabellina* cultures.

10. Biomass quantity examination:

    a) after the incubation is complete, filter the cultures using a KNF LAB Laboport® vacuum pump and Whatman 2 type paper filters of 55 mm diameter.

    b) dry the obtained biomass for 30 min at 95°C until a constant weight is obtained.

11. Perform peroxidation product determination:

    a) Mix 1 g of biomass with 9 ml deionised H_2O and 50 µl BHT (7.2%) in a 50 ml test tube. After homogenisation, take 1 ml, transfer to a new tube, add 2 ml of TBA-TCA solution. Vortex the mixture, heat at 95°C for 30 min, then cool to room temperature (water bath) and centrifuge for 15 min at 2,000 × g. Determination of substances reacting with TBA should be performed at $\lambda = 531$ nm, correction for sugar at $\lambda = 600$ nm.

Analysis of results

Describe the dominant changes in lipid profile under the influence of toxic substances, compare with data on lipid peroxidation.

6.4. Search for biomarkers in industry and medicine

INTRODUCTION

Biomarkers are biological indicators whose examination allows for qualitative or quantitative evaluation of various states, phenomena or biological characteristics. Biomarkers, also called bioindicators, are also used to assess the impact of external factors on the environment and inhabiting organisms. The term "biomarker" is a relatively young term, defining the phenomenon of the existence of measurable biological properties, the value of which is determined by a specific process or physiological or pathological state in the body. The general definition of biomarkers includes many such indicators, including substances (e.g. omics – proteins, nucleic acids, metabolites, etc.) and elements. There are several types of biomarkers:

- biomarkers of exposure – measurable exogenous substances or their metabolites present within the organism or products of interaction between the harmful agent and target cells or molecules;
- biomarkers of effects – measurable biochemical, physiological, behavioural and other changes occurring within the organism which, depending on their size, can be recognised as being associated with health disorders and diseases already present or likely to appear later on;

- biomarkers of sensitivity – indicators of the organism's innate or acquired ability to respond to exposure to a specific harmful agent.

Biomarkers, which are defined as molecular, genetic or biochemical factors for a precise and relatively easy diagnosis of physiological and pathological conditions, play an invaluable role in modern medicine and biotechnology. Biomarkers are an intensely explored area in the search for new therapies, and for many life-science researchers, they are an essential starting point in their research work.

6.4.1. Analytical methods

Biomarkers are subjected to two basic analytical procedures:
- qualitative analysis – linking the presence of a marker to a pathological or physiological image/phenotype;
- quantitative analysis – linking the amount of the biomarker to a pathological or physiological image/phenotype.

The methods used for biomarker analysis can be divided into two basic groups:
- methods of searching for biomarkers (both non-targeted and targeted) – chromatography, electrophoresis (including CE) coupled with MS and MS/MS, NMR, AED detection; confocal and fluorescent microscopy etc.; immunological methods;
- methods used in routine analysis (targeted methods) – mainly immunological, LC-MS/MS and MALDI-TOF/TOF methods.

The results obtained from the analyses are subject to mathematical analysis. At the stage of searching for biomarkers, data clustering methods are used – primarily the statistics of many variables, such as PCA or hierarchical clustering. At further stages, including method validation, mainly statistical methods of one variable are used, such as T-test or ANOVA.

6.4.2. Characteristics and sources of biomarkers

The ideal biomarker should have: high sensitivity, specificity and accuracy and provide predictive information (e.g. during treatment or cleaning of contaminated land). A separate group are screening biomarkers, used to determine the complex states of the examined systems, e.g. confirmation of infection and its stage in relation to the host state and monitoring of other potential threats. Screening biomarkers are characterised by:

- high specificity – minimum number of false positive and false negative measurements;
- clear reflection of the different states of the examined system (especially the early stages of the process);
- easy detection in basic, easily accessible material;
- economic profitability of the analytical method used.

The material for the analysis of biomarkers are all sources of molecules that can reflect the state of the biological system under investigation. First of all, these are: blood, plasma, urine, tissues, respiration, smears, culture media, cells of microorganisms, total or partial biomass from culture, cells with specific morphology or isolated cell elements, water, soil, air or waste gases from bioreactors etc. The discovery and implementation of biomarker packages for routine diagnostics is difficult due to the following problems:

- body fluids and cellular extracts are very complex – there are countless compounds in extracts from biological materials;
- body fluids and cellular extracts have very wide dynamic ranges – from very small to very high concentrations;
- biomarkers are usually at low concentrations;
- very high individual diversity – this applies especially to higher organisms.

Due to the problems mentioned above, most biomarkers will never have clinical relevance, because:

- statistical standards for diagnostic methods are very high;
- the greater the representation of a given biological condition (e.g. disease), the easier it is to discover, identify and analyse a biomarker and subsequently validate and implement a routine analytical procedure.

The biomarker used in routine analysis has a binding meaning only in quantitative analysis, the result of which is related to the applicable regulations, standards, acceptable quantity and their significance, which in total allows for a clear interpretation of the result (Figure 6.4.1). A typical discovery of a biomarker involves examining about 50 samples per type of experimental condition, of which about 10 samples per type of experimental condition usually gives a 90% chance of finding double differences. The method validation usually takes about 1,000 samples. Once the validation process has been successfully completed, the developed analytical method can be used as a basis for implementation in routine analytics.

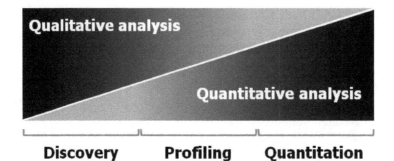

Qualitative analysis

Quantitative analysis

Discovery   Profiling   Quantitation

Fig. 6.4.1. General scheme of the procedure leading to implementation of a biomarker for routine diagnostics.

PRACTICAL PART

Materials and media

LC-MS/MS data previously collected in other laboratories or during previous analyses.

Aim of the exercise
Principal components analysis – learning the principles, calculation methods and bioinformatics tools. Determination of characteristic metabolic changes during the xenobiotic biodegradation process based on PCA analysis.

Procedure

1. Introduction to the functions and operation of MarkerView software (Sciex, USA):
   a) import and formats of imported data;

b) data normalisation;

c) PCA analysis – generation of charts and profiling of variables;

d) PCA analysis – methods of scaling and weighing the input data;

e) PCA analysis – classified and non-classified methods;

f) data analysis with the T-test.

2. Analysis of a simple data series – collect data characterising a group of people, e.g. height, age, hair length, biceps circumference, thigh circumference, weight, etc. Perform PCA analysis in the MarkerView software and identify the most differentiating features of the group, common features and correlated features.

3. Biological data analysis:

a) import data from targeted metabolomic analysis of the process of microbiological distribution of xenobiotics – at least 30 compounds in the LC-MS/MS method, analysed in the control and tested system, at least 5 culture time points, in three repetitions.

b) perform PCA analysis:

- adjust scaling and weighing;
- choose the method of data normalisation (depending on the input data);
- perform non-classified and classified analysis;
- identify the relationships that most differentiate groups at given time points, common features and correlated features.

Analysis of results

1. On the basis of the results obtained in the PCA analysis, determine the effect of the xenobiotic and its biodegradation on the metabolic changes of the examined microorganism – indicate potential biomarkers.

2. On the basis of the results obtained in the PCA analysis and data available in the scientific literature, indicate toxicity, stress, signal markers, etc.

6.5. Quantitative analysis of pesticides – multimethods

INTRODUCTION

The development of technology has made it possible in recent years to create analytical methods involving quantitative analysis of several dozen to even several hundred substances in a sample, the so-called multimethods. The main technique used in such analyses is tandem mass spectrometry coupled with liquid or gas chromatography (LC-MS/MS, GC-MS/MS). As the only analytical method currently used, it allows for selective and reproducible quantitative analysis, with high sensitivity, in a relatively short time and at relatively low cost.

6.5.1. Multimethods

The main application of multimethods is currently the analysis of residues of biologically active substances in food products (both human and animal), in body fluids (e.g. blood, urine) or in water or soil extracts (e.g. sewage, contaminated soil or water). The analyses cover several main classes of compounds, including pesticides and their derivatives, toxins, drugs and designer drugs (and their derivatives), medicines and their derivatives, doping substances, etc. Multimethods are used in medical analysis, the food industry, agriculture, toxicology, forensic science or environmental protection.

The sample preparation is dependent on the material and analytes subject to MS/MS analysis. The procedures used mainly include:
- liquid-liquid extraction (e.g. Quechers);
- extraction to solid phase (SPE);
- concentration or dilution of the sample;
- sample derivatisation (mainly in GC).

The analysis itself is based on the chromatographic separation of the tested compound(s) in the mixture and MS/MS detection in MRM or sMRM scanning mode (scheduled MRM – each MRM pair has a specific retention time and is not scanned during the whole chromatographic separation). Quantitative analysis is carried out on the basis of standard curves drawn usually on the basis of the surface areas of the chromatographic peaks or on the basis of the ratio of the surface areas of the test substance and the internal standard (IS), which is usually a deuterated analogue of the relevant test substances added to the sample at the preparation stage. A general scheme of multimethod preparation is shown in Figure 6.5.1.

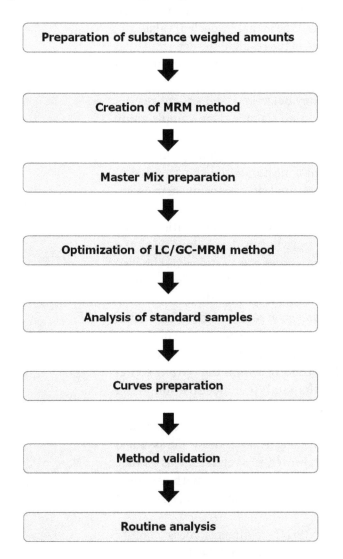

Fig. 6.5.1. General scheme for creating and implementing an analytical multimethod.

6.5.2. Method validation

Method validation is a process of evaluation of an analytical method conducted to ensure compliance with the requirements of the method, defining the method and allowing determination of its usefulness. Validation is a complex process during which all elements of the analytical procedure are checked in turn for the so-called validation parameters. These include:

1. Accuracy –consistency between the obtained measurement result and the actual (expected) value.
2. Trueness – compliance of the result of the determination (calculated on the basis of a series of measurements) with the expected value.
3. Precision –consistency between independent results obtained during the analysis of a given sample using a given analytical procedure:
   - repeatability – precision of results obtained under the same measurement conditions in a short period of time (laboratory data, analyst, measuring instrument, reagents);
   - intermediate precision – long-term deviation of the measurement process, which is determined using the standard deviation of a series of measurements obtained in a given laboratory over a period of several weeks (or by different operators using the same method);
   - reproducibility – precision of the results obtained in different laboratories using a given measuring method.
4. Linearity – the extent of the measurement range of the analytical method in which the output signal is proportional to the analyte concentration being determined:
   - range – range of analyte concentration values in which the error of the measuring device is below the assumed one. Increasingly, however, it is referred to as a range of concentration values determined with assumed precision, accuracy and uncertainty within a specified linearity range.
5. Signal-to-Noise Ratio (S/N) – a dimensionless quantity which determines the ratio of the analytical signal to the average background noise level for a specific sample:

- Limit of Detection (LOD): the smallest amount or concentration of a substance (element, ion, compound) that can be detected by a given method or analytical technique with a certain probability. Usually S/N ≥ 3 (sometimes 5);
- Limit of Quantitation (LOQ): the smallest amount or the smallest concentration of the substance that can be quantified in a given analytical method with assumed accuracy and precision. Its value is always a multiple of the determined value of the limit of detection – most often it is: LOQ = 3 × LOD (but can also be: 2 × LOD or 6 × LOD).
6. Measurement Uncertainty (MU) – parameter related to the measurement result, which defines the distribution range around the average value, in which the expected value may (at the assumed level of materiality) be placed:
   - standard uncertainty – $u(xi)$ – measurement uncertainty presented and calculated as standard deviation;
   - combined standard uncertainty – $uc(y)$ – standard uncertainty of the result y of the measurement, the value of which is calculated on the basis of the uncertainty of the parameters influencing the value of the result of the analysis using the law of propagation of uncertainty;
   - expanded uncertainty – U – quantity defining the interval around the obtained result of the analysis, within which the expected value can occur at an appropriate, accepted level of significance (probability, e.g. 95% confidence interval);
   - coverage factor – k – numerical value used to multiply the total standard uncertainty of measurement to obtain the expanded uncertainty, the value of the coefficient depends on the assumed probability level (e.g. for 95% is 2) and is most often chosen from a range of 2–3.

There are two methods of uncertainty estimation:
- Type A method of uncertainty estimation – a method of uncertainty estimation based on statistical measurements based on the standard deviation of a series of measurements;

- Type B method of uncertainty estimation – a method of uncertainty estimation using other than statistical methods, i.e.:
  - previous experiences;
  - previous results of similar examinations;
  - manufacturer's specifications of the instruments used, the reagents used or e.g. measuring vessels;
  - results taken from previous reports, e.g. on calibrations;
  - uncertainty calculated from the test results for the reference material.

PRACTICAL PART

Materials and media

1. Pesticides – standard substances with minimum 98% purity.
2. Internal Standards (IS) – deuterated analogues of selected pesticides.
3. Reference and pesticide-free matrix material (soil, grain, feed, fruit and vegetables).
4. Reagents for extraction by QUECHERS method.
5. Reagents for LC-MS/MS analysis (ultra pure): water, methanol, acetonitrile (ACN), formic acid, ammonium formate, acetic acid, ammonium acetate.
6. Apparatus: liquid chromatograph Agilent 1200 and mass spectrometer QTRAP 3200 (Sciex).

Aim of the exercise
Getting acquainted with the procedure for the development and quantitative validation of a LC-MS/MS analytical method covering several dozen or more compounds.

Procedure

1. Prepare weights of dozens of commonly occurring pesticides in screwed top glass test tubes. Weigh 2–3 mg of each substance and dissolve in a suitable solvent, so that the final concentration is 1 mg/ml.

2. Create a LC-MS/MS multimethod:

  a) prepare mixtures of several to several dozen or more compounds at a concentration of 1 µg/ml or less of each compound to create an MRM on the LC-MS/MS device. Select the compounds in the mixtures so that they differ in molar mass by at least 1–2 Da;

  b) optimise MS/MS parameters – 2 MRM pairs for each compound and 1 for internal standards;

  c) optimise the ion source for the flow of LC on selected compounds representative of the whole mixture;

  d) make a mixture of all compounds of known concentration (e.g. 5 ng/ml), a so-called Master Mix, at the starting phase of the chromatographic method;

  e) develop a LC gradient method for the separation of the pesticide mixture on a C18 column using a reversed phase system.

3. Plot standard curves:

  a) Prepare the matrix extract (test material, e.g. pre-homogenised feed, soil or grain) using the QUECHERS method;

  b) examine possible contamination of the matrix by the LC-MS/MS method;

  c) make a standard curve using the matrix extract diluted with the initial mobile phase of the chromatographic system, to obtain concentrations of 50, 25, 10, 5, 1, 0.5, 0.1 ng/ml + 25 ng/ml of the internal standard (each point in at least 3 repetitions);

  d) set the analysis sequence to HPLC-MS/MS.

4. Preparation and analysis of samples for method validation:

  a) add a known amount of the pesticides (e.g. 10 µg/kg) to the matrix samples and internal standard in appropriate amounts and extract according to the QUECHERS methodology (6 + 6 repetitions);

  b) to compare the results (100%), use pesticide solutions that are applied in the same amount on a pure matrix extract (3 repetitions);

  c) dilute the samples with the initial mobile phase of the chromatographic system (min. 10 times);

  d) set the analysis sequence to HPLC-MS/MS.

Analysis of results

1. Based on the collected data, validate the basic parameters of the LC-MS/MS method for each analyte:

   a) based on standard curves: limit of detection (LOD), limit of quantification (LOQ), linearity, working range;

   b) based on samples from point 4: repeatability, expanded uncertainty;

   c) develop validation cards;

   d) develop SOP (Standard Operation Procedure) documents.

2. Prepare and analyse the test samples according to a validated procedure, e.g. potentially contaminated feed, soil or grain samples. Describe the results of the analysis in the form of a short summary.

Literature

Books

Bernat, P., 2016. *Lipidomics in studies on adaptation mechanisms of microorganisms to the toxic effects of hazardous compounds*, w: Długoński, J. (red.), *Microbial Biodegradation: From Omics to Function and Application*. Caister Academic Press, Norfolk, s. 85–98.

Bjerrum, J.T., 2015. *Metabonomics: Methods and Protocols*. Springer, New York.

Cutillas, P.R., Timms, J.F. (red.), 2010. *LC-MS/MS in Proteomics: Methods and Applications*. Humana Press, Totowa.

Długoński, J. (red.), 1997. *Biotechnologia mikrobiologiczna – ćwiczenia i pracownie specjalistyczne*. Wydawnictwo Uniwersytetu Łódzkiego, Łódź.

Fan, T.W.M., Lane, A.N., Higashi, R.M. (red.), 2012. *The Handbook of Metabolomics*. Humana Press, Totowa.

Jain, K.K., 2010. *The Handbook of Biomarkers*. Springer, New York.

Joint Committee for Guides in Metrology (JCGM), 2008. *Evaluation of Measurement Data – Guide to the Expression of Uncertainty in Measurement*, JCGM 100:2008, https://www.iso.org/sites/JCGM/GUM-introduction.htm (access: 20.09.2019).

Lundblad, R.L., 2016. *Development and Application of Biomarkers*. CRC Press, Boca Raton.

Martins-de-Souza, D., 2014. *Shotgun Proteomics. Methods in Molecular Biology*. Humana Press, Totowa.

Sussulini, A. (red.), 2017. *Metabolomics: From Fundamentals to Clinical Applications*. Springer, New York.

Swartz, M.E., Krull, I.S. (red.), 2018. *Analytical Method Development and Validation*. CRC Press, Boca Raton.

Vaidya, V.S., Bonventre, J.V. (red.), 2010. *Biomarkers: In Medicine, Drug Discovery, and Environmental Health*. John Wiley and Sons, Hoboken.

Von Hagen, J. (red.), 2008. *Proteomics Sample Preparation*. John Wiley and Sons, Hoboken.

Original scientific papers

Bernat, P., Nykiel-Szymańska, J., Stolarek, P., Słaba, M., Szewczyk, R., Różalska, S., 2018. *2,4-dichlorophenoxyacetic acid-induced oxidative stress: metabolome and membrane modifications in Umbelopsis isabellina, a herbicide degrader*. PLoS ONE 13 (6): e0199677, 1–18.

7
Media, buffers

7.1. Media

1. Wort

Dilute the wort (usually about 16°Blg) with tap water to 12°Blg, adjust the pH to 6.3. Sterilise twice at 121°C for 15 min., and filter after each sterilisation. Pour into flasks, sterilise as before.

°Blg – Balling degree, unit indicating sugar concentration in aqueous solutions.
1°Blg = 1 g of sugar in 100 g of solution, measurement is made with a Balling or Brix aerometer.

2. ZT medium – for culture and storage of microfungi

Wort 12°Blg	500 ml
Yeast extract (Difco)	4 g
Glucose	4 g
Agar	25 g
Distilled water	up to 1,000 ml pH 6.6–7.0

Medium sterilisation at 117°C for 20 min.

3. ZT medium with rose Bengal – selective, for microfungi isolation from natural environments

Wort 12°Blg	500 ml
Yeast extract (Difco)	4 g
Glucose	4 g
Agar	25 g
Distilled water	up to 1,000 ml
rose Bengal	66.7 mg pH 6.6–7.0

Medium sterilisation at 117°C for 20 min.

4. Sabouraud medium – for multiplication and storage of fungi

Neopeptone (Difco)	10 g
Glucose	40 g
Agar	20 g
Distilled water	up to 1,000 ml pH 5.8–6.0

Medium sterilisation at 117°C for 20 min.

5. Agar medium – for multiplication and storage of bacteria

Enrichment broth (POCH)	15 g
Agar	20 g
Distilled water	up to 1,000 ml pH 7.6

Medium sterilisation at 117°C for 20 min.

6. Universal medium – for multiplication and storage of bacteria

Enrichment broth (POCH)	15 g
Yeast extract (Difco)	1.3 g
Glucose	5 g
Gelatin	2 g
Agar	20 g
Distilled water	up to 1,000ml pH 7.5

Medium sterilisation at 117°C for 20 min.

7. Liquid medium PL_2 – for culture of microfungi

Wort 12°Blg	100 ml
Aminobak	5 g
Yeast extract (Difco)	5 g
KH_2PO_4	5 g
Distilled water	up tp 1,000 ml pH 5.8–6.0

Medium sterilisation at 117°C for 20 min.
Separately, prepare 20% glucose solution (sterilisation: 117°C, 15 min). Add to the medium after sterilisation to a final concentration of 2%.

8. Mineral medium – for testing the decomposition of petroleum substances by microorganisms

$MgSO_4 \times 7 H_2O$	0.5 g
Na_2HPO_4	1.0 g
KH_2PO_4	0.5 g
NH_4NO_3	2.5 g
$CaCl_2 \times 2 H_2O$	0.01 g
$Fe_2(SO_4)_3 \times n H_2O$	0.01 g
$MnSO_4 \times H_2O$	0.01 mg
$Co(NO_3)_2 \times 6 H_2O$	0.005 mg
$(NH_4)_6Mo_7O_{24} \times 4 H_2O$	0.1 mg
Agar	20 g
Distilled water	up to 1,000 ml pH 6.8

Medium sterilisation at 117°C for 20 min.
Add anthracene and phenanthrene in a concentration of 0.5 g/l after medium sterilisation.

8.1. Mineral medium – for isolation of mutants able to selectively degrade the side chain of sterols.
Composition of the medium as above, with addition of cholesterol or AD to the final concentration of 0.5 g/l.

9. Enrichment broth

Enrichment broth (POCH)	15 g
Distilled water	up to 1,000 ml pH 6.7

Medium sterilisation at 117°C for 20 min.

10. Media for oxytetracycline production

10.1. IM-I – streaking medium

Corn steep	12 g
Starch	25 g
Soybean flour	25 g
Corn flour	5 g
$(NH_4)_2SO_4$	5 g
Soybean oil	10 g
NaCl	5 g

| CaCO$_3$ | 5 g | |
| Distilled water | up to 1,000 ml | pH 6.8–7.0 |

Medium sterilisation at 121°C for 30 min.

10.2. IM-II – multiplying medium

Corn steep	12 g	
Starch	25 g	
Soybean flour	25 g	
Corn flour	5 g	
(NH$_4$)$_2$SO$_4$	5 g	
Soybean oil	10 g	
CaCO$_3$	5 g	
Distilled water	up to 1,000 ml	pH 6.8–7.0

Medium sterilisation at 121°C for 30 min.

10.3. IM-III – production medium

Corn steep	4 g	
Soybean flour	5 g	
Corn flour	15 g	
Starch	60 g	
(NH$_4$)$_2$SO$_4$	6 g	
CoSO$_4$	20 mg	
CaCO$_3$	7 g	
Soybean oil	20 g	
Distilled water	up to 1,000 ml	pH 7.0

Medium sterilisation at 121°C for 30 min.

11. Agar medium for determining biological activity of oxytetracycline:
   a) compensation layer – 15 ml on a 9 cm diameter Petri dish

| Agar | 15 g | |
| Distilled water | up to 1,000 ml | pH 7.0 |

   b) base layer – 12 ml on a 9 cm diameter Petri dish

| Peptone (Difco) | 6 g |
| Yeast extract (Difco) | 3 g |

Enrichment broth (POCH)   0.9 g
Agar   15 g
Distilled water   up to 1,000 ml   pH 6.5

c) inoculation layer – 4 ml on a 9 cm diameter Petri dish.
Peptone (Difco)   6 g
Yeast extract (Difco)   3 g
Enrichment broth (POCH)   0.9 g
Agar   15 g
Distilled water   up to 1,000 ml   pH 6.5

Medium sterilisation at 117°C for 20 min.

12. Molasses medium

Concentrated molasses   300 g
Phosphoric acid   0.05 ml
Potassium ferrocyanide   400 mg
Tap water   up to 1,000 ml   pH 6.3
Adjust pH to 6.3 with sulfuric acid. Medium sterilisation at 117°C for 20 min.

13. Saccharose medium

Saccharose (POCH)   100 g
NH_4NO_3 3 g
$MgSO_4 \times 7\ H_2O$ 1 g
$KH2PO_4$ 1 g
$FeSO_4$ 0.01 g
Distilled water   up to 1,000 ml   pH 7.0

Medium sterilisation at 117°C for 20 min

14. Liquid medium for dextran biosynthesis

Saccharose (sugar)   100 g
Yeast extract (Difco)   0.4 g
KH_2PO_4 1 g
$NH_4NaHPO_4 \times 4\ H_2O$ 4.5 g
$MgSO_4 \times 7\ H_2O$ 0.2 g
KCl   0.1 g
$MnCl_2 \times 4\ H_2O$ 0.02 g

$(NH_4)_6Mo_7O_{24} \times 4\ H_2O$ 0.005 g

Tap water   up to 1,000 ml   pH 6.7

Adjust pH to 6.7 with sulfuric acid. Medium sterilisation at 117°C for 20 min.

15. Solid medium for *Leuconostoc* culture

$MgSO_4 \times 7\ H_2O$	0.5 g
Na_2HPO_4	1.0 g
KH_2PO_4	0.5 g
NH_4NO_3	2.5 g
$CaCl_2 \times 2\ H_2O$	0.01 g
$Fe_2(SO_4)_3 \times n\ H_2O$	0.01 g
$MnSO_4 \times H_2O$	0.01 mg
$Co(NO_3)_2 \times 6\ H_2O$	0.005 mg
$(NH_4)_6Mo_7O_{24} \times 4\ H_2O$	0.1 mg
Agar	20 g
Distilled water	up to 1,000 ml pH 6.8

Medium sterilisation at 117°C for 20 min.

16. Medium with saccharose – for lactic acid production Saccharose (POCH) 80 g

Malt sprouts (Brewery)	20 g
$CaCO_3$	100 g
Yeast water	100 ml
Enrichment broth (POCH)	1.5 g
Distilled water	up to 1,000 ml pH 6.0–6.5

Medium sterilisation at 117°C for 20 min. Preparation of yeast water: 10 g of yeast spread in 100 ml of tap water, sterilise in the Koch apparatus for 1 h, filter, add to the other components of the medium.

17. Medium with molasses – for lactic acid production Molasses 16°Blg   800 ml

Malt sprouts (Brewery)	20 g
$CaCO_3$	100 g
Yeast water	100 ml

Enrichment broth (POCH)   1.5 g
Distilled water   up to 1,000 ml   pH 6.0–6.5

Medium sterilisation at 117°C for 20 min.

18. Medium for *Gluconobacter* – for ketogenic properties testing

Yeast extract (Difco)   50 g
Glycerol   50 g
Distilled water   up to 1,000 ml   pH 6.0–6.5

Medium sterilisation at 117°C for 20 min.

19. Semi-synthetic medium with glucose

Yeast extract (Difco)   1 g
Glucose   100 g
$(NH_4)_2HPO_4$ 1 g
KH_2PO_4 1 g
$MgSO_4 \times 7 H_2O$ 0.1 g
Distilled water   up to 1,000 ml   pH 4.8–5.0

Medium sterilisation at 117°C for 20 min.

20. Semi-synthetic medium with sorbitol

Yeast extract (Difco)   5 g
Sorbitol   100 g
$(NH_4)_2HPO_4$ 2 g
KH_2PO_4 1 g
$MgSO_4 \times 7 H_2O$ 0.25 g
Distilled water   up to 1,000 ml   pH 4.8–5.0

Medium sterilisation at 117°C for 20 min.

21. Medium 1a – peptone-yeast

Yeast extract (Difco)   5 g
Meat extract (POCH)   1.5 g
Peptone (Difco)   6 g
Glucose   1 g
Distilled water   up to 1,000 ml   pH 6.8–7.0

Medium sterilisation at 117°C for 20 min.

22. Medium EG, bars with osmotic stabiliser – for preparation of casting plates with protoplasts

Yeast extract (Difco)	5 g
Glucose	20 g
KCl	44.7 g
Agar	7 g
Distilled water	up to 1,000 ml pH 6.5

Medium sterilisation at 117°C for 20 min.

23. Medium EG, liquid with osmotic stabiliser

Yeast extract (Difco)	5 g
Glucose	20 g
KCl	44.7 g
Distilled water	up to 1,000 ml pH 6.5

Medium sterilisation at 117°C for 20 min.

24. Czapek-Dox medium, bars with osmotic stabiliser

Ready-made medium from Difco, with an addition of 44.9 g KCl and 7 g agar per 1 l medium.
Medium sterilisation at 117°C for 20 min.

25. Medium for the regeneration of protoplasts, with osmotic stabiliser

Mannitol	140 g
Acidic casein hydrolyzate (POCH)	1 g
Yeast extract (Difco)	1 g
Glucose	4 g
Sorbose	8 g
Distilled water	up to 1,000 ml pH 6.5

Medium sterilisation at 117°C for 20 min.

26. Medium for *Acetobacter* culture with glucose

Yeast extract (Difco)	1 g
Glucose	50 g

(NH$_4$)$_2$HPO$_4$	2 g	
KH$_2$PO$_4$	1 g	
MgSO$_4$ × 7 H$_2$O	0.5 g	
Distilled water	up to 1,000 ml	pH 4.5–5.0

Medium sterilisation at 117°C for 20 min.

27. Medium for *Acetobacter* culture with sorbitol

Yeast extract (Difco)	1 g	
Sorbitol	50 g	
(NH$_4$)$_2$HPO$_4$	2 g	
KH$_2$PO$_4$	1 g	
MgSO$_4$ × 7 H$_2$O	0.5 g	
Distilled water	up to 1,000 ml	pH 4.5–5.0

Medium sterilisation at 117°C for 20 min.

28. Medium for *Trichoderma*

KH$_2$PO$_4$	2 g	
(NH$_4$)$_2$SO$_4$	1.4 g	
MgSO$_4$ × 7 H$_2$O	0.3 g	
CaCl$_2$	0.3 g	
Metal traces	1 ml concentrated solution	
Glucose	3 g	
Mycelium	100 g wet mass	
Distilled water	up to 1,000 ml	pH 4.5–4.8

Medium sterilisation at 117°C for 20 min.

Prepare glucose separately, e.g. 20% solution. Sterilise at 117°C for 15 min.

Solution (concentrated) of metal traces, prepared separately, contains (in 100 ml of distilled water):

FeSO$_4$ × 7 H$_2$O	0.5 g
MnSO$_4$ × H$_2$O	0.156 g
ZnCl$_2$	0.167 g
CaCl$_2$	0.2 g
HCl – 19%	1 ml

Store the solution at 4°C, without sterilisation.

Preparation of *Cunninghamella echinulata* mycelium:
Rinse the mycelium (100 g wet weight), obtained from an 18 to
24 h culture on PL2 liquid medium, twice with distilled water.
Suspend in 700 ml distilled water and sterilise at 117°C for 15
min. After cooling, add mineral salts, adjust pH to 4.5–4.8,
make up to 985 ml with distilled water and sterilise at 117°C
for 15 min. Immediately before using for *Trichoderma* culture,
add 15 ml of 20% sterile glucose solution.

29. Skimmed milk

Pour 10 ml of freshly skimmed milk (by centrifugation 3,000
× g, 10 min) into test tubes.
Medium sterilisation at 117°C for 20 min.

30. Potato slants – for the multiplication of Actinomycetes

Potato dextrose agar (Difco)	30 g	
Agar	5 g	
Distilled water	up to 1,000 ml	pH 6.7–6.8

Medium sterilisation at 117°C for 20 min.

31. LB medium – liquid – for the multiplication of *E. coli*

Bacto tryptone (Difco)	10 g	
Yeast extract (Difco)	22 g	
NaCl	10 g	
Distilled water	up to 1,000 ml	pH 7.0

Medium sterilisation at 117°C for 20 min.

32. Medium for biosynthesis of hybrid protein

Bacto tryptone (Difco)	11 g	
Yeast extract (Difco)	22 g	
Glycerol/glucose/maltose	4 ml	
1M K_2HPO_4	51 ml	
1M KH_2PO_4	15.7 ml	
Ampicillin	100 mg	
Distilled water	up to 1,000 ml	pH 7.0

Medium sterilisation at 117°C for 20 min. Add the ampicillin to the sterilised medium.

33. Protective medium – for filamentous fungi lyophilisation

Food grade gelatin	100 g
Distilled water	up to 1,000 ml

Sterilise three times at 117°C for 30 min. After sterilisation, combine with 10% glucose in a 1:1 ratio.

34. Solid wort medium for yeast breeding

Wort 12°Blg	1,000 ml	
Agar	25 g	
Distilled water	up to 1,000 ml	pH 6.5

Medium sterilisation at 117°C for 20 min.

35. Sabouraud medium with chloramphenicol

Yeast extract (Difco)	2 g	
Peptone (Difco)	3 g	
Peptone SP	3 g	
Peptone K	3 g	
Glucose	19 g	
Dipotassium hydrogen phosphate	0.5 g	
Potassium dihydrogen phosphate	0.5 g	
Chloramphenicol	0.5 g	
Agar	13 g	
Distilled water	up to 1,000 ml	pH 6.4–6.5

Medium sterilisation at 117°C for 20 min.

36. Liquid medium WHI3

Peptone (Difco)	5 g	
Yeast extract (Difco)	5 g	
KH_2PO_4	5 g	
Glucose	20 ml	
Soybean oil	0.25 ml	
Distilled water	up to 1,000 ml	pH 5.4

Medium sterilisation at 117°C for 20 min. Aqueous glucose solution (20%) sterilised separately.

37. Medium with milk – isolation of microorganisms with proteolytic properties

Glucose	1 g	
Broth	15 g	
Agar	20 g	
Distilled water	900 ml	pH 7.5

Medium sterilisation at 117°C for 20 min.

After sterilisation, add 100 ml of sterile, skimmed milk to the medium (50°C).

38. Medium with starch – isolation of microorganisms with amylolytic properties

Starch	10 g	
Broth	15 g	
Agar	20 g	
Distilled water	up to 1,000 ml	pH 7.5–7.6

Medium sterilisation at 117°C for 20 min.

39. Medium with Tween 80 – isolation of microorganisms with lipolytic properties

Bacto peptone (Difco)	10 g	
NaCl	5 g	
$CaCl_2$	0.1 g	
Agar	20 g	
Distilled water	900 ml	pH 7.4

Medium sterilisation at 117°C for 20 min.
After sterilisation, add 100 ml of sterile 10% Tween 80 solution to the medium (50°C).

40. Blood medium

Agar	25 g
Broth	15 g

Fresh sheep blood   50 g
Distilled water   up to 1,000 ml   pH 7.6

Medium sterilisation at 117°C for 20 min. After sterilisation, add 50 ml of blood to the medium (50°C).

41. Cake medium

Rapeseed cake   50 g
Distilled water   1,000 ml   pH 6.8

Cook for 30 min, then filter, make up to 1,000 ml with water. Medium sterilisation at 117°C for 20 min.

42. Mineral medium according to Lobos

K_2HPO_4 4.35 g
KH_2PO_4 1.7 g
NH_4Cl 2.7 g
$MgSO_4$ 0.2 g
$MnSO_4$ 0.05 g
$FeSO_4 \times 7\ H_2O$ 0.01 g
$CaCl_2 \times 2\ H_2O$ 0.03 g
Distilled water   up to 1,000 ml   pH 6.8

Medium sterilisation at 117°C for 20 min.
Medium with different amounts of glucose (1 or 2%), which is added separately, after sterilisation.

43. Middlebrook 7H9 medium with 0.05% Tween 80

7H9   5.23 g
Tween 80   0.5 g
Deionised water   up to 1,000 ml   pH 6.6–6.8

Medium sterilisation at 117°C for 20 min.

44. Synthetic medium with 4% glucose

K_2HPO_4 4.36 g
KH_2PO_4 1.7 g
NH_4Cl 2.1 g
$MgSO_4 \times 7\ H_2O$ 0.2 g

MnSO$_4$	0.05 g
FeSO$_4$ × 7 H$_2$O	0.01 g
CaCl$_2$ × 2 H$_2$O	0.03 g
Glucose	40 g
Distilled water	up to 1,000 ml pH 5.6

Medium sterilisation at 117°C for 20 min.

45. Broth enriched with tryptophan

Peptone K	10 g
NaCl	5 g
DL-tryptophan	1 g
Distilled water	up to 1,000 ml pH 7.5

Medium sterilisation at 117°C for 20 min.

46. McClung medium

Enzymatic hydrolyzate of animal tissue	40 g (Proteose Peptone)
Na$_2$HPO$_4$	5 g
KH$_2$PO$_4$	1 g
NaCl	2 g
MgSO$_4$ × 7 H$_2$O	0.1 g
Agar	25 g
Glucose	2 g
Distilled water	up to 1,000 ml pH 7.4

Suspension of chicken egg yolk in saline (1:1).
Add 10 ml of yolk in saline for every 100 ml of cooled agar medium (50°C). Medium sterilisation at 117°C for 20 min.

47. Maltose medium

Yeast extract (Difco)	4 g
Malt extract	15 g
Glycerin	4 ml
Agar	20 g
Distilled water	up to 1,000ml pH 7.5

Medium sterilisation at 117°C for 20 min.

48. Broth medium

Peptone (Difco)	4 g
Meat extract (POCH)	0.4 g
Enzymatic casein hydrolyzate	5.4 g
Yeast extract (Difco)	1.7 g
NaCl	3.5 g
Distilled water	up to 1,000ml pH 7.5

Medium sterilisation at 117°C for 20 min.

49. Synthetic substrate by Fred (with cellulose strips) – for testing of cellulolytic properties.

$NaNO_3$	2 g
KH_2PO_4	1 g
KCl	0.5 g
$MgSO_4$	0.5 g
$FeSO_4$	0.01 g
Distilled water	up to 1,000 ml pH 4.8–5.0

Medium sterilisation at 117°C for 20 min.
Place 3 strips of blotting paper (previously sterilised) on the plates with the medium (solidified).

50. Czapek-Dox medium

$NaNO_3$	3 g
KH_2PO_4	1 g
KCl	0.5 g
$MgSO_4 \times 7 H_2O$	0.5 g
$FeSO_4 \times 7 H_2O$	0.01 g
Glucose	40 g
Distilled water	up to 1,000 ml pH 6.7

Medium sterilisation at 117°C for 20 min. Glucose (sterile) should be added separately, after sterilisation.

51. LB selective medium with ampicillin

Tryptone	10 g
Yeast extract (Difco)	5 g

| NaCl | 5 g | |
| Distilled water | up to 1,000 ml | pH 7.0 |

Medium sterilisation at 117°C for 20 min.
Add ampicillin (100 µg ml–1) after medium sterilisation.

52. LB medium with ampicillin, X-Gal and IPTG (Blue/White Screening)

LB – ready-made medium	150 ml
Ampicillin (solution 100 µl/ml)	150 µl
X-Gal (5-Bromo-4-chloro-3-indolyl β- -D-galactopyranoside)	
(2% solution in DMF)	300 µl
IPTG (Isopropyl β- -D-1-thiogalactopyranoside)	
(1M aqueous solution)	75 µl

Medium sterilisation at 117°C for 20 min. Add ampicillin, X-Gal, IPTG after medium sterilisation

53. MRS medium (MRS agar)

Yeast extract (Difco)	4 g	
Meat extract (POCH)	8 g	
Peptone (Difco)	10 g	
Glucose	20 g	
Sodium acetate	5 g	
Tri-base ammonium citrate	2 g	
K_2HPO_4	2 g	
$MgSO_4 \times 7 H_2O$	0.2 g	
$MnSO_4 \times 4 H_2O$	0.05 g	
Tween 80	1 ml	
Agar	15 g	
Distilled water	up to 1,000 ml	pH 6.2

Medium sterilisation at 117°C for 20 min.

54. Whey medium

Sterilise the whey (117°C, 20 min), then centrifuge (3,000 × g for 25 min) and sterilise the obtained supernatant (pH 4.5) at 117°C, 20 min.

55. WSM medium (carrot peel extract)

Aqueous carrot peel extract*	up to 1,000 ml	
Tryptone	2.5 g	pH 7.0

Medium sterilisation at 117°C for 20 min.

* Add 1,000 ml of sterile distilled water to 200 g of carrot (*Daucus carota* L.) waste (peel about 4 mm thick) and cook for 1 h. Then filter the suspension through a filter paper, add 2 g tryptone, set the pH (pH 7) and sterilise (117°C, 10 min.). The medium is used to prepare the culture of appropriate bacterial strains.

56. Solid medium with cationic complex CTAB-MB (Cetyltrimethylammonium bromie-Methylene Blue)

$MgSO_4 \times 7 H_2O$	0.5 g	
Na_2HPO_4	1.0 g	
KH_2PO_4	0.5 g	
NH_4NO_3	2.5 g	
$CaCl_2 \times 2 H_2O$	0.01 g	
$Fe_2(SO_4)_3 \times n H_2O$	0.01 g	
$MnSO_4 \times H_2O$	0.01 mg	
$Co(NO_3)_2 \times 6 H_2O$	0.005 mg	
$(NH_4)_6Mo_7O_{24} \times 4 H_2O$	0.1 mg	
Agar	20 g	
CTAB	5 mg	
MB	2 mg	
Distilled water	up to 1,000 ml	pH 6.8

Medium sterilisation at 117°C for 20 min.

57. Krik and Farell medium (PMM)

Glucose	2 g
Ammonium tartrate	2 g
Maltose extract	2 g
KH_2PO_4	0.26 g
Na_2HPO_4	0.26 g
$MgSO_4 \times 7 H_2O$	0.5 g

$CuSO_4 \times 5\,H_2O$	0.01 g
$CaCl_2 \times 2\,H_2O$	0.0066 g
$FeSO_4$	0.005 g
$ZnSO_4 \times 7\,H_2O$	0.0005 g
$Na_2MoO_4 \times 7\,H_2O$	0.02 mg
$MnCl_2 \times 4\,H_2O$	0.09 mg
$H_3BO_3 \times 4\,H_2O$	0.07 mg
Agar	20 g
ABTS	0.35 g
Distilled water	up to 1,000 ml pH 5.5

Medium sterilisation at 117°C for 20 min.

58. Mueller-Hinton medium

Beef extract	2 g
Acidic casein hydrolyzate (POCH)	17.5 g
Starch	1.5 g
Distilled water	up to 1,000 ml pH 7.4

Medium sterilisation at 117°C for 20 min.

59. Medium with urea by Christensen

Enzymatic hydrolyzate of animal tissue (Proteose Peptone)	1 g
NaCl	5 g
KH2PO$_4$	2 g
Cresol red (0.2% solution in 50% ethanol)	4.8 ml
Urea (20% aqueous solution)	100 ml
Distilled water	up to 1,000 ml pH 6.7

Medium sterilisation at 117°C for 20 min.

60. Casein medium

K_2HPO_4	1 g
$(NH_4)_2SO_4$	0.5 g
$MgSO_4 \times 7\,H_2O$	0.25 g
Glucose	1 g
Yeast extract (Difco)	7 g
Casein	1 g

Agar	20 g	
Distilled water	up to 1,000 ml	pH 7.0

Medium sterilisation at 117°C for 20 min.

61. Medium with casein hydrolyzate

K_2HPO_4	1 g	
$(NH_4)_2SO_4$	0.5 g	
$MgSO_4 \times 7 H_2O$	0.25 g	
Glucose	1g	
Yeast extract (Difco)	7 g	
Enzymatic casein hydrolyzate	1 g	
Agar	20 g	
Distilled water	up to 1,000 ml	pH 7.0

Medium sterilisation at 117°C for 20 min.

62. Medium with yeast extract

KH_2PO_4	1 g	
$(NH_4)_2SO_4$	0.5 g	
$MgSO_4 \times 7 H_2O$	0.25 g	
Glucose	4g	
Yeast extract (Difco)	12 g	
Enzymatic casein hydrolyzate	1 g	
Agar	20 g	
Distilled water	up to 1,000 ml	pH 7.0

Medium sterilisation at 117°C for 20 min.

63. SOC medium

Bacto peptone (Difco)	2%	
Yeast extract (Difco)	0.5%	
NaCl	8.5 mM	
Glucose	20 mM	
$MgCl_2 \times 6 H_2O$	10 mM	
$MgSO_4 \times 7 H_2O$	10 mM	
Distilled water	up to 1,000 ml	pH 7.0

Medium sterilisation at 117°C for 20 min.

64. Synthetic medium for antibiotics production

Glucose	20 g
Yeast extract (Difco)	5 g
$(NH_4)_2SO_4$	3 g
KH_2PO_4	2 g
K_2HPO_4	1 g
$MgSO_4 \times 7\ H_2O$	0.5 g
Distilled water	up to 1,000 ml pH 6.5

Medium sterilisation at 117°C for 20 min.

65. Liquid medium 1a – for multiplication of *Rhodococcus* sp. IM58 and for steroid biotransformation

Corn seep	20 g
Saccharose (POCH)	30 g
$(NH_4)_2SO_4$	2 g
$CaCO_3$	7 g
Distilled water	up to 1,000 ml pH 6.8–7.0

Medium sterilisation at 117°C for 20 min.

7.2. Buffers

1. Phosphate buffer

KH_2PO_4	4.54 g	
$Na_2HPO_4 \times 2\ H_2O$	5.94 g	
Distilled water	1,000 ml	pH 6.4

2. TE buffer

Tris-HCl 1 M	100 ml	
EDTA 0.5 M	20 ml	
Distilled water	up to 1,000 ml	pH 8.0

3. Electrode buffer

Tris-HCl	30.3 g
Glycine	144.42 g
SDS	10 g
Distilled water	up to 1,000 ml

4. SB buffer

1 M Tris-HCl buffer	2.5 ml
20% aqueous SDS solution	4 ml
Glycerol	3 ml
2- mercaptoethanol	0.5 ml
Bromphenol	3–5 crystals

Mix the suspension thoroughly and place it in aboiling bath for 1 min.

5. Load buffer

Glycerin	1 ml
1 M Tris	0.5 ml
10% (w/v) SDS	0.5 ml
β-mercaptoethanol	0.5 ml
1% (w/v) bromophenol blue	0.5 ml

6. SSSB-thio buffer

7 M urea	420 g
2 M thiourea	152 g
4% (w/v) CHAPS	40 g
40 mM Tris	4.84 g
0.2% ampholytes (stock 40%)	50 ml

7. HEPES buffer

KH_2PO_4	0,418 g
HEPES	4.77 g
Distilled water	1,000 ml

8. TAE (50×) buffer

Tris base	42 g
Glacial acetic acid	57.1 ml
0.5 M EDTA	100 ml
Complete with deionised water	up to 1,000 ml pH 8.0

9. TBE (10×) buffer

Tris base	108 g
Boric acid	55 g

| 0.5 M EDTA | 40 ml | |
| Complete with deionised water | up to 1,000 ml | pH 8.0 |

10. Mc Ilvaine buffer

| 0.2 M sodium hydrogen phosphate solution | 8.8 ml | |
| 0.1 M citric acid solution | 11.2 ml | pH 4.5 |

11. Sodium malonate buffer (50 mM solution)

1 M malonic acid solution	5 ml	
1 M sodium malonate solution	5 ml	
Complete with distilled water	up to 100 ml	pH 4.5

12. Sodium tartrate buffer (10 mM solution)

7.5 M tartaric acid solution	1.13 g	
2.5 mM sodium tartrate dihydrate solution	0.58 g	
Distilled water	up to 1,000 ml	pH 4.5

Literature

Books

Atlas, R.M., 2010. *Handbook of Microbiological Media*, 4th ed. CRC Press, Taylor and Francis Group, Boca Raton–London–New York.

Burbianka, M., Pliszka, A., Burzyńska, H., 1983. *Mikrobiologia żywności*, wyd. 5 popr. i unowocześnione, Państwowy Zakład Wydawnictw Lekarskich, Warszawa.

Długoński, J. (red.), 1997. *Biotechnologia mikrobiologiczna – ćwiczenia i pracownie specjalistyczne*. Wydawnictwo Uniwersytetu Łódzkiego, Łódź.

8
Macroscopic* and microscopic images of fungal strains applied at the Department of Industrial Microbiology and Biotechnology of the University of Łódź

---

* The fungi were cultured on Petri dishes with a diameter of 9 cm.

8.1. Photograph of fungi cultured under laboratory conditions

8.1.1. *Aureobasidium pullulans*

A

Aureobasidium pullulans mycelium obtained after 3 days of culture on ZT medium (A)

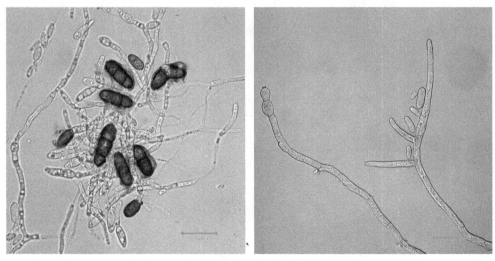

B   **C**

Hyphae and spores of *A. pullulans* mycelium after 5 days of culture on ZT
medium (B) and PL-2 medium (C). The scale corresponds to 20 μm

8.1.2. *Ashbya gossypii*

A
Ashbya gossypii mycelium obtained after 3 days of culture on ZT medium (A)

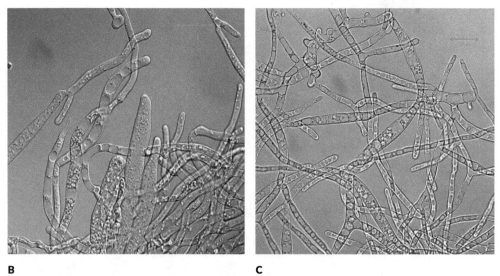

B   **C**
Hyphae and spores of A. *gossypii* mycelium after 5 days of culture on ZT
medium (B) and PL-2 medium (C). The scale corresponds to 20 μm

8.1.3. *Aspergillus niger*

A   **B**

Aspergillus niger mycelium obtained after 4 days of culture on ZT (A)
and Sabouraud medium (B)

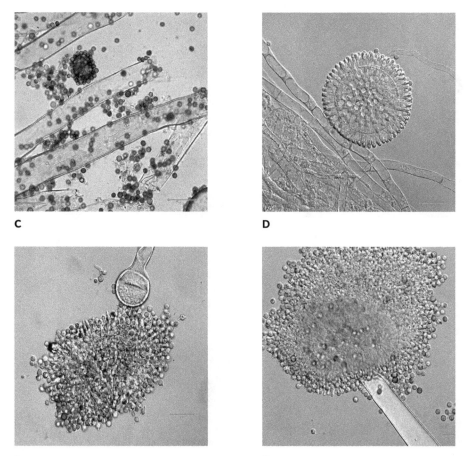

C   **D**

E   **F**

Mycelium, conidiophores and spores of *A. niger* after 5–6 days of culture on
ZT medium (C–F). The scale corresponds to 20 μm

8.1.4. *Aspergillus versicolor* IM2161

 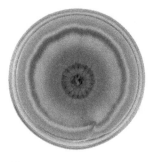

A   **B**

Aspergillus versicolor IM2161 mycelium obtained after 5–6 days of culture on
ZT (A) and Sabouraud medium (B)

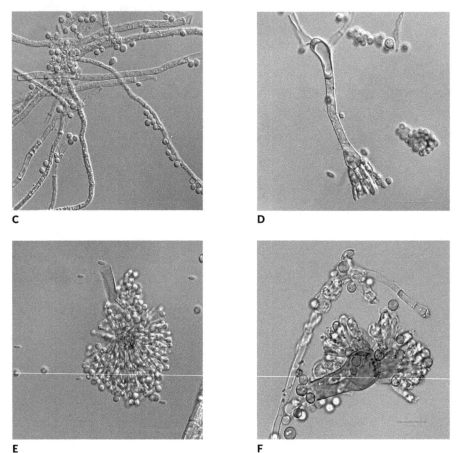

C   **D**

E   **F**

Mycelium, hyphae and spores of *A. versicolor* IM2161 obtained on ZT medium
(C–F). The scale corresponds to 20 μm (C–E) and 10 μm (F)

8.1.5. *Chaetomium globosum*

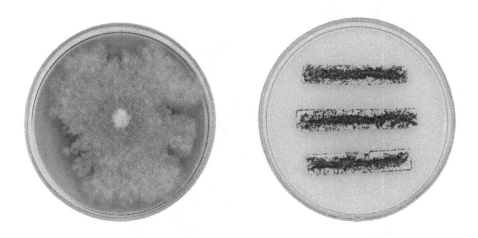

A   **B**

Chaetomium globosum mycelium on solid ZT medium (A) and perithecium type fruiting bodies on mineral medium with cellulose strips (B)

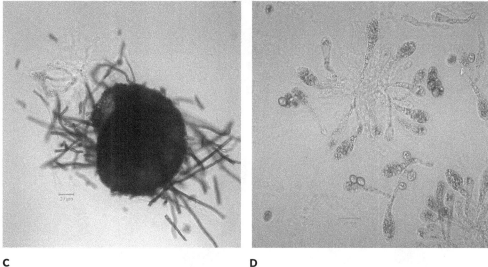

C   **D**

Microscopic images: perithecium of *Chaetomium globosum* (C) and asci with ascospores (D). The scale corresponds to 20 μm

8.1.6. *Cunninghamella echinulata* IM1785 21Gp (previously *C. elegans*)

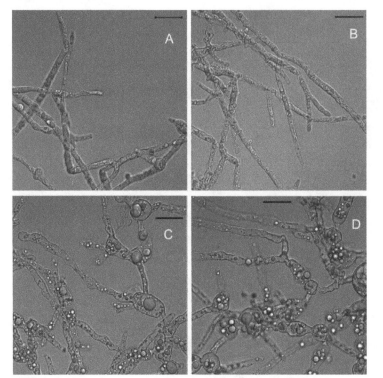

Microscopic observations of morphology of *Cunninghamella echinulata* cultured without additives (A), with cortisolone (B), with TBT (C) and with cortisolone and TBT (D). The scale corresponds to 20 μm

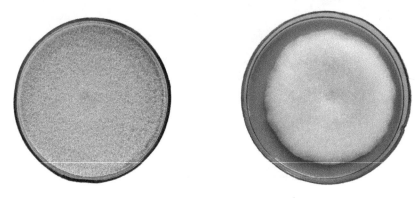

E   **F**

C. echinulata IM1785 21Gp mycelium obtained after 3–4 days of culture on solid ZT (E) and Sabouraud medium (F)

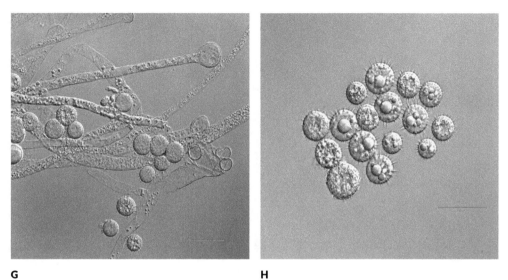

G   **H**
Sporangiophores (G) and unicellular sporangioles (H) of *C. echinulata* IM1785
21Gp on ZT medium. The scale corresponds to 20 µm

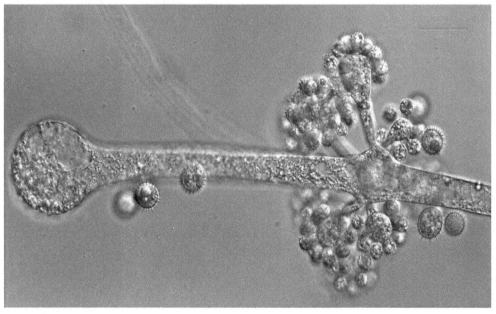

I
Sporangiophre of *C. echinulata* IM1785 21Gp with a cluster of lateral hyphae
and spherical sporangioles on ZT medium (I). The scale corresponds to 20 µm

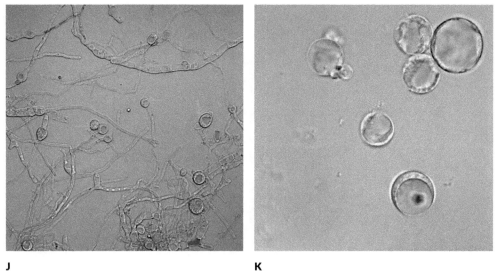

J   **K**

Formation (J) and release of protoplasts (K) from *C. echinulata* IM1785
21Gp mycelium suspended in 0.8 M MgSO$_4$ after 7 h of digestion with lytic
enzymes produced by *Trichoderma viride*. Explanation: (→)the remains of
hyphae from which the protoplasts were released; (→) released protoplasts.
The scale corresponds to 10 μm

8.1.7. *Curvularia lunata* IM2901

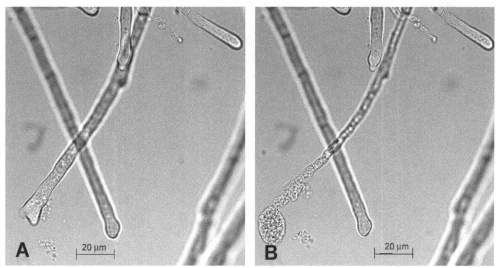

Effect of Ni ions on *Curvularia lunata* IM 2901 mycelium (loss of cell integrity due to cracking of top hyphae) observed after 5 h of mycelium culture in PL_2 medium with addition of Ni ions at an initial concentration of 5 mM. Figures A and B were made at an interval of 2 min. The scale corresponds to 20 μm

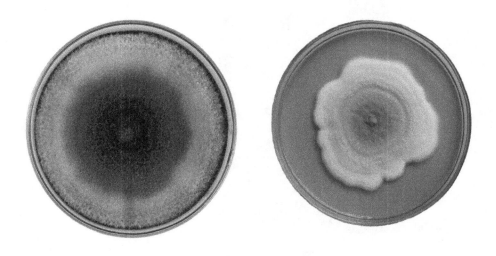

C   **D**

Curvularia lunata IM2901 mycelium obtained after 4–5 days of culture on ZT (C) and Sabouraud medium (D)

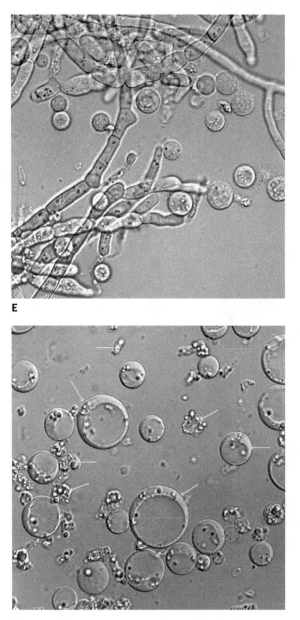

E

F
Release of protoplasts from hyphae of *Curvularia lunata* IM2901 suspended in an osmotic stabiliser (0.8 M MgSO$_4$ in phosphate and citrate buffer with pH 5.4) with addition of lytic enzymes produced by *Trichoderma viride* 1131 CBS 354–33 (E); protoplasts of *Curvularia lunata* IM2901 suspended in an osmotic stabiliser (F). Explanations: (⟶)cytoplasm located on the edge of the protoplast; (⟶)residues of protoplasts after their disintegration, (→)protoplast diameter

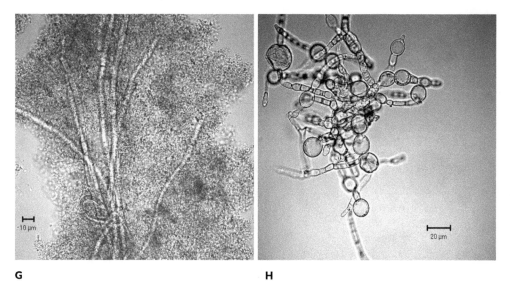

G   **H**

Hyphae of *C. lunata* IM2901 directly after the introduction into the PL_2 medium with 10 mM of lead acetate (G) and after 48 h of culture (H). The scale corresponds to 10 μm (G) and 20 μm (H)

8.1.8. *Curvularia lunata* IM4417

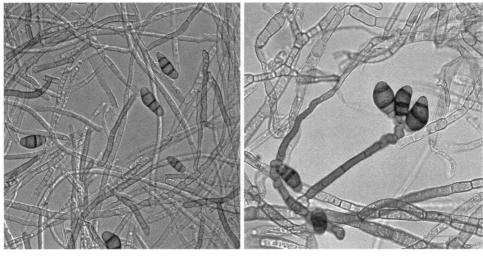

A   **B**

Curvularia lunata IM4417 mycelium obtained after 5–6 days of culture on ZT
(A) and Sabouraud medium. (B)

C   **D**

Mycelium and spores of *Curvularia lunata* IM4417 cultured on ZT medium (C).
Conidiophore with apically formed fragmoconidia containing three transverse
septa (D). The scale corresponds to 20 μm

8.1.9. *Exophiala* sp.

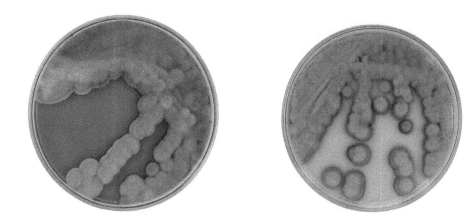

A   **B**

Exophiala sp. mycelium obtained after 5–6 days of growth on ZT (A) and
Sabouraud medium (B)

8.1.10. *Kluyveromyces marxianus*

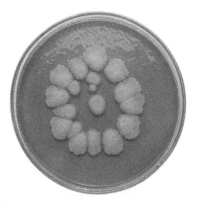

A
Kluyveromyces marxianus growth after 2–3 days of culture on ZT medium (A)

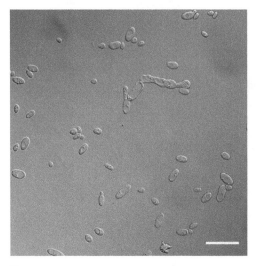

B
Microscopic image of cells of *Kluyveromyces marxianus* from culture on ZT medium (B). The scale corresponds to 10 μm

8.1.11. *Metarhizium robertsii* IM2358

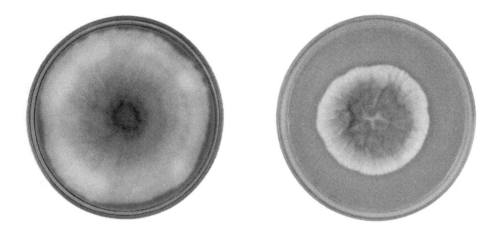

A   **B**

Metarhizium robertsii IM2358 mycelium obtained after 4–5 days of culture on
ZT (A) and Sabouraud medium. (B)

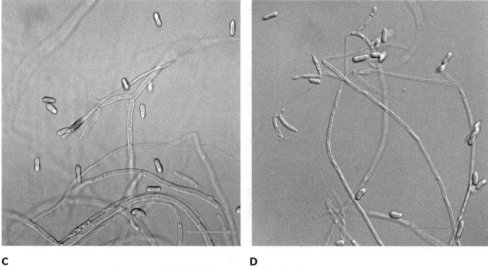

C   **D**

Mycelium and spores of *M. robertsii* IM2358 (ZT medium) (C, D). The scale
corresponds to 20 µm

8.1.12. *Mucor ramosissimus* IM6203

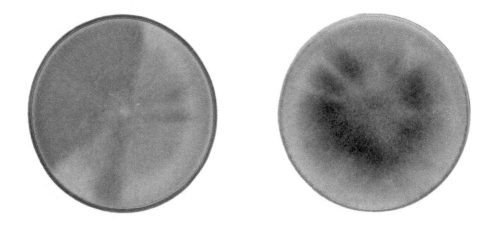

A   **B**

Mucor ramosissimus IM6203 mycelium obtained after 4–6 days of culture on ZT (A) and Sabouraud medium (B)

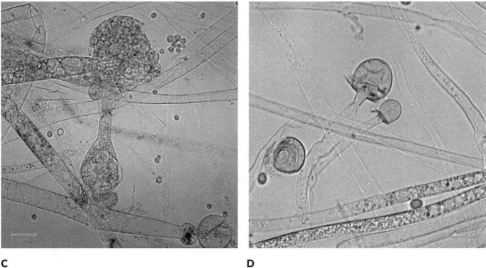

C   **D**

Mycelium and spores of *M. ramosissimus* IM6203 obtained in culture on ZT (C) medium. The large sporangia are made up of a central columella and a shield (connected to the columella at its base) (D). The scale corresponds to 20 μm (C) and 10 μm (D)

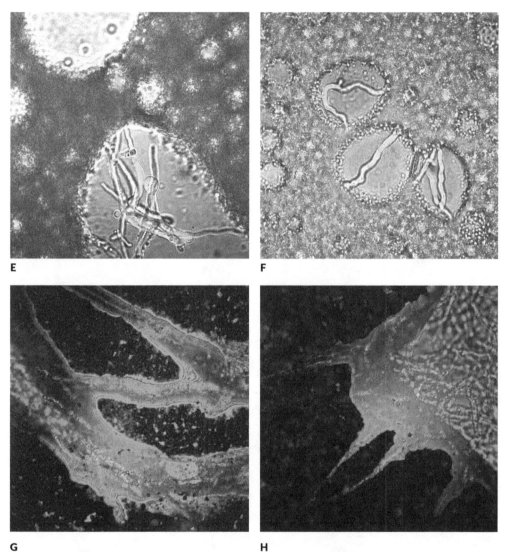

E

F

G

H

Growth of *M. ramosissimus* IM6203 on OX medium with added
pentachlorophenol (PCP) at a concentration of 10 mg/l. A – zones produced
after 5 and more days of incubation (E-F), growth in oil on 1–3 days of
incubation (G–H)

8.1.13. *Myrothecium roridum* IM6482

A
Myrothecium roridum IM6482 mycelium obtained after 4 days of culture on ZT medium (A)

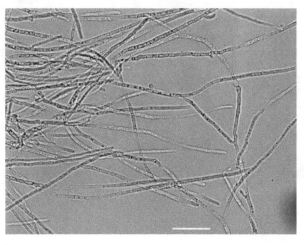

B
Hyphae in 24 h culture of *M. roridum* IM6482 on Czapek-Dox III medium (B). The scale corresponds to 20 μm

8.1.14. *Nectriella pironii* IM6443

A
Nectriella pironii IM6443 mycelium obtained after 4 days of culture on ZT medium (A)

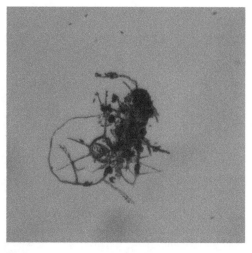

B
Hyphae and spores of *N. pironii* IM6443 (ZT medium) (B). Magnification x 125

8.1.15. *Paecilomyces marquandii* IM6003
(current name *Metarhizium marquandii*)

A   **B**

Metarhizium marquandii IM6003 mycelium obtained after 4–6 days of culture on ZT (A) and Sabouraud medium (B)

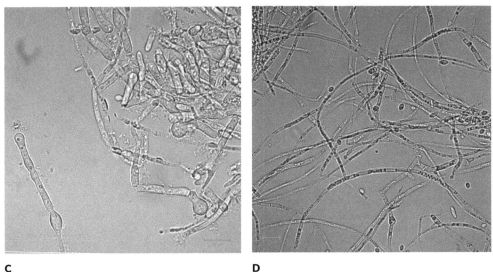

C   **D**

Microscopic images of 168 h culture of *M. marquandii* IM6003 on Sabouraud medium with 10 mM Zn and 50 mg/l alachlor (C). Liquid cultures on Sabouraud medium (D). The scale corresponds to 10 μm

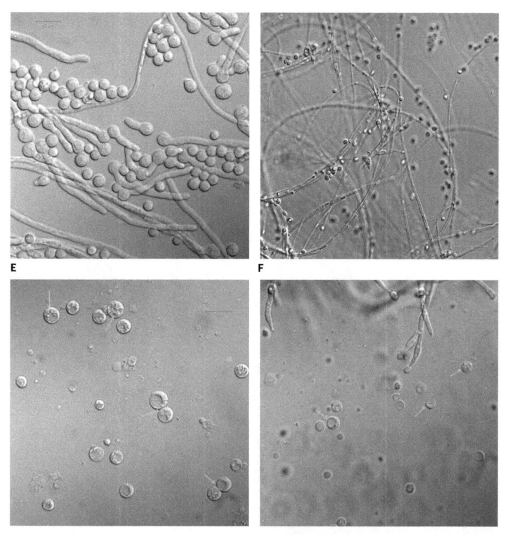

G   **H**

Microscopic images of *M. marquandii* IM6003: germinating spores after 20 h of incubation (E), mycelium from ZT medium (F), protoplasts released from mycelium in an environment of 0.8 M KCl (G) and 0.8 M MgSO$_4$ (H) after 6 h of digestion with a complex of lytic enzymes obtained from *Trichoderma viride*. Explanations: ($\longrightarrow$) released protoplasts. The scale corresponds to 10 μm (E, G, H) and 20 μm (F)

8.1.16. *Phanerochaete chrysosporium* DSM1556

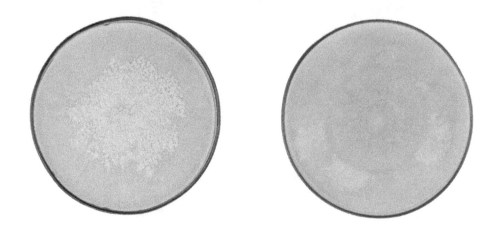

A   **B**

Phanerochaete chrysosporium DSM1556 mycelium obtained after 4–6 days of culture on ZT (A) and Sabouraud medium (B)

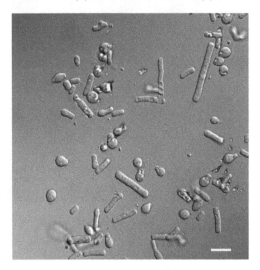

C

Hyphae and spores of *P. chrysosporium* DSM1556 (ZT medium) (C). The scale corresponds to 10 μm

8.1.17. *Schizosacharomyces pombe*

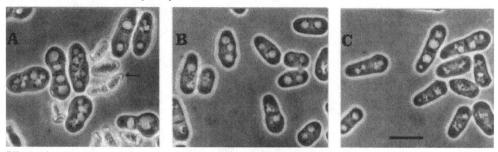

Effects of androgens on microscopic morphology of *Schizosaccharomyces pombe*. Cells cultured with androstenedione (AD) (A). Presence of abnormal swollen cells against dead cells (→). Cells cultured with testosterone (TS) (B). Cells are almost unchanged compared to control culture without steroids (C). The scale corresponds to 10 µm

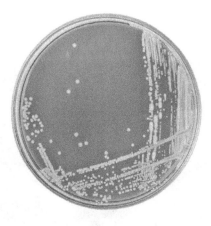

D
S. pombe growth after 2–3 days of culture on ZT medium (D)

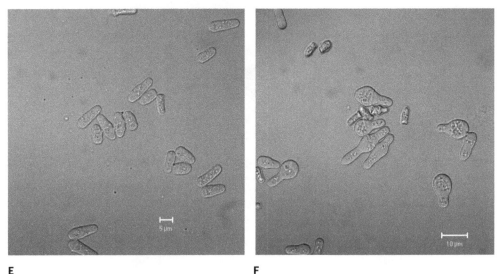

E   **F**

Microscopic image of cells of *S. pombe* 972 h– after 72 h of culture with addition of androgens. Control (E) (The scale corresponds to 5 μm), androstadiendione ADD (F) The scale corresponds to 10 μm)

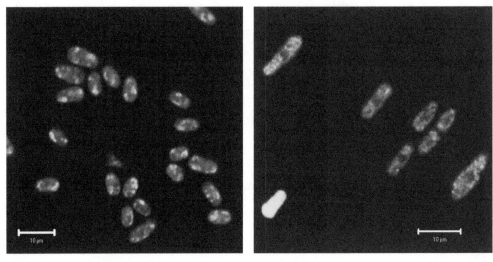

G   **H**

Microscopic image of 24 h culture of *S. pombe* with addition of androgens. Cells stained with a set of dyes FUN-1 (Molecular Probes). Control (G), androstadienone (ADD) (H). The scale corresponds to 10 μm

8.1.18. *Serpula himantioides* DSM6419

A

Serpula himantioides DSM6419 mycelium obtained after 5–6 days of culture on maltose medium (A)

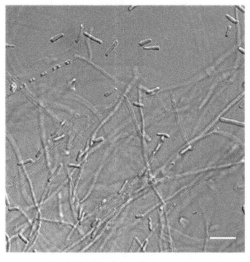

B

Microscopic image: hyphae *S. himantioides* DSM6419 from culture on Sabouraud medium (B). The scale corresponds to 20 μm.

8.1.19. *Stachybotrys chartarum* DSM2144

A   **B**

Stachybotrys chartarum DSM2144 mycelium obtained after 4–6 days of culture on ZT (A) and Sabouraud medium (B).

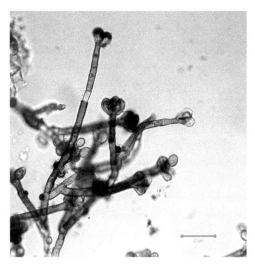

C

Microscopic image: hyphae, conidiophores and phialides of *S. chartarum* DSM2144 from culture on ZT medium (C). The scale corresponds to 20 µm

8.1.20. *Trametes versicolor*

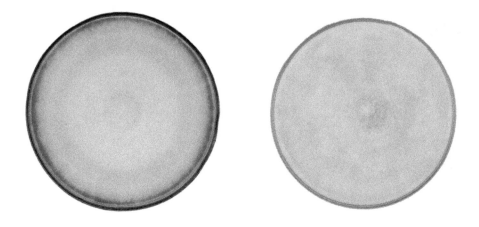

A   **B**

Mycelium of *T. versicolor* after 7–8 days of culture on ZT (A) and on a maltose medium (B).

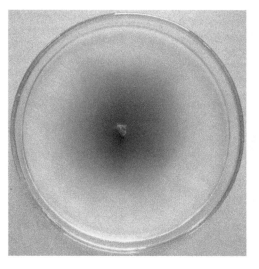

C

Green colouring around the colony indicating ABTS oxidation by laccase *T. versicolor* (C).

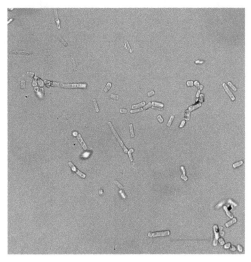

D

Mycelium and spores of *T. versicolor* (ZT medium). The scale corresponds to 20 µm (D).

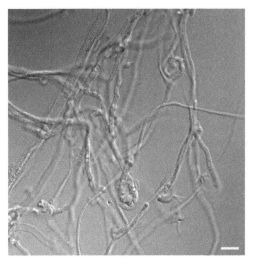

E

Culture of *T. versicolor* on liquid ZT medium after 72 h of incubation (E). Explanation: (→) clamp cennection. The scale corresponds to 10 µm.

8.1.21. *Trichoderma harzianum* QF10

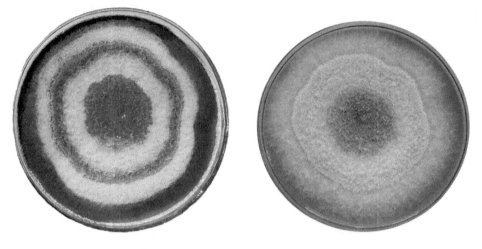

A   **B**

Trichoderma harzianum QF10 mycelium obtained after 5–6 days of culture on
ZT (A) and Sabouraud medium (B).

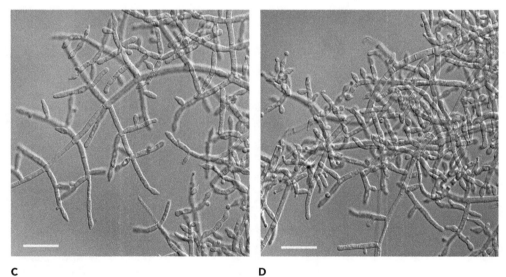

C   **D**

Multiple vertically branched conidiophores with phialides in *T. harzianum*
QF10 on ZT medium (C–D). The scale corresponds to 20 μm.

8.1.22. *Trichoderma viride* IM6325

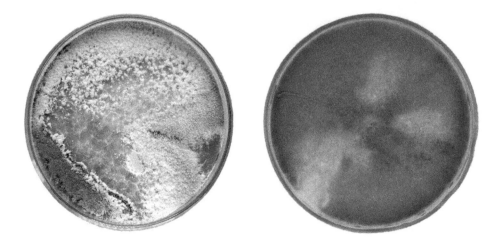

A   **B**

Trichoderma viride IM6325 mycelium obtained after 5–6 days of culture on ZT (A) and Sabouraud medium (B).

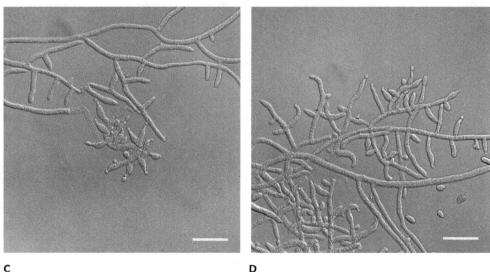

C   **D**

Branched, vertical conidiophores with clusters of *T. viride* IM6325 cultivated on ZT medium (C–D). The scale corresponds to 20 μm.

8.1.23. *Umbelopsis ramanniana* IM833

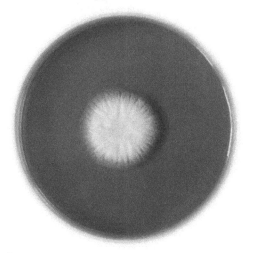

A

Umbelopsis ramanniana IM 833 mycelium obtained after 3 days of culture on ZT medium (A).

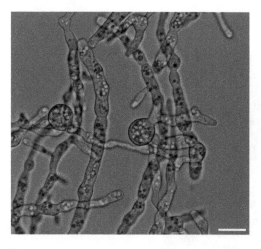

B

Hyphae of *U. ramanniana* IM 833 obtained after 24 h of culture in Lobos medium (B). The scale corresponds to 10 µm.

8.2. Photographs of trees and wood infected by ligninolytic fungi

8.2.1. *Pleurotus ostreatus* (oyster mushroom)

Photographs 8.2.1–8.2.12 taken by A. Długoński in the period from February to June 2020 in parks and the Łagiewnicki Forest in Łódź (disclaimer: copying of the illustration only with the author's consent).

A
Tree trunk infested with *Pleurotus ostreatus* (A)

B
Fruiting bodies of *P. ostreatus* fungus (B).

8.2.2 *Trametes versicolor* (turkey tail)

A   **B**

Stems of trees affected by *Trametes versicolor*. Visible fruiting bodies of the fungus (A–B).

8.2.3. *Heterobasidion annosum* (annosum root rot)

Heterobasidion annosum (annosum root rot). It attacks mainly coniferous trees, but also some deciduous ones, causing soft rot. Fruits grow low – in the ground part of trunks.

8.2.4. Brown wood rot

A

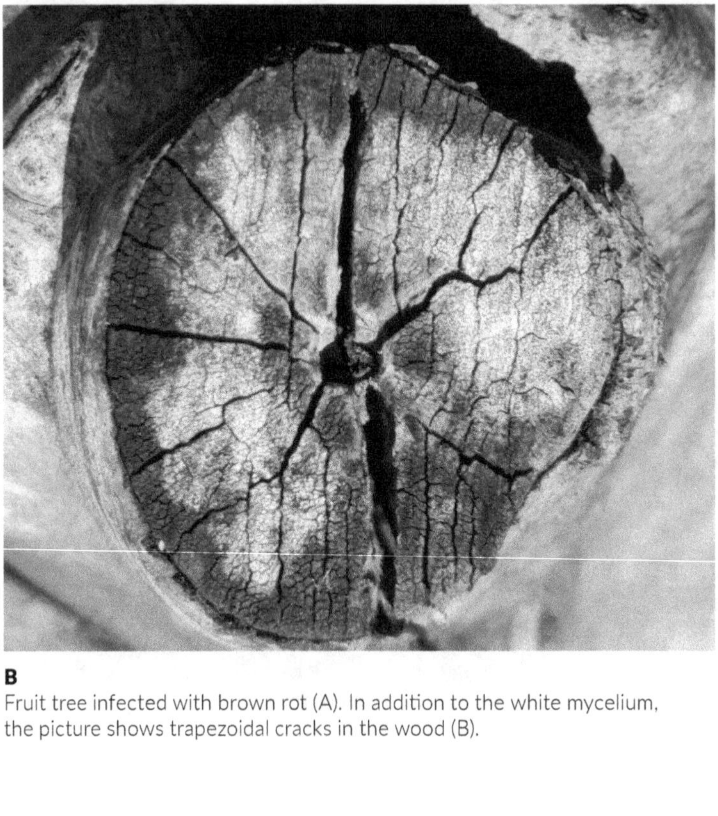

B

Fruit tree infected with brown rot (A). In addition to the white mycelium, the picture shows trapezoidal cracks in the wood (B).

C
A fragment of a board affected by brown wood rot. Visible white mycelium (→) and characteristic trapezoidal cracks of the wood (C).

8.2.5. White wood rot

White wood rot. White mycelium is visible on the surface of the decayed board.

8.2.6. Soft wood rot

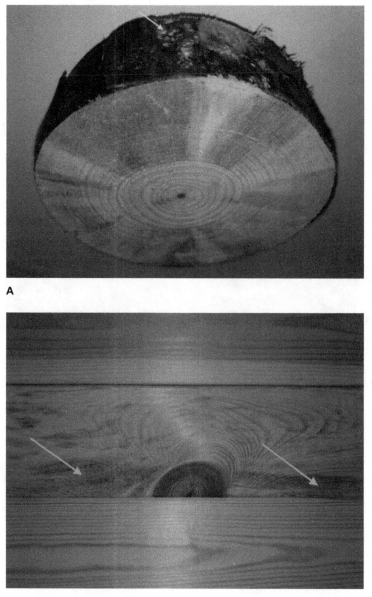

A

B

Bluish streaks (→) caused by fungi of soft wood rot visible in the sapwood part of the log (B) and on the left and right side of the log. White mycelium is visible on the surface of the bark (A) (→)

8.2.7. *Tremella mesenterica* (yellow brain)

A

It causes white wood rot. At high humidity, it forms bright yellow fruiting bodies with a gelatinous consistency (A). During the rain-free period, the fruiting bodies dry out, forming a yellow-orange membrane (B) adjacent to the branches and swell again in the presence of water. (C).

B

C

8.2.8. *Phellinus pomaceus*

It causes white wood rot. It attacks mainly fruit trees, especially plums.
Perennial fruiting bodies, of various shapes. Adjoining fruiting bodies often
grow together or they are tile-like one above the other in the form of large
clusters. The colour of the fruits: from sand- ochre in young to maroon and
grey-black in older ones.

8.2.9. *Piptorus betulinus* (birch polypore)

It causes brown wood rot. The fruiting bodies extracts of the fungus show bactericidal and antiviral effects, inhibit the growth of cancer cells and induce interferon synthesis. They also stimulate the growth of moulds attacking trees and proliferation of yeast cells. Used in folk medicine to treat various diseases.

8.2.10. *Fomes fometorius* (tinder fungus)

Silver maple infected by Fomes fometorius (tinder fungus) causing white wood rot. F. fometorius has the ability to decompose all the components of the wood, which initially becomes brown, later it becomes spongy, soft and white-yellowish, with dark brown narrow regions.

8.2.11. Interaction of pathogens causing the rot of trees

Birch infested by *Piptorus betulinus* (birch polypore) – beige fruiting body and
Fomes fometorius (tinder fungus) – grey fruiting body visible on the right side.

8.1.12. *Schizophyllum commune* (split gill)

It grows in damaged areas of coniferous and deciduous trees, as well as on dead wood. It causes white wood rot. Fruits of white or grey colour contain schizophylate (β-1,3 beta-glucan with β-1,6 branches) with immunostimulating and anticancer properties.

Literature

Original scientific papers

Bernat P., Długoński J., 2012. *Comparative study of fatty acids composition during cortexolone hydroxylation and tributyltin chloride (TBT) degradation in the filamentous fungus Cunninghamella elegans.* International Biodeterioration & Biodegradation 74, 1–6.

Długoński J. Wilmańska D., 1998. *Deleterious effects of androstenedione on growth and cell morphology of Schizosaccharomyces pombe.* Antonie van Leeuwenhoek 73, 189–194.

Paraszkiewicz K., Bernat P., Długoński J., 2009. *Effect of nickel, copper, and zinc on emulsifier production and saturation of cellular fatty acids in the filamentous fungus Curvularia lunata.* International Biodeterioration & Biodegradation 63, 100–105.

Authors

Przemysław Bernat – dr hab., prof. UŁ, University of Łódź, Faculty of Biology and Environmental Protection, Institute of Microbiology, Biotechnology and Immunology, Department of Industrial Microbiology and Biotechnology, e-mail: przemyslaw.bernat@biol.uni.lodz.pl.

Scientific interests: metabolomics of microorganisms with particular emphasis on lipidomics, degradation ability of microorganisms in relation to toxic substances, interactions between plants and microfungi.

Andrzej Długoński – dr inż. arch. kraj., Stefan Wyszyński University in Warsaw, Faculty of Biology and Environmental Sciences, Institute of Biological Sciences, e-mail: a.dlugonski@ uksw. edu.pl.

Scientific interests: influence of anthropopressure (emission of gas and dust pollution) on the properties of natural environment, revitalisation of post-industrial areas, remote sensing in landscape architecture, green infrastructure of cities, ecology of the city, natural photography, landscape engineering, digital techniques in public space design.

Jerzy Długoński – prof. dr hab., University of Łódź, Faculty of Biology and Environmental Protection, Institute of Microbiology, Biotechnology and Immunology, Department of Industrial Microbiology and Biotechnology, e-mail: jerzy.dlugonski@biol. uni.lodz.pl.

Scientific interests: metabolic diversity of microorganisms, microbiological degradation and detoxification of environmental pollutants, bioconversion of steroids with the use of bacteria and fungi, protoplasts of filamentous fungi and their use in the production of steroid hormones.

Hobby: genealogy, ornithology, gardening, tourism.

Aleksandra Felczak – dr, University of Łódź, Faculty of Biology and Environmental Protection, Institute of Microbiology, Biotechnology and Immunology, Department of Industrial Microbiology and Biotechnology, e-mail: aleksandra.felczak@biol.uni.lodz.pl.

Scientific interests: microbiological degradation and detoxification of N-heterocyclic compounds, biodegradation of fluoroquinol antibiotics by filamentous microfungi, evaluation of antibacterial, antifungal and anticancer activity of newly synthesised N-heterocyclic derivatives.

Hobby: tourism.

Aleksandra Góralczyk-Bińkowska – mgr, University of Łódź, Faculty of Biology and Environmental Protection, Institute of Microbiology, Biotechnology and Immunology, Department of Industrial Microbiology and Biotechnology, e-mail: aleksandra.goralczyk@biol.uni.lodz.pl.

Scientific interests: biosynthesis of enzymes by filamentous fungi, microbiological degradation of dyes using laccases, protein purification and characterisation.

Anna Jasińska – dr, University of Łódź, Faculty of Biology and Environmental Protection, Institute of Microbiology, Biotechnology and Immunology, Department of Industrial Microbiology and Biotechnology, e-mail: anna.jasinska@biol.uni.lodz.pl.

Scientific interests: filamentous fungi in environmental protection, fungal oxidative enzymes, fungal proteomics.

Mariusz Krupiński – dr, University of Łódź, Faculty of Biology and Environmental Protection, Institute of Microbiology, Biotechnology and Immunology, Department of Industrial Microbiology and Biotechnology, e-mail: mariusz.krupinski@biol.uni.lodz.pl.

Scientific interests: ecotoxicological studies in the processes of microbiological degradation of xenobiotics, analysis of environmental risk caused by chemical substances and their

mixtures, assessment of toxic effects of endocrine disruptors, microbiological decomposition and mineralisation of toxic compounds present in personal and household chemicals.

Hobby: military tourism, searching for and visiting abandoned places that are forgotten cultural heritage, board games, collecting and listening to vinyl records.

Katarzyna Lisowska – prof. dr hab., University of Łódź, Faculty of Biology and Environmental Protection, Institute of Microbiology, Biotechnology and Immunology, Department of Industrial Microbiology and Biotechnology, e-mail: katarzyna. lisowska@biol.uni. lodz.pl.

Scientific interests: biodegradation and detoxification of environmental pollutants, including pharmacologically active compounds, microbiological synthesis of nanoparticles, antimicrobial activity of nanomaterials, bioconversion of steroids.

Krystyna Milczarek – mgr, University of Łódź, Faculty of Biology and Environmental Protection, Institute of Microbiology, Biotechnology and Immunology, Department of Industrial Microbiology and Biotechnology, retired.

Scientific interests: physiology of microorganisms, microbiological transformation of steroidal compounds, fungi protoplasts, confocal microscopy.

Hobby: photography, travel, politics, gardening.

Justyna Nykiel-Szymańska – mgr, University of Łódź, Faculty of Biology and Environmental Protection, Institute of Microbiology, Biotechnology and Immunology, Department of Industrial Microbiology and Biotechnology, e-mail: justyna. nykiel@biol.uni.lodz.pl.

Scientific interests: microbiological degradation and detoxification of herbicides, influence of Trichoderma genus fungi on plant growth, adaptation changes of filamentous fungi in response to toxic xenobiotics.

Katarzyna Paraszkiewicz – dr hab., prof. UŁ, University of Łódź, Faculty of Biology and Environmental Protection, Institute of Microbiology, Biotechnology and Immunology, Department of Industrial Microbiology and Biotechnology, e-mail: katarzyna.paraszkiewicz@biol.uni.lodz.pl.

Scientific interests: biosurfactants of microorganisms, microbiological degradation and detoxification of anthropogenic contamination, bioconversion of steroids using microfungi.

Hobby: gardening, tourism.

Sylwia Różalska – dr hab., prof. UŁ, University of Łódź, Faculty of Biology and Environmental Protection, Institute of Microbiology, Biotechnology and Immunology, Department of Industrial Microbiology and Biotechnology, e-mail: sylwia.rozalska@biol.uni.lodz.pl.

Scientific interests: conventional and unconventional methods of entomopathogenic fungi application, confocal microscopy, mycology, microbiological degradation and detoxification of anthropogenic contaminants, metabolomics, secondary metabolites of filamentous fungi.

Mirosława Słaba – dr hab., prof. UŁ, University of Łódź, Faculty of Biology and Environmental Protection, Institute of Microbiology, Biotechnology and Immunology, Department of Industrial Microbiology and Biotechnology, e-mail: miroslawa.slaba@biol.uni.lodz.pl.

Scientific interests: interaction of microorganisms with heavy metals and toxic xenobiotics, microbiological degradation and detoxification of anthropogenic environmental contaminants, filamentous fungi promoting plant growth.

Rafał Szewczyk – dr, University of Łódź, Faculty of Biology and Environmental Protection, Institute of Microbiology, Biotechnology and Immunology, Department of Industrial Microbiology and Biotechnology (until September 30, 2019).

Scientific interests: proteomics and metabolomics of biological processes; search, identification and profiling of various biomarkers; bioinformatics, modern microbiological diagnostics and biodegradation of xenobiotics using mass spectrometry and various separation techniques.

Aleksandra Anna Walaszczyk – mgr, graduate of full-time studies of the second degree in Microbiology Faculty of Biology and Environmental Protection, University of Łódź, e-mail: a.walaszczyk96@gmail.com.

Scientific interests: immunology, infection diagnosis, microbial biosurfactants.

Hobby: science fiction films and literature, guitar playing, tourism.

Stanisław Walisch – dr inż., University of Łódź, Faculty of Biology and Environmental Protection, Institute of Microbiology, Biotechnology and Immunology, Department of Industrial Microbiology and Biotechnology, retired.

Scientific interests: nitrogen metabolism of industrial acetic acid bacteria species, aerobic processes of organic acid biosynthesis using immobilised microorganisms, optimisation of biotechnological processes using microorganisms in food and pharmaceutical industry, storage of industrial strains of microorganisms.

Hobby: alpine tourism, philately – research in this field.

Danuta Wilmańska – dr, University of Łódź, Faculty of Biology and Environmental Protection, Institute of Microbiology, Biotechnology and Immunology, Department of Industrial Microbiology and Biotechnology, retired.

Scientific interests: various applications of protoplasts of filamentous fungi, microbiological bioconversion of steroids, bioconversion of androgens by yeast, mycology.

Natalia Wrońska – dr, University of Łódź, Faculty of Biology and Environmental Protection, Institute of Microbiology, Biotechnology and Immunology, Department of Industrial Microbiology and Biotechnology, e-mail: natalia.wronska@biol.uni.lodz.pl.

Scientific interests: nanotechnology, antibacterial nanomaterials, biopolymers, nanocomposites, bioconversion of steroids using bacteria.

Hobby: dance, tourism, theatre.

Katarzyna Zawadzka – dr, University of Łódź, Faculty of Biology and Environmental Protection, Institute of Microbiology, Biotechnology and Immunology, Department of Industrial Microbiology and Biotechnology, e-mail: katarzyna.zawadzka@biol.uni.lodz.pl.

Scientific interests: microbiological synthesis and properties of nanoparticles, mechanisms of antibacterial action of nanomaterials, synthesis of new compounds with antimicrobial activity, microbiological degradation and detoxification of environmental pollutants.

REWIEVERS
Grażyna Płaza, Jerzy Falandysz, Dominik Kopeć, Alwyn Fernandes

INITIATING EDITOR
Beata Koźniewska

TECHNICAL EDITOR
Leonora Gralka

TYPESETTING
Munda – Maciej Torz

COVER DESIGN
krzysztof de mianiuk

Descriptions and sources of the photos used on the cover are listed on p. 552

Published under the patronage of:
Committee of Biotechnology of The Polish Academy of Sciences

and The Polish Mycological Society

Edition of the book has been financed by: The Rector of The University of Lodz
and The Dean of The Faculty of Biology and Environmental Protection
The publication of this book has been partially sponsored by The Polish Mycological Society

First edition. W.10000.20.0.S

Publisher's sheets 24.0; printing sheets 34.5

Descriptions and sources of photos used on the cover

ZEISS Primo Star light microscope with a camera, photo by Rafał Szewczyk, taken in the microscopic laboratory of the Department of Industrial Microbiology and Biotechnology, University of Łódź, Poland

Hyphae of the microscopic fungus *Aureobasidium pullulans* used, inter alia, in for the production of pullulan polysaccharide and hydrolytic enzymes, photo by Krystyna Milczarek, taken during research at the he Department of Industrial Microbiology and Biotechnology, University of Łódź, Poland

Drone used to evaluate (using cameras and sensors) the plant health algorithm and detect plant diseases caused by microbes © Depositphotos.com/cboswell

Hyphae of the microscopic mushroom *Ashbya gossypii* used for the production of riboflavin (vitamin B2), photo by Krystyna Milczarek, taken during research at the Department of Industrial Microbiology and Biotechnology, University of Łódź, Poland

Observations of biological material samples using the ZEISS LSM 510 Meta confocal microscope (at the microscope, Dr. Sylwia Różalska, prof. University of Łódź), photo by Rafał Szewczyk, made in the microscopic laboratory of the Department of Microbiology Industrial and Biotechnology of the University of Łódź, Poland

Thermostatic bioreactor for the cultivation of microorganisms © Depositphotos.com/PTHamilton

The fruiting body of *Tremella mesenterica* (orange-yellow scrapie) is formed on the surface of decaying wood and fruiting bodies of the *Trametes versicolor* fungus used for the production of ligninolytic enzymes, photos by Andrzej Długoński, taken in the Łagiewniki Forest in Łódź, Poland

The post-glacial lake, Małe Łąki near Sianów (Kashubian Swiónowò), in the buffer zone of the Kashubian Landscape Park, Poland, photo by Jerzy Długoński, taken from the lake's pier

Sub-cultivation of the microscopic fungus *Mucor ramosissimus* capable of simultaneous utilization of used engine oil and degradation of pentachlorophenol and a fragment of the chromatography laboratory with Agilent 1200 liquid chromatograph and QTRAP 3200 mass spectrometer visible in the depth, photos by Rafał Szewczyk, taken in the Department of Industrial Microbiology and Biotechnology, University of Łódź, Poland